Modern Medical Genetics and Genomics

Edited by Israel Gomy

Published in London, United Kingdom

IntechOpen

Supporting open minds since 2005

Modern Medical Genetics and Genomics
http://dx.doi.org/10.5772/intechopen.76597
Edited by Israel Gomy

Contributors
Esther Lopez-Bayghen, Almena Lopez-Luna, Dinorah Hernandez-Melchor, Leticia Ramirez, Rodney Scott, Tom Nolis, Isam Al Jashi, Al Jashi Cristina Gladys, Leonardo Porchia, Elizbeth Schaeffer, Israel Gomy

First published in London, United Kingdom, 2019 by IntechOpen
IntechOpen is the global imprint of INTECHOPEN LIMITED, registered in England and Wales, registration number: 11086078, 7th floor, 10 Lower Thames Street, London, EC3R 6AF, United Kingdom
Printed in Croatia

British Library Cataloguing-in-Publication Data
A catalogue record for this book is available from the British Library

Additional hard and PDF copies can be obtained from orders@intechopen.com

Modern Medical Genetics and Genomics
Edited by Israel Gomy
p. cm.
Print ISBN 978-1-83968-142-4
Online ISBN 978-1-83968-143-1
eBook (PDF) ISBN 978-1-83968-144-8

For EU product safety concerns:
IN TECH d.o.o., Prolaz Marije Krucifikse Kozulić 3, 51000 Rijeka, Croatia, info@intechopen.com or visit our website at intechopen.com.

Meet the editor

Dr. Israel Gomy graduated in Medicine at Pontificia Universidade Católica do Paraná, Curitiba, Brazil. He received his Master's degree in Biomedical Research at Universidade de São Paulo, Ribeirão Preto, Brazil, and his PhD in Oncology at Fundação Antônio Prudente, São Paulo, Brazil. Dr. Gomy is an assistant professor at Faculdades Pequeno Príncipe, Curitiba, Brazil, and an assistant physician at Hospital de Clínicas da Universidade Federal do Paraná, Curitiba, Brazil. He is also a postdoctoral researcher in medical oncology at Dana-Farber Cancer Institute, Boston, USA.

Contents

Preface

The field of medical genetics and genomics is undergoing a constant revolution based on new breakthroughs that bring increasing insights into the etiology of rare and common diseases, which help improve the health care of individuals with these disorders.

The development of high-throughput technologies such as next-generation sequencing (NGS) has enabled the detection of inherited susceptibilities and actionable mutational profiles of tumors for tailored therapies and management. Additionally, the generation of such massive amounts of data has prompted advances in the field of bioinformatics, with regard to complex tools and numerous databases to guide the interpretation of such big data, and to deal with noninformative results such as the notorious variants of uncertain significance. Besides these technical challenges, important ethical issues must also be considered: for example, the disclosure of results from prenatal and newborn screening through NGS, and findings from population-level screening of asymptomatic individuals. Addressing such challenges is not easy. We propose that investigators must provide tentative answers for one vital question: What is the analytic validity, clinical validity, and clinical utility of genetic testing?

This book aims to depict some of the myriad applications of genetics and genomics in the practice of medicine. We present the material in three sections: (1) an introductory landscape of medical genetics and genomics and portraits of personalized genomic medicine; (2) the field of reproductive genetics and prenatal screening: the first two chapters of this section deal with the most critical factors involved in the success of embryo implantation through assisted reproductive technology, including endometrial receptivity and rates of embryonic aneuploidy; the third chapter evaluates the accuracy of invasive and noninvasive prenatal screening for predicting the risk of fetal malformations; and (3) the area of hereditary cancer genetics, wherein is presented an updated review of the clinical utility of genetic testing, as exemplified by the description of hereditary breast and ovarian cancers.

Even though the information presented here offers a very limited look at the modern-day developments and breakthroughs of the centenarian field of medical genetics, this book nevertheless provides an important glimpse of the fascinating and challenging era of personalized or precision medicine.

Israel Gomy
Universidade Federal do Paraná,
Curitiba, Brazil

Section 1

Introduction

Chapter 1

Modern Medical Genetics and Genomics in the Era of Personalized/Precision Medicine

Israel Gomy

1. Introduction

Modern medical genetics as a well-defined field of medicine has developed so fast since its origins half a century ago that we cannot bear in mind how long time ago its roots and beginnings came from.

It can be argued that genetics overall was based on measuring the problems of human hereditary features and inherited diseases, since before the twentieth-century acceptance of Mendelian laws of heredity. Thus, medical genetics, viewed from its broadest perspective, is perhaps the oldest area of genetics and not a recent area that it is sometimes believed.

2. Modern genomics in medicine

During the years, with developments in cellular and molecular biology, the field of medical genetics expanded from a small clinical subspecialty focused at describing a few rare hereditary disorders to a recognized medical specialty whose principles and approaches are essential parts of the diagnosis and management of many disorders, both common and rare. These genetic concepts and approaches are not restricted to any one medical specialty or subspecialty, as they permeate many, and perhaps all areas of medicine.

The medical geneticist is usually a physician who works as part of a team of clinical providers, including many other physicians, nurses, and genetic counselors, to evaluate patients and their relatives for possible hereditary diseases. They characterize the patient (or proband) through analyses of personal and family history and physical examination; assess risk and possible modes of inheritance; indicate diagnostic testing; manage prevention, treatment, and surveillance; and participate in communicating to other family members at risk for the disorder.

During the twentieth century, it gradually became clear that hereditary factors were implicated in many conditions and that different genetic mechanisms were involved. Virtually, any disease is the result of the interaction of genes and environment, but the relative influence of the genetic component may be large or small. Traditionally, genetic conditions have been classified into three categories: monogenic, chromosomal, and multifactorial disorders. However, it is becoming increasingly evident that the interplay of different genes (polygenic inheritance) is essential in disease and that an additional category—acquired somatic genetic disease—should also be included.

Improvements in all areas of medicine, mainly public health and therapeutics, resulted in modifying patterns of diseases, with improving recognition of the role of genetic factors for most of common disorders or even for the susceptibility to infectious diseases. For complex chronic degenerative diseases of adult onset, their overall contribution of heritability has been identified, as life expectancy increases and high-throughput technologies improve. These provide the opportunity for understanding the interactions between the genetic and environmental factors of diseases such as cancer, Alzheimer's disease, diabetes mellitus, macular degeneration, and cardiomyopathy.

In recent years, we have been facing the applications of modern genomics to the practice of medicine. With robust molecular biology technologies, one can identify the actionable mutations present in a tumor and establish the profile of its pattern of RNA expression, which are currently being used for determining prognosis and choosing appropriate targeted therapies for individual cancer patients. Another application is how modern genomic approaches are increasing our abilities in risk assessment and helping provide more accurate genetic counseling to patients and families affected with hereditary diseases as well as advances in prenatal diagnosis.

There are further examples of applications of genomics to individualized health care: screening asymptomatic individuals for genetic predisposition to various diseases in order to improve health care, population-based newborn screening for preventable and treatable genetic diseases, identifying couples that are carriers for autosomal recessive or X-linked diseases that could affect their children before conception, prenatal screening of the fetus for aneuploidy by maternal cell-free DNA, and applications of pharmacogenomics on the detection of individual variation affecting drug therapy, which can be used to improve therapeutic efficacy and reduce adverse events.

At the beginning of the twenty-first century, the Human Genome Project (HGP) provided a virtually complete sequence of human genome from which now derives the efforts to catalog all human genes, understand their structure and regulation, determine the extent of their variation in different populations, and uncover how genetic variation contributes to susceptibility. The whole genome of any individual can now be sequenced rather than sequencing one gene at a time. These achievements are making possible the practice of genomic medicine, which aims at applying a wide analysis of the human genome and its products, including the epigenetic regulation of gene expression, gene variation, and their interactions with the environment, to medical care.

The HGP has now been succeeded by the Human Variome Project, which seeks to compile and share the huge variation in human DNA sequence worldwide. This is potentially possible since whole exome sequencing (WES) and whole genome sequencing (WGS) have been increasingly performed in several population studies. For example, the 100,000 Genomes Project in the United Kingdom have recently achieved its goal of 100,000 genomes sequenced [1].

The study of genetics and its role in causing human disease is now widely acknowledged as being among the most exciting and influential areas of medical research. Certainly, their valuable discoveries have benefited patients and families dramatically, but this achievement will be measured by translating them into both treatment and prevention of disease.

3. Diagnostic advances

Advances over the past few years in mutation identification have provided many improvements in risk assessment, carrier detection, and prenatal diagnosis, allowing the detection of particular mutations with almost 100% accuracy. Laboratory testing for pathogenic (disease-causing) mutations is available for more than 4,500 genes associated to over 11,000 genetic conditions [2]. The better knowledge of the genes involved in hereditary disease and the rapidly cheaper cost of DNA sequencing have permitted the identification of mutations in a patient or family member, and, therefore, the molecular diagnosis has become the standard of care for many conditions. DNA samples for testing are available not only from readily accessible tissues, such as a buccal scraping or blood sample, but also from tissues obtained by more invasive testing, such as chorionic villus sampling or amniocentesis.

For many hereditary disorders (including retinal degeneration, deafness, hereditary breast and ovarian cancer, Lynch syndrome, congenital myopathy, mitochondrial disorders, and hypertrophic or dilated cardiomyopathies), there is a substantial locus heterogeneity, that is, numerous genes are known to be mutated in different families with these disorders. When a patient with one of these highly heterogeneous disorders seeks for testing, recent advances in DNA sequencing make it possible to analyze large panels of dozens to more than 100 genes simultaneously and cost-effectively for mutations in every gene in which mutations have been seen previously to cause the disorder.

In those conditions for which even a large panel of relevant genes cannot be formulated for a particular phenotypically defined disorder, or for those genetically heterogeneous entities (e.g. intellectual disability, autism spectrum disorder) genetic diagnosis still can be achieved by analyzing all the coding exons of every gene (by WES) or by sequencing the entire genome (by WGS) to identify pathogenic mutations.

The use of large gene panels and, even more so, WGS or WES raises special issues for sequence interpretation and risk assessment. As there are more genes being tested, the number of sequence differences between a patient's DNA and that of an arbitrary reference sequence also increases; consequently, many previously undescribed variants will be found whose pathogenicity is unknown. These are so-called variants of uncertain significance (VUSs). This is the case, for example, of a missense mutation that results in the substitution of one amino acid for another in the encoded protein.

Thus, the interpretation of variants is a challenging and demanding area for all professional geneticists who provide molecular diagnosis. The American College of Medical Genetics and Genomics has recommended that variants be categorized into one of five classes: pathogenic, likely pathogenic, of uncertain significance, likely benign, and benign variants [3] Specialists in molecular diagnostics, human genomics, and bioinformatics have developed a series of criteria for addressing the mutation status. In most cases, none of these criteria is absolutely definitive but must be considered together to provide an overall assessment of the pathogenicity of a variant. Only those variants with a high probability of being pathogenic are communicated to the health professional and patient. It is arguable whether the testing laboratory should disclosure all VUSs, at the same time remaining available for updating as new information allow reclassification as either benign or pathogenic.

Despite all the time and effort put into interpretation, it is still impossible to ensure any clinical significance to the vast majority of all variants found through next-generation sequencing (NGS). There is a general concern that individuals and their clinical providers, when confronted with VUSs, will require additional unnecessary testing, with the potential for finding results with even more VUSs, thereby increasing patient's uncertainty and anxiety. Moreover, there is the additional concern that even when a variant is known to be pathogenic and highly penetrant in families with multiple affected individuals, the true penetrance of a variant when it is found in individuals with a negative family history may be much lower.

Therefore, risk assessment and genetic counseling in this context are challenging processes and depend on continually addressing recently available information and communicating it properly to healthcare providers and patients.

4. Ethical dilemmas

Each new advance in genetic technology has simultaneously brought new ethical concerns and raised new dilemmas about how science will be applied and utilized in medicine. At the center of this is the recognition that an individual's genetic background is fundamental to both their identity and possible disease susceptibility. The most controversial field is prenatal genetics and reproductive choice, though national legal frameworks and cultural practices vary widely worldwide. The debate surrounding the early ability to perform prenatal diagnosis for Down syndrome through invasive procedures in the mid-1960s is compared to the new technology that makes it possible to perform detailed genetic screening of the fetus on cell-free fetal DNA in the maternal blood or on embryos created through in vitro fertilization for preimplantation genetic diagnosis (PGD). Great controversy has taken place, and will go on, regarding the disclosure of unexpected but significant "incidental findings" from WES or WGS carried out for specific clinical purposes [4]. Furthermore today there is the technical feasibility of all newborns having their genome sequenced and screened for either childhood disorders or adult-onset conditions [5].

Advances in genetics attract great media attention, and this has brought the ethical debate to a wide public scenario. Issues about insurance, forensic science and DNA databases, patenting, gene therapy, population screening, cloning, stem cell research, and hybrids are considered to have major societal, commercial, and political importance and therefore impact clinical and laboratory practice in medical genetics. On a global scale, it is essential to safeguard fundamental principles such as privacy, confidentiality, and respect for human life at all stages and ages.

Many of the questions raised do not have easy or definitive answers, which means that there will be a great need for both public awareness and properly trained clinicians and counselors to balance the needs of their patients and families with these ethical challenges for the foreseeable future.

5. Personalized and precision medicine

The aim of personalized or individualized medicine is to use knowledge of an individual's genetic (or genomic) background relevant to the maintenance of health, prevention, and treatment of diseases as a routine part of medical care.

During the past 10 years, many examples of stratified medicine have blossomed, where the therapy of a particular disease is dependent on the germline or somatic variants patients may carry. These examples include monogenic rare diseases where a different treatment is recommended for patients with some types of germline mutations in a specific gene, such as *CFTR* gene mutations that cause cystic fibrosis, and the molecular therapy targeted on an actionable somatic mutation of a specific tumor, such as the *BRAF* V600E mutation in malignant melanoma. The genetic (or genomic) diagnosis is therefore a crucial step toward the most appropriate treatment and/or prevention, what is so-called clinical utility or actionability [6, 7]. Recent initiatives are focusing on improving health outcomes through precision medicine especially in Oncology [8] (**Table 1**). This is a multidisciplinary integrated approach that analyzes human samples and personal data to improve health care through increased precision in the knowledge of mechanisms of both disease and drug response.

Personalized genomic medicine (PGM) is only one component of precision medicine, which means, in a broadest sense, it requires clinical care providers to combine genomic information with other types of information, such as biochemical or physiological testing results, neurodevelopmental history, environmental exposures, and psychosocial experiences. The most important goal is provide more

Tumor(s)	Genes	Clinical application(s)
Melanoma	*BRAF, CDKN2A, KIT*	Therapy; risk assessment; inherited susceptibility
Breast and ovarian cancer	*BRCA1, BRCA2*	Therapy; inherited susceptibility
Colorectal cancer	*MLH1, MSH2, MSH6, PMS2, EPCAM, BRAF, KRAS, NRAS*	Therapy; inherited susceptibility
Non-small-cell lung cancer	*EGFR, BRAF, KRAS, ERBB2, RET, MET, ALK, ROS1*	Therapy
Myelodisplastic syndrome	*TP53, GATA2, JAK2, ASXL1, ETV6, RUNX1, SF3B1, EZH2*	Diagnosis; risk stratification
Acute myeloid leukemia	*KIT, CEBPA, FLT3, NPM1*	Risk stratification

Table 1.
Genetic/genomic profiling for selected malignancies (adapted from [8]).

precise diagnosis, genetic counseling, management, prevention, and therapy. This effort has already got started, but plenty of work still needs to be done before PGM becomes integrated into medicine itself.

Conflict of interest

The author declares no conflict of interest.

Author details

Israel Gomy
Universidade Federal do Paraná, Curitiba, Brazil

*Address all correspondence to: israel.gomy@hc.ufpr.br

References

[1] Genomics England. Available from: https://www.genomicsengland.co.uk. [Accessed 01/31/19]

[2] Genetic testing registry. Available from: https://www.ncbi.nlm.nih.gov/gtr [Accessed: 01/31/19]

[3] Richards S, et al. Standards and guidelines for the interpretation of sequence variants: A joint consensus recommendation of the American College of Medical Genetics and Genomics and the Association for Molecular Pathology. Genetics in Medicine. 2015;**17**:405-424. DOI: 10.1038/gim.2015.30

[4] Kalia SS, Adelman K, Bale SJ, et al. Recommendations for reporting of secondary findings in clinical exome and genome sequencing, 2016 updated (ACMG SF v2.0): a A policy statement of the American College of Medical Genetics and Genomics. Genetics in Medicine. 2017;**19**:249-255. DOI: 10.1038/gim.2016.190

[5] Anderson, J. et al. Parents perspectives on whole genome sequencing for their children: qualified Qualified enthusiasm? Journal of Medical Ethics. 2017;**43**(8):535-539. DOI: 10.1136/medethics-2016-103564

[6] Chakravaty D et al. OncoKB: A precision oncology knowledge base JCO precision Oncology. 2017. Available from: https://oncokg.org. DOI: 10.1200/PO.17.00011. [Accessed: 01/31/19]

[7] ClinGen. Actionability Working Group. Evidence-based Summaries. Available from: https://clinicalgenome.org/working-groups/actionability/projects-initiatives/actionability-evidence-based-summaries [Accessed: 01/31/19]

[8] National Comprehensive Cancer Network. Biomarker compendium. Available from: https://www.nccn.org/professionals/content/ [Accessed: 01/31/19]

Section 2

Reproductive and Prenatal Genetics

Chapter 2

The Genetic and Biochemical Blueprint of Endometrial Receptivity: Past, Present, and Future Factors Involved in Embryo Implantation Success

Almena López-Luna, Dinorah Hernández-Melchor, Leticia Ramírez-Martínez and Esther López-Bayghen

Abstract

In the field of assisted reproductive technology, endometrial receptivity is a crucial aspect that affects implantation rates in *in-vitro* fertilization procedures; in fact, impaired endometrial receptivity has been identified as the rate-limiting step for favorable pregnancy outcomes once factors regarding embryo quality have been optimized. The endometrium is a dynamic tissue that undergoes proliferative and secretory changes in each menstrual cycle, acquiring a short and transient period of embryo receptivity known as the Window of Implantation. Precise embryo-endometrial synchrony is necessary to achieve a successful pregnancy, and it involves complex and multifactorial processes related to morphological, biochemical, and genetic changes. On that behalf, defining the receptive window of each patient for personalized embryo transfer is a current goal. Here, we review different indicators of endometrial receptivity throughout the menstrual cycle, spotlighting the opening of the window of implantation: classical histological and biochemical markers, genetic factors, leading-edge transcriptomic signatures and miRNA profiles, and novel features such as the microbiome and secretome. Understanding the molecular mechanisms behind endometrial receptivity will facilitate the optimization and improvement of infertility treatments.

Keywords: endometrial receptivity, embryo implantation, menstrual cycle, window of implantation, decidualization

1. Introduction

The field of assisted reproductive technology (ART) has grown significantly in use and understanding over the past few decades, nevertheless, the rates of successful pregnancies in *in-vitro* fertilization (IVF) procedures are still relatively low. Impaired endometrial receptivity (ER) has been identified as the rate-limiting step for favorable pregnancy outcomes once all other factors, including the acquisition and selection of the best quality embryo(s), have been optimized. Correct and synchronized maturation of the endometrial tissue is essential for embryo

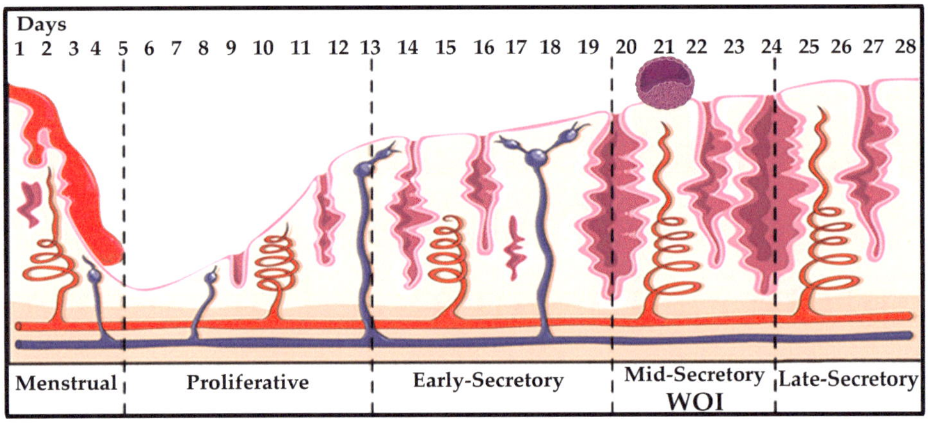

Figure 1.
Endometrial dynamics throughout the menstrual cycle. This picture was modified from Servier Medical Art under the Creative Commons License 2018.

implantation [1, 2]. The endometrium is a dynamic tissue that undergoes proliferative and secretory changes in each menstrual cycle. Throughout most of this cycle, the endometrium remains "non-adhesive" to embryos and it only acquires a short and transient period of embryo receptivity known as the "window of implantation" (WOI) [3, 4]. In humans, during a natural cycle, the endometrium becomes receptive 6 to 8 days after ovulation and it remains receptive for approximately 24–48 h, this time is assumed to occur between days 20 and 24 of a regular menstrual cycle [2, 5] (**Figure 1**). The cyclic fluctuations of the endometrium are coordinated by the ovarian hormones estrogen (E2) and progesterone (P4); a finely balanced signaling process mediated by these hormones defines the WOI. The menstrual cycle is divided in two phases: proliferative or follicular, and secretory or luteal [6]. During the proliferative phase (PP), E2 from the maturing follicle allows the elongation of the spiral arteries and the proliferation of endometrial stromal cells (EnSCs) and glands [7]. Afterwards, during the secretory phase (SP), P4 from the corpus luteum induces secretory changes; the endometrium thickens and it acquires a receptive phenotype that will support blastocyst attachment [8–10]. During the Mid-Secretory Phase (MSP) circulating P4 induces EnSCs to undergo decidualization [11]. Decidualization is the transformation of the endometrial stroma into a dense cellular matrix known as the decidua, this process initiates during the SP in the stroma and, if pregnancy occurs, it progresses into the development of the decidua which will in turn form the maternal placenta [12]. The optimization of the endometrium to support embryo implantation is a complex and multifactorial process that involves morphological, biochemical, and genetic changes [13]. ER is a key aspect that affects implantation rates in IVF procedures considering that a precise embryo-endometrial synchrony is completely necessary to achieve a successful pregnancy [4, 14]. Thus, understanding the molecular mechanisms behind ER will facilitate the optimization and improvement of infertility treatments.

2. Factors involved in endometrial receptivity

2.1 Evaluation of endometrial morphology for receptivity assessment

Morphological changes during the endometrial cycle generate markers that have been used over decades to assess receptivity, such as histological evaluation of a biopsy and ultrasound examination of the endometrium. Endometrial biopsies are now considered to provide little clinically relevant information [15, 16]. Additionally, the formation of pinopodes was thought to show potential as a clinical marker to assess ER [17]. However, the presence of pinopodes was demonstrated not only during the WOI but also in the post-receptive endometrium, precluding in this way its use as a marker of ER [17, 18]. On the other hand, ultrasound examination is a routinely used technique in IVF procedures [9, 19]. This non-invasive technique is based on the interpretation of a medical ultrasound of the endometrium. Various ultrasonographic parameters have been proposed as pregnancy predictors, such as endometrial thickness, volume, and blood flow patterns. The most commonly used is endometrial thickness [20, 21]. Due to differences in stimulation protocols, sonographic approaches, and difficulties in obtaining a standard sagittal view of the uterus, discrepancies in the cut-off value of endometrial thickness to achieve pregnancy arise [22]. Generally, it is considered that a minimum of 6–8 mm in endometrial thickness is necessary for a successful pregnancy [23–25]. Nevertheless, case reports have described pregnancy establishment despite an endometrial thickness of no more than 4 mm [26, 27]. Three-dimensional (3D) sonography assesses ER by considering endometrial thickness, volume, and angiogenic dynamics. The endometrial volume of fewer than 2 ml has been shown to decrease pregnancy rates significantly [19, 28, 29]. Another evaluated criterion is an endometrial pattern, which can be classified as triple-line, intermediate, or homogenous [30]. Among these, the triple-line pattern has been suggested to reflect ER [24, 31] broadly. Finally, the impact of ovarian stimulation on ER has yet to be determined. Abnormal hormone concentrations the due to stimulation protocols during IVF might affect endometrial morphology and thereby ER [32]. Comprehensively, although morphological elements are important components of receptivity, there is still no consensus on the extent in which they can be used as WOI predictors.

2.2 Genetic factors involved in ER

Endometrial genetic abnormalities can lead to implantation failure due to dysregulation of critical processes such as trophoblast invasion and angiogenesis. Here, we discuss common genetic abnormalities that have been analyzed to determine their role in implantation failure (**Table 1**). Parental chromosomal abnormalities such as mutations and translocations should be considered relevant in the efficacy of improving reproductive outcome.

2.2.1 Angiogenetic factors: vascular endothelial growth factor (VEGF), endothelial nitric oxide synthase (eNOS), TP53 tumor suppressor (TP53), murine double minute 2 (MDM2), herpes virus-associated ubiquitin-specific protease (HAUSP)

Successful pregnancy is dependent on adequate placental circulation and fetal vasculature. The development of a normal vascular network during implantation, embryo development, and placentation requires cooperation between different cell types and various growth factors. *VEGF* is a potent angiogenic factor that plays an essential role in embryo implantation/development. Four *VEGF* polymorphisms have been reported to affect VEGF activity and expression increasing aberrations in vascular formation and/or function. The polymorphism −1154G/A located in the promoter region has been associated with RPL [33], RSA [34, 35], and RIF [36, 37]. Moreover, a meta-analysis showed that genotypes −2578C/A, −634G/C, and 936C/T increase the risk of RSA as well [38]. Furthermore, *eNOS*, which is expressed in the terminal chorionic villous vessels, is important for vascular nitric oxide (NO) production to supply nutrients to the fetus. Only the eNOS Glu298Asp polymorphism has been shown to be significantly associated with RPL [39]. Additionally, successful trophoblast invasion requires the induction of paracrine apoptotic reactions to

Gene	Polymorphism	rs code	Relevance	References
APOE			Heterozygous genotype is more frequent in women with RPL	[33]
eNOS	VNTR (B/A)		Association to the risk of RPL	[77]
			No associated	[78]
	Glu298Asp		Homozygote genotype T/T is associated with risk of IRM	[39]
ESR1	IVS1-397T>C	rs2234693	Related to unknown thin endometrium in which P allele may be the risk and X allele its guard factor	[44]
	IVS1-351A>G	rs9340799		
F2	G20210A	rs1799963	No association	[58, 60, 62, 63]
			Heterozygous genotype is more frequent in women with RSA in the first trimester [57] and women with RPL [42]	[42, 57]
F5	G1691A, Leiden	rs6025	No association	[57, 58, 62, 63]
			More frequent in women with RIF	[60]
	H1299R (R2)		No association	[58, 62, 63]
	Y1702C		No association	[58, 62, 63]
F8	V34L		More prevalent in women with RPL	[42, 58]
			No association	[62, 63]
FGB	G-455A	rs1800790	No association	[57, 58, 62, 63]
GPIIIa	C1565T		No association	[57]
HAUSP	rs1529916 G/A	rs1529916	Allele A is associated with RIF	[40]
HPA1	HPA1 a/b (L33P)		No association	[58, 62, 63]
LIF	C715A		No associated	[72]
	G3400A		More frequent in nulligravid women	[72, 73]
	G3424A		No associated	[72]
	T1414G	rs929271	G/G genotype is associated with RIF	[40, 74]
MDM2	T309G	rs2279744	Allele G is associated with RIF	[40]
MTHFR	A1298C	rs1801131	No association	[58, 62–64]
	C677T	rs1801133	More frequent in women with unexplained infertility [64], RPL [42, 58], and RSA [60]	[42, 58–60]
			No association	[57, 62, 63]
MUC1	VNTR		Women with unexplained infertility might have susceptibility to implantation failure due to small *MUC1* allele size	[67]
			No association	[68, 69]
MUC4	VNTR		No association	[70]
PAI-1	4G/5G	rs1799889	More prevalent in women with RIF [63] and RPL [42, 58]	[42, 58, 63]
			No association	[62]

Gene	Polymorphism	rs code	Relevance	References
PR	H770H-C/T		No association	[54]
	G/T—Val660Leu	rs1042838	More prevalent in women with unexplained infertility	[65]
	V660L		No association	[54]
PT53	Codon 72 Pro	rs1042522	Homozygote genotype is associated with RPL [41, 75], IRM [76] and RIF [37, 40]	[37, 40, 41, 75, 76]
	Codon 72 Arg	rs1042522	Homozygote genotype is associated with RIF	[75]
PTGS2	G-765C	rs20417	Association with implantation failure susceptibility	[55]
TFF3	rs225361 A/G	rs11701143	Homozygous genotype is associated with less live births before their first spontaneous abortion	[61]
	rs225361 A/G	rs11701143	No association	[66]
	rs11701143 T/C	rs225361	Associated with idiopathic RSA	[61]
	rs11701143 T/C	rs225361	No association	[66]
	rs225439 G/A	rs225439	No association	[61, 66]
	ros533093 C/T	rs533093	No association	[61, 66]
	rs77436142 G/C	rs7743614	No association	[61, 66]
VEGF	G-1154A	rs1570360	Homozygote A/A genotype associated with RSA [34, 35], RPL [33], and RIF [36, 37]	[33–37]
			No associated	[71]
	C-2578A	rs699947	No associated	[34]
	G-634C	rs2010963	No associated	[34]
	C936T	rs3025039	No associated	[34]

Probably implicated mutations in implantation failure, studies are listed even when no association was found, rs code is mentioned whenever it is reported.

Table 1.
Genetic abnormalities involved in implantation failure.

secrete proteases capable of digesting the endometrial extracellular matrix (ECM) [36]. TP53 is a potent regulator of apoptosis, cell cycle, angiogenesis, and embryonic development. A TP53 polymorphism at codon 72, encoding either proline or arginine, was reported to alter the TP53 activity and affect human fertility [40]. The Arg72 variant has been shown to induce higher apoptotic activity than Pro72. Therefore, Pro72 variant might cause inadequate trophoblast invasion, increasing the risk of RPL [41] and RIF [37, 40]. In this manner, women with Pro/Pro genotype have a higher risk of RPL than women with the Arg/Arg or Arg/Pro genotypes. Following, MDM2 and HAUSP regulate TP53. MDM2 binds to TP53 to degrade it through poly-ubiquitination, blocking its ability to function as a transcription factor. The *MDM2* SNP309 is a functional SNP that increases MDM2 expression levels and attenuates TP53 pathway. HAUSP, on the other hand, acts as a specific deubiquitinase for TP53, the A allele has a significant association with infertility in young patients (<35 years) but not in the older patients, similarly to the *MDM2* SNP309 G allele. Those observations suggest that *MDM2* and *HAUSP* may be involved in the regulation of human fertility through the regulation of TP53 [40].

2.2.2 Apolipoprotein E (APOE)

Due to the increase in total cholesterol levels during pregnancy, *APOE* plays a crucial role in lipid metabolism. *APOE* has three alleles in the long arm of chromosome 19 at position 13.2: ε2, ε3, and ε4. Individuals harboring the allele ε4 have higher cholesterol levels than the ones carrying the ε3/ε3 allele, whereas levels in those with the ε2 allele are lower [42].

2.2.3 Estrogen receptor α (ESR1)

ESR1 is a ligand-activated transcription factor essential for sexual development and reproductive function; its dysregulation leads to the development of various diseases such as cancer, cardiovascular disease, and inflammation, among others [43]. Due to alternative splicing of mRNA, it possesses three isoforms: ERαΔ3, ERα36, and ERα46 [43]; in a study performed by Yuan and Le [44], the polymorphisms rs2234693 and rs9340799 were related to the uncommonly thin endometrium.

2.2.4 Leukemia inhibitor factor (LIF)

LIF is an important implantation factor that promotes proliferation, invasion, and differentiation; its expression is regulated by the transcription factor tumor protein TP53 (TP53). Few studies have found a correlation between LIF gene polymorphism and reproductive capacity, Kang et al. demonstrated that SNP in the 3'UTR of the LIF (rs929271) gene is associated with infertility [40].

2.2.5 Mucin 1 (MUC1), Mucin 4 (MUC4)

MUC1 is an anti-adhesion molecule secreted by human endometrial epithelium, it has been suggested that its expression prevents the adherence of blastocyst to the endometrium. Interestingly, MUC1 must be locally removed in a paracrine fashion at the implantation site during the WOI to allow contact between the embryo and the endometrium, making it an important factor in determining ER [45–48]. *MUC1* is a highly polymorphic gene that differs in the size of the region carrying the O-glycosylation sites: the variable number tandem repeat region (VNTR), which can go from 20 to 125 repeats [49]. Similarly, MUC4 is a greatly expressed mucin in endometrial epithelium [50], its gene is highly polymorphic and it contains a VNTR region that can go from 145 to 395 repeats [51]. Although its role in human infertility has not been fully explored, studies in other species have suggested that it plays a role in embryo implantation [52, 53].

2.2.6 Progesterone receptor (PR)

The PROGINS complex are three mutations in the *PR* gene that may be associated with unexplained infertility and implantation failure: a 306 bp insert in intron G of the dT2 allele in *PR*, the mutated alleleV660L, and guanine to thymine substitution in exon 4, resulting in a valine to leucine change in the hinge region of PR, and a cytosine to thymine substitution at exon 5 [54].

2.2.7 Prostaglandin-endoperoxide synthase 2 (PTGS2)

PTGS2 is a key enzyme involved in the conversion of arachidonic acid to prostaglandins (PGs). The −765G>C SNP mutation in the promoter region of *PTGS2* upstream the transcriptional start site in the putative Sp1 site can cause alterations in Sp1 binding

[55]. Accordingly, the hypermethylation of the NF-IL6 site within the *PTGS2* promoter results in elevated gene expression in eutopic endometrium in endometriosis [56].

2.2.8 Thrombolytic factors: coagulation factor II (F2), coagulation factor V (F5), coagulation factor XIII a chain (F13A1), methylenetetrahydrofolate reductase (MTHFR), plasminogen activator inhibitor-1 (PAI-1)

Thrombophilia, the predisposition for thrombosis, has been shown to be a risk factor for successful pregnancy due to impaired vascularization at the time of implantation. Therefore, the possible association between early pregnancy loss and polymorphisms at coagulation factors and thrombolytic genes responsible for inherited or acquired thrombophilia has been investigated. The coagulation factor II SNP G20210A in the 3′-untranslated region of *F2* causes elevated prothrombin in plasma, leading to enhanced blood coagulation [57]. Furthermore, factor V Leiden mutation, G1691A, is a single nucleotide substitution in the *F5* gene that results in reduced clearance of factor Va due to its blocked inactivation by activated protein C, increasing the risk of thrombosis [42, 57]. Also, factor XIII V34L polymorphism is a guanine to thymine substitution in exon 2 of *F13A1* that leads to a valine for leucine change in residue 34; this SNP leads to reduced susceptibility to fibrinolysis and influences fibrin degradation [57]. Moreover, MTHFR is the rate-limiting enzyme in the methyl cycle. C677T polymorphism causes a substitution of valine for alanine, resulting in a thermolabile variant of the enzyme with reduced catalytic activity; combined with the SNP AC1298C, it is associated with hyper-homocysteinemia, a risk factor for venous and arterial thrombosis [42, 58–60]. Additionally, PAI-1 is a key regulatory element in the fibrinolysis cascade, it is believed to control proteolysis and remodeling of maternal tissue during trophoblast invasion. The 4G/5G polymorphism is located 657 bp upstream from the start site of transcription within the *PAI-1* promoter and results in an allele with decreased transcriptional activity [42].

2.2.9 Trefoil factor 3 (TFF3)

TFF3 is a mucin-associated peptide co-expressed in mucus cells that acts as a mitogen to promote epithelial cell migrations and mediates epithelial repair after damage. The SNP rs11701143 is located in the promoter region of *TFF3* within the regulatory region of the transcription binding site, whereas rs225361 is an intron variant located within a regulatory region. The exact function of both SNPs remains to be elucidated [61].

2.3 Immunological factors contributing to ER

The immune system plays a major role in the process of implantation and pregnancy maintenance [62]. During decidualization, endocrine processes transform uterine fibroblasts into cells that can produce hormones, growth factors, and matrix components to support embryo implantation [63, 64]. Furthermore, tolerance of the immune maternal system is required in pregnancy to avoid rejection of the semi-allograft or allograft embryo and for its successful implantation [65]. The decidua is a privileged site for immune tolerance; a large number of molecules and immune cell types participate in this process, leukocytes, macrophages, T lymphocytes, and dendritic cells comprise around 30 to 40% of the cells within the decidual stroma in early pregnancy. Among leukocytes, uterine natural killer (uNK) cells are activated and they significantly increase during decidualization (65–70%) [66, 67]. Increases in uNK cells denote three main functions in the endometrium: regulation of placental and trophoblast growth by cytokines [68, 69], local immunomodulation [70, 71], and control of trophoblast invasion [69]. Furthermore, trophoblast cells play a major role in immune tolerance since these cells do not express major histocompatibility complex (MHC) class I (HLA-A and HLA-B) or class II molecules; ensuring that maternal T cells with αβ receptors cannot mount a classic cytotoxic attack against fetal paternal alloantigens. The trophoblast also protects itself by expressing Fas ligand (Fas L), which is important in the elimination of maternal reactive T cells by apoptosis induction [72–74]. Other important component of this process is T-regulatory (Treg) cells; these cells are essential for immunosuppression, prevention of autoimmunity, and maternal tolerance to the fetus [75–78]. Treg cells have been shown to be locally enriched in decidua during early normal pregnancy [79]. Furthermore, Forkhead box P3 (Foxp3) is a master regulator of Treg cell development, function, and differentiation [80]. Expression of *FOXP3* was reduced approximately two-fold in endometrial biopsies of infertile women, implicating that the impaired differentiation of uterine T-cells into the Treg phenotype is a key determinant of fertility [81, 82]. On the other hand, helper T cells (CD4+) facilitate embryo implantation by regulating endometrial differentiation; they secrete interleukins and interferons that establish the implantation microenvironment. Successful pregnancy is dependent upon Th1/Th2 balance [83]. While Th1 cytokines are harmful for pregnancy, Th2 cytokines favor fetal growth and regulate uterine expression of fatty acid amide hydrolase (FAAH), LIF, and trophoblast release of human chorionic gonadotropin (hCG), which are known to play important roles during implantation [84–86]. Piccinni et al. demonstrated that T cells from decidua of women with a miscarriage show predominantly Th1-type cytokines with decreased Th2-type [84, 87, 88]. Finally, P4 and E2 mediate the downregulation of the maternal immune system [89]. P4 stimulates decidual proliferation; therefore, pregnancy results in an upregulation of P4 receptors on activated lymphocytes among placental cells and decidual CD56+ cells. In the presence of sufficient P4, these cells express progesterone-induced blocking factor (PIBF), a mediator that exerts substantial anti-abortive activities.

2.4 Biochemical markers involved in ER

We review molecular markers involved in the decidualization process that could be suitable as makers to assess ER.

2.4.1 Homeobox A10 (HOXA10)

HOXA10 is a transcription factor member of the homeobox family, known to be involved in the genetic control of embryonic development and in the regulation of the adult female reproductive tract [12, 90]. HOXA10 regulates downstream target genes that lead to endometrium development and receptivity acquisition [91], such as insulin-like growth factor binding protein 1 (*IGFBP1*) [12, 92, 93], genes of the Wnt pathway (reviewed by Sonderegger et al. [94], integrin β3 (*ITGB3*) [12, 90, 92], and empty spiracles homolog 2 (*EMX2*) [12, 95]. HOXA10 is regulated by P4 in a dose-dependent manner [12, 90, 91]. The expression of HOXA10 is low during the PP and it rapidly increases in the MSP [12, 90]. The diminished expression of HOXA10 in endometria of women with recurrent pregnancy loss (RPL) [96], adenomyosis [97], endometriosis [92], polycystic ovary syndrome (PCOS) [98], and idiopathic infertility [99] indicates that this gene could be essential for fertility [12].

2.4.2 Heparin-binding epidermal growth factor-like growth factor (HB-EGF)

HB-EGF is a member of the epidermal growth factor (EGF) family, it is expressed in the human uterus at the time of implantation and its expression is under steroidal hormone control [100–103]. The transmembrane form is associated with cell adhesion and migration, it allows communication with the blastocyst by acting as a chemoattractant [101, 104]. HB-EGF expression is low in the PP and increases in the ESP immediately prior to the WOI, after which its levels decrease [101–103, 105]. Also, mRNA levels are low in pregnant endometrium and high in placental tissues at an early stage of development, suggesting that the HB-EGF ligand not only potentiates the health and survival of the peri-implantation embryo, but also induces the progression of its development [104]. HB-EGF stimulates epithelial expression of key endometrial proteins that are important biomarkers of the WOI, including LIF, HOXA10, and ITGB3 [100].

2.4.3 Leukemia inhibitory factor (LIF)

LIF is a member of a cytokine family with functional redundancy that includes interleukin 6 (IL6), oncostatin (OSM), ciliary neurotrophic factor (CNTF), and cardiotrophin 1 (CT1). They regulate proliferation, differentiation, and cell survival in different cellular systems [106]. LIF acts on cells by binding to the heterodimeric LIF receptor (LIFR), which consists of two transmembrane proteins, LIFR and glycoprotein 130 (gp130). LIFR activates several signaling pathways including the JAK/STAT, MAPK, and P13-kinase pathways, whereas gp130 participates in the activation of STAT1, STAT3, and STAT5B [107]. LIF induces the expression of cytokines and other regulatory molecules that could serve to regulate preimplantation development and embryo implantation [106–108]. LIF is one of the most important cytokines for receptivity during the WOI, the expression of LIF and LIFR reaches its highest level during the WOI in the MSP, LSP, and in early pregnancy in both surface and glandular epithelial cells under the influence of P4 [106, 107, 109, 110]. LIF can also be detected in decidual leukocytes, which are abundant at the implantation site; interestingly, LIF expression is low in women with unexplained infertility [106, 107, 111]. LIF also plays a crucial role in the regulation of fetal-maternal interactions during pregnancy, this cytokine mediates uterine receptivity through autocrine/paracrine interactions by binding to LIFR on the luminal epithelium to permit blastocyst attachment [106], but also regulates trophoblast function and vascular formation in the placenta [109].

2.4.4 Integrin β3 (ITGB3)

Integrins are ubiquitous cell adhesion molecules involved in maintaining normal tissue morphology and participate in cell–cell and cell-substrate interactions [100, 110, 112, 113]. In the human endometrium, integrins are involved in early embryo-endometrial interactions [90]. ITGB3 subunit is present after the ESP and its expression extend into the pregnancy [112, 114]. It has been reported that healthy fertile women show higher ITGB3 expression than patients with unexplained infertility [46, 96, 113–115]. Moreover, its dysregulation appears to characterize two distinct pathophysiological conditions that involve distinct mechanisms of defective ER: Type I and Type II. Type I defect is an out-of-phase endometrium with negative ITGB3 subunit expression, portrays a shifted WOI, and hormonal inadequacy or responsiveness is implicated, on the other hand, Type II defect is an "in-phase" endometrium with negative ITGB3 subunit expression and connotes the complete loss of the WOI. Furthermore, ITGB3 is expressed in EnSCs and endometrial glands with the highest levels in the MSP to LSP, suggesting a role in the regulation of endometrial function and implantation [115, 116]. Due to its temporal distribution and the effects of implantation when it is not present, ITGB3 is a useful molecular marker to assess ER. ITGB3 is regulated in the endometrium through a molecular mechanism via sex steroid signaling where HOXA10 acts as an intermediary [90, 96].

Other identified markers that are important for decidualization in the human endometrium include *PR*, particularly its encoded isoform progesterone receptor A (PR-A), homeobox A11 (*HOXA11*), *PTGS2*, *MUC1,* and interleukin 11 receptor (*IL11R*) [12].

2.5 Transcriptomic signature to determine the WOI

Microarray technology has been widely used to determine the transcriptomic profile of the endometrium by analyzing the expression of large batches of genes at different stages of the menstrual cycle. The most representative and commercially available test in this regard is perhaps the Endometrial Receptivity Array (ERA), developed in 2009 by Diaz-Gimeno et al., this test identifies the unique transcriptomic signature of the receptive endometrium by analyzing 238 differentially expressed genes, predicting the WOI for personalized embryo transfer (pET) [108]. Various research groups have analyzed changes in gene expression during the different phases of the endometrial cycle using microarray-based technologies [117–120], however, due to differences on results, unanimity about the main genes to be analyzed to determine the WOI has not been reached. Factors that contribute to the disagreement among studies results include differences on experimental design, utilized probes, sample acquisition day, sample size, collection method, and the application of distinct statistical analyses. Nevertheless, some genes have been reported to be expressed similarly in more than one work, here, we present a compilation of the expression profiles of those candidate genes in the human endometrium (**Table 2**).

2.6 miRNAs involved in ER

Micro-RNAs (miRNAs) are small, single-stranded, non-protein-coding RNA sequences of ~18–25 nucleotides in length that play an important post-transcriptional regulatory role in gene expression [121, 122] by targeting mRNAs for cleavage or transcriptional repression [123]. More than two decades have passed since the initial discovery of miRNAs in *Caenorhabditis elegans* by Lee et al. [124]; since then, great progress has been made in the understanding of miRNAs: what they are, how are synthesized, how regulate gene expression, and how they are involved in the formation and progression of pathological disorders. Extracellular miRNAs have been ubiquitously detected in body fluids [125]. Therefore, the presence and stability of miRNAs in biological fluids have advocated their potential as non-invasive biomarkers. Nevertheless, the identification of reliable miRNA biomarkers with reproducible profiles has been a challenge, and their diagnostic promise has remained a work in progress since they have still not entered the clinical field [126]. Nonetheless, given that miRNAs are differentially expressed in the endometrium across the menstrual cycle [127–131], several studies have been conducted to explore their role in ER [131–136]. **Table 3** presents a summary of these studies.

2.7 The endometrial secretome as a potential tool to ascertain ER

The aim to develop alternative non-invasive strategies to provide accurate receptivity assessment has drawn assiduity to the endometrial secretome, which is based in the identification of factors secreted by cells or tissues at a particular time in either physiological states or pathological conditions [137], including proteins, lipids, and metabolites. Therefore, the analysis of differentially present molecules in the uterine cavity at different time points of the menstrual cycle could potentially help to identify the WOI and to diagnose uterine pathologies. Sample collection of endometrial fluid (EF) collection in the peri-implantation period is an easy procedure performed with minimally invasive tools that could easily be implemented

Gene	Functional category	Expression profile		References
Annexin 4 [*ANX4*]	Apoptosis	PP to MSP ESP to MSP MSP to LSP	↑ ↑ ↓	[73, 86, 91–95]
Apolipoprotein D [*APOD*]	Cholesterol transport and trafficking	PP to ESP ESP to MSP	↑ ↑	[85, 92, 96]
Claudin 4 [*CLDN4*]	Cell adhesion	ESP to MSP	↑	[73, 85, 93, 94, 96, 97]
Decay-accelerated factor [DAF]	Immune modulators/ cytokines	ESP to MSP	↑	[85, 93, 94, 98]
Dickkopf-1 [*DKK1*]	Regulation of WNT signaling pathway	PP to ESP ESP to MSP ESP to LSP	↑ ↑ ↓	[73, 85, 92, 94, 96, 97, 99]
Endothelin 3 [*EDN3*]	Vasoactive substances	ESP to MSP	↓	[73, 92, 93, 100]
Growth arrest and DNA-damage-inducible protein [*GADD45*]	Cell cycle	PP to MSP ESP to MSP MSP to LSP	↑ ↑ ↑	[73, 92–94, 97, 101]
Glutathione peroxidase 3 [*GPX3*]	Response to stress and oxidoreductase activity	ESP to MSP	↑	[73, 86, 96]
Homeobox A10 [*HOXA10*]	Transcription factor	ESP to MSP MSP to LSP	↑ ↑	[55, 64, 94, 102, 103]
Inhibitor of DNA binding 4, dominant negative helix–loop helix protein [*ID4*]	Anatomical structure development	ESP to MSP	↑	[73, 92–94, 104]
Insulin-like growth factor binding protein 1 [IGFBP-1]	Anatomical structure development	ESP to MSP MSP to LSP	↑ ↑	[73, 94]
IL15 precursor [*IL15*]	Immune response	ESP to MSP	↑	[73, 85, 92–94, 96]
Mitogen-activated protein kinase kinase kinase 5 [*MAPKKK5*]	Signal transduction	PP to ESP	↑	[92, 101]
Matrix metalloproteinase 26 [*DAMMP26*]	Tissue remodeling and blastocyst invasion	MP to LPP PP to ESP MSP to LSP	↑ ↑ ↓	[87, 94] [73]
Msh homeobox homologs 1,2 [*MSX1*]	Anatomical structure development	ESP to MSP	↓	[73, 88, 91, 93, 96]
Natural cytotoxicity-triggering receptor 3 [*NCR3*]	Immune response	ESP to MSP MSP to LSP LSP to MP MP to PP	↑ ↑ ↑ ↓	[86, 87]
Olfactomedin 1 [*OLFM1*]	Anatomical structure development	PP to ESP ESP to MSP	↓ ↓	[73, 88, 92, 93]
Sex-determining region Y-box 4 [*SOX4*]	Apoptotic pathways	PP to ESP MSP to LSP MP to LPP	↓ ↑ ↑	[87, 88]
Osteopontin [*SPP1*]	Cell adhesion	PP to SP ESP to MSP	↑ ↑	[73, 85, 91–93, 96, 97, 101]
Tissue inhibitor metalloproteinase 3 [*TIMP-3*]	Degradation of the extracellular matrix	PP to ESP ESP to MSP MSP to LSP	↑ ↑ ↑	[73, 94, 95]

Genes expressed in human endometrium and its expression profiles at the different phases of the endometrial cycle. A compilation of the genes exhibiting the same expression profile in more than one work is presented, regardless of the differences among studies.

Table 2.
Expression profiles of genes involved in ER.

Reference and relevance	Methodology	Results	Targets
[132] Shows distinct miRNA profiles in LPP and MSP	**Samples:** Endometrial biopsies LPP [n = 4] and MSP [n = 4] **Technique:** Microarray Technology	49 differentially expressed miRNAs: 12 ↑ in the MSP [miR-214, 503, 134, 450, 382, 376A, 369-5p, 222, 370, 542-3p, 105, and 127] and 12 ↓ in LPP [miR-210, 193a-3p, 345, 29b, 29c, 30b, 204, 203, 582-5p, 30d, 200c, and 31]	**Predicted:** Cell cycle pathways in the LPP. Wnt signaling pathway in the PP. **Validated:** Decreased transcripts of predicted targets [cyclins, CDKs, and E2F3]
[134] Evaluates differentially expressed miRNAs in RIF-IVF patients	**Samples:** Secretory endometrium; 12 fertile women versus 11 women with RIF **Technique:** TaqMan miRNA array cards	13 differentially expressed miRNAs: 10 ↑ [miR-23b, I45, 99a, 27b, 652, 139-5p, 195, 342-3p, 150, and 374b] and 3 ↓ [miR-32, 628-5p, and 874]	**Predicted:** Wnt signaling, adherents junctions, p53 signaling, cell adhesion, and cell cycle pathways. **Validated:** Decreased transcripts of N-cadherin, H2AFX, NTN4, and SFRP4. Wnt and cell cycle pathways ↓ in RIF-IVF
[154] Makes comparison between natural vs. stimulated cycles. Suggests that ovarian stimulation may shift the WOI	**Samples:** Infertile women. Receptive [LH + 7, n = 5] vs. prereceptive [LH + 2, n = 5] in natural cycles. Receptive [hCG + 7, n = 5] vs. prereceptive [hCG + 4, n = 5] in stimulated IVF cycles **Technique:** Next-gen seq., 626 miRNAs evaluated	20 differentially expressed miRNAs in natural cycles: 8 ↑ [miR-30d, 30b, 30b, 31, 21, 193a-5p, 193a3p, 203] and 12 ↓ [miR-33a, 452, 125b, 455-3p, 455-5p, 483-5p, 143, 100, 504, 424, 424, 503]. 22 differentially expressed miRNAs in stimulated cycles: 19 ↑ [miR-187, 708, 433, 320a, 320b, 34c-5p, 320c, 320d, 485-5p, 574-5p, 375, 23b, 423-5p, 193b, 34b, 503, 424, 455-5p, 483-5p] and 3 ↓ [miR-886-5p, let-7f, let-7a]	**Predicted:** Cell cycle, transport, cell adhesion, cell death, and metabolism
[136] Provides miRNA signature of EnSCs during decidualization *in vitro*	**Samples:** Endometrial samples on oocyte retrieval day from healthy ovum donors [n = 50]. EnSCs [n = 20] isolated and cultured **Technique:** miRNA PCR array, 704 miRNAs + specific miRNA-200b primers evaluated	43 differentially expressed miRNAs: 26 ↑ [miR-95, 888, 936, 1185, 518f, 548 k, 593, 486-5p, 29c, 449b, 300, 371-5p, 1224-3p, 891a, 365, 541, 409-5p, 33b, 154, 376a, 133a, 218–2, 22, 614, 369-3p, 185] and 17 ↓ [miR-146a, 155, 181b, 181a, 135b, 181d, 200c, 141, 182, 429, 483-3p, 200a, 96, 183, 9, 30a, 126] miR-95 ↑ by P4 and E2 + P4; miR-96 and miR-135b both ↓ by E2 or P4	**Predicted:** Growth factors, interleukins, ECM remodeling enzymes. • Top pathways by ↓ miRNAs: axon guidance, adherents junction, actin cytoskeleton regulation, ErbB [EGFR] signaling, and renal cell carcinoma. • Top pathways by ↑ miRNAs: actin cytoskeleton regulation, adherent junction, axon guidance, Wnt signaling, and MAPK
[133] Provides miRNA signature of fertile human endometria: receptive vs. prereceptive	**Samples:** Receptive MSP [LH + 7, n = 4] vs. prereceptive ESP [LH + 2, n = 5] endometrial biopsies from 9 healthy fertile women **Technique:** Microarray, 723 human and 76 human viral miRNAs evaluated	4 significantly expressed different miRNAs in receptive samples: 2 ↑ [hsa-miR-30b and 30d] and 2 ↓ [hsa-miR-494 and 923] Suggests 12 genes that could serve as a new panel for ER: CAST, CFTR, DPYSL2, FI IR, FGFR2, LIF, MTFI, NPAS2, P4HA2, PPARGCIA, TACC2, RAB40B	**Predicted:** Transcription, cell proliferation, and apoptosis. Involvement in pathways such as axon guidance, Wnt/β-catenin, ERK/MAPK, TGF-β, p53 and leukocyte extravasation. They identified *SEPT7, CRMP1, SLC44A1, HES1, FXR2,* and *TNF144B* as genes that interact with genes *MIR30B* and *MIR30D*

Reference and relevance	Methodology	Results	Targets
[130] Differential miRNAs across cycle. Release of miRNAs into the EF and its uptake by the embryo	**Samples:** EF [n = 20] at different phases of the menstrual cycle of healthy women. Timing of sample collection: EPP, LPP, ESP, WOI, and LSP **Technique:** Microarray, 866 miRNAs evaluated	Compared with the WOI, 9 differentially expressed miRNAs were identified in the EPP, 8 in the LPP, 6 in the ESP and 4 in the LSP. MiR-30d was the most differentially secreted maternal miRNA in the EF during the WOI	**Predicted:** Cell cycle and endocrine processes
[135] Shows a significant different expression of miRNAs in the WOI of RIF patients that may contribute to impaired ER	**Samples:** Endometrial biopsies from the WOI [5–7 days after ovulation]: 7 from RIF group and 5 from control group [infertile patients that delivered after one transfer attempt] **Technique:** Microarray Technology	With a 2-fold threshold: 105 miRNAs were differentially expressed: 93 ↑ and 12 ↓. After raw signal value correction, 15 were found to be significantly different. 10 ↑ [hsa-miR-374a-5p, 145-5p, 30b-5p, 196b-5p, 199a-5p, 199b-5p, 449a, 424-5p, 125b-5p, 21-5p] and 5 ↓ [hsa-miR-1207-5p, 4306, 572, 5739, 6088]	**Predicted:** TAM analysis: miR-30 family, human embryonic stem cell regulation, epithelial-mesenchymal transition, and miRNA tumor suppressors. Network regulatory analysis: 176 miRNA-mRNA interactions. The top core mRNA were ABP1, AQP3, ASS1, and TIMP3, the top core miRNAs were has-miR-4668-5p, 429, and 5088

Studies conducted to analyze miRNA expression profiles during the menstrual cycle as potential biomarkers of the WOI. A summary of the differentially expressed miRNAs and predicted targets found in recent studies is presented. Its relevance regarding the role of miRNAs is also addressed.

Table 3.
Studies of miRNA-profiling during the WOI.

in ART procedures. Proteomics of EF has already rendered valuable information regarding ER; Casado-Vela et al. identified 803 proteins in EF aspirates using three different proteomic strategies [138]. Additionally, Boomsma et al. [139] analyzed endometrial secretions prior to embryo transfer from 210 women undergoing IVF to determine differences in cytokine profiling at the time of implantation, finding a negative and a positive association of monocyte chemo-attractant protein-1 (MCP-1) and IFN-γ-inducible 10 kDa protein (IP-10) levels and implantation, respectively. Lipidomics, on the other hand, seems to have rendered slight information on receptivity [140], nevertheless, a study performed by Berlanga et al. [141] and followed by Vilella et al. [142, 143] carried out lipidomic analyses of EF from patients at different stages of their menstrual cycle, they determined a significant increase in Prostaglandin E2 (PGE2) and Prostaglandin F2α (PGF2α) between days 19 and 21, coincident with the WOI. In a recent study performed by Durairaj, Aberkane et al. [144], the contribution of EnSCs to failed implantation was examined by analyzing the secretome profile of EnSCs cultures *in-vitro*. From there, they encountered that secretome profiles of pregnant women are less divergent in implantation-positive-cultures particularly in Day 0 (undifferentiated cells), suggesting that endometrial defects linked to reproductive failure could be more prominent in the PP, a phase that is commonly thought to be not relevant for ER studies. This research group also demonstrated that the secretome of undifferentiated EnSCs compromises blastocyst development. Finally, they determined that a deficiency of endometrial mesenchymal stem-like cells (MSCs) could lead to aberrant EnSC function and implantation failure. Overall, this study remarks the importance of progenitor cell populations in the endometrium that supports the acquisition of receptivity and raises the prospect of screening the endometrium before the initiation of an ART procedure.

2.8 The microbiome as a novel aspect of ER

Historically, the uterus was assumed to be free of bacteria as the fetal environment was considered to be physiologically sterile [145], this notion implies that the neonate's microbiome is acquired only during and after birth. Although recent research still supports this conception [146], others have characterized upper genital tract microbiota [147–150], suggesting that the endometrial and vaginal microbiota not be identical [151]. What is more, the study by Moreno et al. [147, 152], which defined the microbiota in the EF as *Lactobacillus*-dominated (LD) or non-*Lactobacillus*-dominated (NLD), suggested that the presence of an NLD microbiota in a receptive endometrium was associated with a significant decrease in implantation, pregnancy, ongoing pregnancy, and live birth rates. Nevertheless, they acknowledge that in the absence of pathological signs, an NLD microbiota could be considered normal since *Lactobacillus*-deficient communities have been identified in the genital tract of otherwise healthy asymptomatic women [153]. This conception sets the stage for further research of the human microbiome, expands the possibilities to assess individualized receptivity based on the endometrial microbiome, and opens the door to explore targeted therapies for an altered endometrial microbial habitat.

3. Remarks

This review encloses different aspects of ER, spotlighting the opening of the WOI. Altogether, this compilation could aid in the development of new clinical practices that define an individual's receptive window for pET to improve ART results ultimately.

Nowadays, ER is the rate-limiting step in successful ART procedures that end up in pregnancy and child delivery. The endometrial tissue is a ponderous element in fertility; it constitutes the soil in which a viable embryo will implant to achieve progeny. The attainment of ER involves an extensive assortment of genetic and biochemical mechanisms that must integrate in a parallel manner. Understanding of this process as an entity is still insufficient; nevertheless, the surge of new technologies is contributing in the deciphering of receptivity mechanisms and in search of novel biomarkers that could serve to detect the WOI. Notably, although most studies focus on individual genetic mutations, a more comprehensive view of the parental genetics is needed to determine whether an endometrium is adequate for embryo transfer before the initiation of an ART procedure. Due to ambiguous or non-conclusive results in search of genetic predispositions of endometrial-associated infertility, it would be controversial to provide genetic counseling currently. Perhaps, in the future, massive sequencing could help to provide insights into the importance of single and multiple genetic mutations to establish a receptive or non-receptive profile. If assertive, this profile could be applied as an endometrial pre-implantation parental test to improve the rates of healthy pregnancies and live births. With this in mind, in our opinion, the best short-term approaches towards detecting or improving ER are the transcriptome, microbiome, and miRNA signatures, all achievable using the power of NGS. This review encloses different aspects of ER, spotlighting the opening of the WOI. Altogether, this compilation could aid in the development of new clinical practices that define an individual's receptive window for pET to improve ART results ultimately.

Abbreviations and nomenclatures

3D	three-dimensional
ApoE	apolipoprotein E
ART	assisted reproductive tech
CNTF	ciliary neurotrophic factor
CT1	cardiotrophin 1
E2	estrogen
ECM	extracellular matrix
EF	endometrial fluid
EGF	epithermal growth factor
EMX2	empty spiracles homolog 2
eNOS	endometrial nitric-oxide synthase
EnSCs	endometrial stromal cells
EPP	early-proliferative phase
ER	endometrial receptivity
ERA	endometrial receptivity array
ESP	early-secretory phase
ESR1	estrogen receptor 1
F13A1	coagulation factor XIII A chain
F2	coagulation factor II
F5	coagulation factor V
F8	coagulation factor 8
FAAH	fatty acid amide hydrolase
Fas L	Fas ligand
FGB	β fibrinogen
Foxp3	Forkhead box P3
gp130	glycoprotein 130
GPIIIa	glycoprotein IIIa
HAUSP	herpesvirus-associated ubiquitin-specific protease
HB-EGF	heparin-binding epithermal growth factor-like growth factor
hCG	human chorionic gonadotropin
HOXA10	homeobox A10
HOXA11	homeobox A11
HPA1	human platelet alloantigens 1
HSCORE	histological score
IGFBP1	insulin-like growth factor binding protein 1
IL11R	interleukin 11 receptor
IL6	interleukin 6
IP-10	IFN-γ-inducible 10 kDa protein
ITGB3	integrin β3
IVF	in-vitro fertilization
LD	lactobacillus-dominated
LH	luteinizing hormone
LIF	leukemia inhibitory factor
LIFR	leukemia inhibitory factor receptor
LPP	late-proliferative phase
LSP	late-secretory phase
MCP-1	monocyte chemo-attractant protein-1
MDM2	murine double minute 2
MHC	major histocompatibility complex
MiRNA	microRNA
MP	menstrual phase

MSCs	mesenchymal stem-like cells
MSP	mid-secretory phase
MTHFR	methylenetetrahydrofolate reductase
MUC1	Mucin 1
MUC4	Mucin 4
NGS	next generation sequencing
NLD	non-lactobacillus-dominated
OSM	oncostatin
P4	progesterone
TP53	tumor protein 53
PAI-1	plasminogen activator inhibitor-1
PCOS	polycystic ovarian syndrome
PCR	polymerase chain reaction
pET	personalized embryo transfer
PGE2	prostaglandin E2
PGF2α	prostaglandin F2α
PGs	prostaglandins
PIBF	progesterone-induced blocking factor
PP	proliferative phase
PR	progesterone receptor
PR-A	progesterone receptor A
RIF	repeated implantation failure
RPL	recurrent pregnancy loss
RSA	recurrent spontaneous abortion
SNP	single-nucleotide polymorphism
SP	secretory phase
TAM	tool annotations human miRNAs
TFF3	trefoil factor 3
Treg	T-regulatory cells
uNK	uterine natural killer
VEGF	vascular endothelial growth factor
VNTR	variable number tandem repeats
WOI	window of implantation

Author details

Almena López-Luna[1,2], Dinorah Hernández-Melchor[1,2], Leticia Ramírez-Martínez[2] and Esther López-Bayghen[2]*

1 Laboratorio de Investigación y Diagnóstico Molecular, Instituto de Infertilidad y Genética SC, INGENES, Mexico

2 Departamento de Toxicología, Cinvestav-IPN, Ciudad de México, Mexico

*Address all correspondence to: ebayghen@cinvestav.mx

References

[1] Lessey BA. The role of the endometrium during embryo implantation. Human Reproduction. 2000;**15**(Suppl 6):39-50

[2] Diedrich K et al. The role of the endometrium and embryo in human implantation. Human Reproduction Update. 2007;**13**(4):365-377

[3] Strowitzki T et al. The human endometrium as a fertility-determining factor. Human Reproduction Update. 2006;**12**(5):617-630

[4] Teh WT, McBain J, Rogers P. What is the contribution of embryo-endometrial asynchrony to implantation failure? Journal of Assisted Reproduction and Genetics. 2016;**33**(11):1419-1430

[5] Achache H, Revel A. Endometrial receptivity markers, the journey to successful embryo implantation. Human Reproduction Update. 2006;**12**(6):731-746

[6] Reed BG, Carr BR. The normal menstrual cycle and the control of ovulation. In: De Groot LJ et al., editors. South Dartmouth (MA): Endotext; 2000

[7] Cha J, Sun X, Dey SK. Mechanisms of implantation: Strategies for successful pregnancy. Nature Medicine. 2012;**18**(12):1754-1767

[8] Mahajan N. Endometrial receptivity array: Clinical application. Journal of Human Reproductive Sciences. 2015;**8**(3):121-129

[9] Heger A, Sator M, Pietrowski D. Endometrial receptivity and its predictive value for IVF/ICSI-outcome. Geburtshilfe und Frauenheilkunde. 2012;**72**(8):710-715

[10] Beier HM, Beier-Hellwig K. Molecular and cellular aspects of endometrial receptivity. Human Reproduction Update. 1998;**4**(5):448-458

[11] Lockwood CJ et al. The role of decidualization in regulating endometrial hemostasis during the menstrual cycle, gestation, and in pathological states. Seminars in Thrombosis and Hemostasis. 2007;**33**(1):111-117

[12] Lu Z, Hardt J, Kim JJ. Global analysis of genes regulated by HOXA10 in decidualization reveals a role in cell proliferation. Molecular Human Reproduction. 2008;**14**(6):357-366

[13] Jabbour HN et al. Endocrine regulation of menstruation. Endocrine Reviews. 2006;**27**(1):17-46

[14] Fox C et al. Local and systemic factors and implantation: What is the evidence? Fertility and Sterility. 2016;**105**(4):873-884

[15] Murray MJ et al. A critical analysis of the accuracy, reproducibility, and clinical utility of histologic endometrial dating in fertile women. Fertility and Sterility. 2004;**81**(5):1333-1343

[16] Coutifaris C et al. Histological dating of timed endometrial biopsy tissue is not related to fertility status. Fertility and Sterility. 2004;**82**(5):1264-1272

[17] Nikas G. Endometrial receptivity: Changes in cell-surface morphology. Seminars in Reproductive Medicine. 2000;**18**(3):229-235

[18] Quinn CE, Casper RF. Pinopodes: A questionable role in endometrial receptivity. Human Reproduction Update. 2009;**15**(2):229-236

[19] Zhang T et al. The role of three-dimensional power Doppler ultrasound parameters measured on hCG day in the prediction of pregnancy during in vitro

fertilization treatment. European Journal of Obstetrics, Gynecology, and Reproductive Biology. 2016;**203**:66-71

[20] Abdalla HI et al. Endometrial thickness: A predictor of implantation in ovum recipients? Human Reproduction. 1994;**9**(2):363-365

[21] Yuval Y et al. The relationships between endometrial thickness, and blood flow and pregnancy rates in in-vitro fertilization. Human Reproduction. 1999;**14**(4):1067-1071

[22] Schild RL et al. Endometrial receptivity in an in vitro fertilization program as assessed by spiral artery blood flow, endometrial thickness, endometrial volume, and uterine artery blood flow. Fertility and Sterility. 2001;**75**(2):361-366

[23] Noyes N et al. Factors useful in predicting the success of oocyte donation: A 3-year retrospective analysis. Fertility and Sterility. 2001;**76**(1):92-97

[24] Friedler S et al. The role of ultrasonography in the evaluation of endometrial receptivity following assisted reproductive treatments: A critical review. Human Reproduction Update. 1996;**2**(4):323-335

[25] Khalifa E et al. Sonographic appearance of the endometrium: The predictive value for the outcome of in-vitro fertilization in stimulated cycles. Human Reproduction. 1992;**7**(5):677-680

[26] Sundström P. Establishment of a successful pregnancy following in-vitro fertilization with an endometrial thickness of no more than 4 mm. Human Reproduction (Oxford, England). 1998;**13**(6):1550-1552

[27] Remohi J et al. Endometrial thickness and serum oestradiol concentrations as predictors of outcome in oocyte donation. Human Reproduction (Oxford, England). 1997;**12**(10):2271-2276

[28] Mercé LT. Ultrasound markers of implantation. The Ultrasound Review of Obstetrics and Gynecology. 2002;**2**(2):110-123

[29] Raga F et al. Assessment of endometrial volume by three-dimensional ultrasound prior to embryo transfer: Clues to endometrial receptivity. Human Reproduction. 1999;**14**(11):2851-2854

[30] Zhao J, Zhang Q, Li Y. The effect of endometrial thickness and pattern measured by ultrasonography on pregnancy outcomes during IVF-ET cycles. Reproductive Biology and Endocrinology. 2012;**10**:100

[31] Zhao J et al. Endometrial pattern, thickness and growth in predicting pregnancy outcome following 3319 IVF cycle. Reproductive Biomedicine Online. 2014;**29**(3):291-298

[32] Fatemi HM, Popovic-Todorovic B. Implantation in assisted reproduction: A look at endometrial receptivity. Reproductive Biomedicine Online. 2013;**27**(5):530-538

[33] Coulam CB, Jeyendran RS. Vascular endothelial growth factor gene polymorphisms and recurrent pregnancy loss. American Journal of Reproductive Immunology. 2008;**59**(4):301-305

[34] Papazoglou D et al. Vascular endothelial growth factor gene polymorphisms and idiopathic recurrent pregnancy loss. Fertility and Sterility. 2005;**83**(4):959-963

[35] Lee HH et al. Association study of vascular endothelial growth factor polymorphisms with the risk of recurrent spontaneous abortion. Fertility and Sterility. 2010;**93**(4):1244-1247

[36] Goodman C, Jeyendran RS, Coulam CB. Vascular endothelial growth factor gene polymorphism and implantation failure. Reproductive Biomedicine Online. 2008;**16**(5):720-723

[37] Goodman C, Jeyendran RS, Coulam CB. P53 tumor suppressor factor, plasminogen activator inhibitor, and vascular endothelial growth factor gene polymorphisms and recurrent implantation failure. Fertility and Sterility. 2009;**92**(2):494-498

[38] Xu X et al. Association of VEGF genetic polymorphisms with recurrent spontaneous abortion risk: A systematic review and meta-analysis. PLoS One. 2015;**10**(4):e0123696

[39] Su MT, Lin SH, Chen YC. Genetic association studies of angiogenesis- and vasoconstriction-related genes in women with recurrent pregnancy loss: A systematic review and meta-analysis. Human Reproduction Update. 2011;**17**(6):803-812

[40] Kang HJ et al. Single-nucleotide polymorphisms in the p53 pathway regulate fertility in humans. Proceedings of the National Academy of Sciences of the United States of America. 2009;**106**(24):9761-9766

[41] Coulam CB, Kay C, Jeyendran RS. Role of p53 codon 72 polymorphism in recurrent pregnancy loss. Reproductive Biomedicine Online. 2006;**12**(3):378-382

[42] Yenicesu GI et al. A prospective case-control study analyzes 12 thrombophilic gene mutations in Turkish couples with recurrent pregnancy loss. American Journal of Reproductive Immunology. 2010;**63**(2):126-136

[43] Jia M, Dahlman-Wright K, Gustafsson JA. Estrogen receptor alpha and beta in health and disease. Best Practice & Research. Clinical Endocrinology & Metabolism. 2015;**29**(4):557-568

[44] Yuan R, Le AW. A study on the estrogen receptor alpha gene polymorphism and its expression in thin endometrium of unknown etiology. Gynecologic and Obstetric Investigation. 2012;**74**(1):13-20

[45] Hey NA et al. The polymorphic epithelial mucin MUC1 in human endometrium is regulated with maximal expression in the implantation phase. The Journal of Clinical Endocrinology and Metabolism. 1994;**78**(2):337-342

[46] Xu B et al. Pinopodes, leukemia inhibitory factor, integrin-beta3, and mucin-1 expression in the peri-implantation endometrium of women with unexplained recurrent pregnancy loss. Fertility and Sterility. 2012;**98**(2):389-395

[47] Meseguer M et al. Human endometrial mucin MUC1 is up-regulated by progesterone and down-regulated in vitro by the human blastocyst. Biology of Reproduction. 2001;**64**(2):590-601

[48] Brayman M, Thathiah A, Carson DD. MUC1: A multifunctional cell surface component of reproductive tissue epithelia. Reproductive Biology and Endocrinology. 2004;**2**:4

[49] Carson DD, DeSouza MM, Regisford EG. Mucin and proteoglycan functions in embryo implantation. BioEssays. 1998;**20**(7):577-583

[50] Audie JP et al. Mucin gene expression in the human endocervix. Human Reproduction. 1995;**10**(1):98-102

[51] Nollet S et al. Human mucin gene MUC4: Organization of its 5′-region and polymorphism of its central tandem repeat array. The Biochemical Journal. 1998;**332**(Pt 3):739-748

[52] Ferrell AD et al. Sialomucin complex (Muc4) expression in porcine endometrium during the oestrous cycle

and early pregnancy. Reproduction in Domestic Animals. 2003;**38**(1):63-65

[53] Carraway KL, Idris N. Regulation of sialomucin complex/Muc4 in the female rat reproductive tract. Biochemical Society Transactions. 2001;**29**(Pt 2):162-166

[54] Coulam CB, Jeyendran R, Roussev R. Association of progesterone receptor polymorphisms with recurrent implantation failure after in vitro fertilization and embryo transfer. Journal of Assisted Reproduction and Genetics. 2008;**25**(4):119-122

[55] Salazar LA et al. Association of −765G>C polymorphism of the COX-2 gene with recurrent embryo implantation failure in Southern Chilean women. Clinica Chimica Acta. 2010;**411**(21-22):1822-1824

[56] Wang D et al. DNA hypomethylation of the COX-2 gene promoter is associated with up-regulation of its mRNA expression in eutopic endometrium of endometriosis. European Journal of Medical Research. 2012;**17**:12

[57] Pihusch R et al. Thrombophilic gene mutations and recurrent spontaneous abortion: Prothrombin mutation increases the risk in the first trimester. American Journal of Reproductive Immunology. 2001;**46**(2):124-131

[58] Goodman CS et al. Which thrombophilic gene mutations are risk factors for recurrent pregnancy loss? American Journal of Reproductive Immunology. 2006;**56**(4):230-236

[59] Vacquier VD. Oogenesis: Developmental biology. Science. 1985;**229**(4718):1078-1079

[60] Qublan HS et al. Acquired and inherited thrombophilia: Implication in recurrent IVF and embryo transfer failure. Human Reproduction. 2006;**21**(10):2694-2698

[61] Haroun S et al. Association between trefoil factor 3 gene variants and idiopathic recurrent spontaneous abortion. Reproductive Biomedicine Online. 2014;**29**(6):737-744

[62] Singh M, Chaudhry P, Asselin E. Bridging endometrial receptivity and implantation: Network of hormones, cytokines, and growth factors. The Journal of Endocrinology. 2011;**210**(1):5-14

[63] Croy BA et al. Decidual natural killer cells: Key regulators of placental development (a review). Journal of Reproductive Immunology. 2002;**57**(1-2):151-168

[64] Abrahamsohn PA, Zorn TM. Implantation and decidualization in rodents. The Journal of Experimental Zoology. 1993;**266**(6):603-628

[65] Gleicher N, Kushnir VA, Barad DH. Redirecting reproductive immunology research toward pregnancy as a period of temporary immune tolerance. Journal of Assisted Reproduction and Genetics. 2017;**34**(4):425-430

[66] Lash GE, Robson SC, Bulmer JN. Review: Functional role of uterine natural killer (uNK) cells in human early pregnancy decidua. Placenta. 2010;**31**(Suppl):S87-S92

[67] Bulmer JN, Williams PJ, Lash GE. Immune cells in the placental bed. The International Journal of Developmental Biology. 2010;**54**(2-3):281-294

[68] King A, Loke YW. Effect of IFN-gamma and IFN-alpha on killing of human trophoblast by decidual LAK cells. Journal of Reproductive Immunology. 1993;**23**(1):51-62

[69] Dosiou C, Giudice LC. Natural killer cells in pregnancy and recurrent pregnancy loss: Endocrine and immunologic perspectives. Endocrine Reviews. 2005;**26**(1):44-62

[70] Koopman LA et al. Human decidual natural killer cells are a unique NK cell subset with immunomodulatory potential. The Journal of Experimental Medicine. 2003;**198**(8):1201-1212

[71] Rachmilewitz J et al. Negative regulation of T cell activation by placental protein 14 is mediated by the tyrosine phosphatase receptor CD45. The Journal of Biological Chemistry. 2003;**278**(16):14059-14065

[72] Poole JA, Claman HN. Immunology of pregnancy. Clinical Reviews in Allergy & Immunology. 2004;**26**(3):161-170

[73] Runić R et al. Apoptosis and Fas expression in human fetal Membranes1. The Journal of Clinical Endocrinology & Metabolism. 1998;**83**(2):660-666

[74] Aschkenazi S et al. Differential regulation and function of the Fas/Fas ligand system in human trophoblast Cells1. Biology of Reproduction. 2002;**66**(6):1853-1861

[75] Saito S, Sasaki Y, Sakai M. CD4(+) CD25high regulatory T cells in human pregnancy. Journal of Reproductive Immunology. 2005;**65**(2):111-120

[76] La Rocca C et al. The immunology of pregnancy: Regulatory T cells control maternal immune tolerance toward the fetus. Immunology Letters. 2014;**162** (1 Pt A):41-48

[77] Guerin LR, Prins JR, Robertson SA. Regulatory T-cells and immune tolerance in pregnancy: A new target for infertility treatment? Human Reproduction Update. 2009;**15**(5):517-535

[78] Heitmann RJ et al. Maternal T regulatory cell depletion impairs embryo implantation which can be corrected with adoptive T regulatory cell transfer. Reproductive Sciences. 2017;**24**(7):1014-1024

[79] Dimova T et al. Maternal Foxp3 expressing CD4+ CD25+ and CD4+ CD25- regulatory T-cell populations are enriched in human early normal pregnancy decidua: A phenotypic study of paired decidual and peripheral blood samples. American Journal of Reproductive Immunology. 2011;**66**(Suppl 1):44-56

[80] Haiqi H, Yong Z, Yi L. Transcriptional regulation of Foxp3 in regulatory T cells. Immunobiology. 2011;**216**(6):678-685

[81] Jasper MJ, Tremellen KP, Robertson SA. Primary unexplained infertility is associated with reduced expression of the T-regulatory cell transcription factor Foxp3 in endometrial tissue. Molecular Human Reproduction. 2006;**12**(5):301-308

[82] Betz AG. Tolerating pregnancy. Nature. 2012;**490**:47

[83] Weetman AP. The immunology of pregnancy. Thyroid. 1999;**9**(7):643-646

[84] Clark DA, Croitoru K. TH1/TH2,3 imbalance due to cytokine-producing NK, gammadelta T and NK-gammadelta T cells in murine pregnancy decidua in success or failure of pregnancy. American Journal of Reproductive Immunology. 2001;**45**(5):257-265

[85] Saito S. Cytokine network at the feto-maternal interface. Journal of Reproductive Immunology. 2000;**47**(2):87-103

[86] Szekeres-Bartho J. Immunological relationship between the mother and the fetus. International Reviews of Immunology. 2002;**21**(6):471-495

[87] Piccinni MP, Romagnani S. Regulation of fetal allograft survival by a hormone-controlled Th1- and Th2-type cytokines. Immunologic Research. 1996;**15**(2):141-150

[88] Piccinni MP et al. Role of hormone-controlled Th1- and Th2-type cytokines

in successful pregnancy. Journal of Neuroimmunology. 2000;**109**(1):30-33

[89] Szekeres-Bartho J, Wegmann TG. A progesterone-dependent immunomodulatory protein alters the Th1Th2 balance. Journal of Reproductive Immunology. 1996;**31**(1):81-95

[90] Daftary GS et al. Direct regulation of beta3-integrin subunit gene expression by HOXA10 in endometrial cells. Molecular Endocrinology. 2002;**16**(3):571-579

[91] Taylor HS et al. HOXA10 is expressed in response to sex steroids at the time of implantation in the human endometrium. The Journal of Clinical Investigation. 1998;**101**(7):1379-1384

[92] Kim JJ et al. Altered expression of HOXA10 in endometriosis: Potential role in decidualization. Molecular Human Reproduction. 2007;**13**(5):323-332

[93] Ruiz-Alonso M, Blesa D, Simon C. The genomics of the human endometrium. Biochimica et Biophysica Acta. 2012;**1822**(12):1931-1942

[94] Sonderegger S, Pollheimer J, Knofler M. Wnt signalling in implantation, decidualisation and placental differentiation--review. Placenta. 2010;**31**(10):839-847

[95] Troy PJ et al. Transcriptional repression of peri-implantation EMX2 expression in mammalian reproduction by HOXA10. Molecular and Cellular Biology. 2003;**23**(1):1-13

[96] Germeyer A et al. Endometrial beta3 integrin profile reflects endometrial receptivity defects in women with unexplained recurrent pregnancy loss. Reproductive Biology and Endocrinology. 2014;**12**:53

[97] Fischer CP, Kayisili U, Taylor HS. HOXA10 expression is decreased in endometrium of women with adenomyosis. Fertility and Sterility. 2011;**95**(3):1133-1136

[98] Cermik D, Selam B, Taylor HS. Regulation of HOXA-10 expression by testosterone in vitro and in the endometrium of patients with polycystic ovary syndrome. The Journal of Clinical Endocrinology and Metabolism. 2003;**88**(1):238-243

[99] Szczepanska M et al. Expression of HOXA-10 and HOXA-11 in the endometria of women with idiopathic infertility. Folia Histochemica et Cytobiologica. 2011;**49**(1):111-118

[100] Lessey BA et al. Regulated expression of heparin-binding EGF-like growth factor (HB-EGF) in the human endometrium: A potential paracrine role during implantation. Molecular Reproduction and Development. 2002;**62**(4):446-455

[101] Stavreus-Evers A et al. Co-existence of heparin-binding epidermal growth factor-like growth factor and pinopodes in human endometrium at the time of implantation. Molecular Human Reproduction. 2002;**8**(8):765-769

[102] Yoo HJ, Barlow DH, Mardon HJ. Temporal and spatial regulation of expression of heparin-binding epidermal growth factor-like growth factor in the human endometrium: A possible role in blastocyst implantation. Developmental Genetics. 1997;**21**(1):102-108

[103] Birdsall MA et al. Expression of heparin-binding epidermal growth factor messenger RNA in the human endometrium. Molecular Human Reproduction. 1996;**2**(1):31-34

[104] Chobotova K et al. Heparin-binding epidermal growth factor and its receptor ErbB4 mediate implantation of the human blastocyst. Mechanisms of Development. 2002;**119**(2):137-144

[105] Leach RE et al. Multiple roles for heparin-binding epidermal growth factor-like growth factor are suggested by its cell-specific expression during the human endometrial cycle and early placentation. The Journal of Clinical Endocrinology and Metabolism. 1999;**84**(9):3355-3363

[106] Cullinan EB et al. Leukemia inhibitory factor (LIF) and LIF receptor expression in human endometrium suggests a potential autocrine/paracrine function in regulating embryo implantation. Proceedings of the National Academy of Sciences of the United States of America. 1996;**93**(7):3115-3120

[107] Aghajanova L et al. Coexpression of pinopodes and leukemia inhibitory factor, as well as its receptor, in human endometrium. Fertility and Sterility. 2003;**79**(Suppl 1):808-814

[108] Diaz-Gimeno P et al. A genomic diagnostic tool for human endometrial receptivity based on the transcriptomic signature. Fertility and Sterility. 2011;**95**(1):50-60 60e1-15

[109] Guzeloglu-Kayisli O, Kayisli UA, Taylor HS. The role of growth factors and cytokines during implantation: Endocrine and paracrine interactions. Seminars in Reproductive Medicine. 2009;**27**(1):62-79

[110] Lessey BA et al. Distribution of integrin cell adhesion molecules in endometrial cancer. The American Journal of Pathology. 1995;**146**(3):717-726

[111] Steck T et al. Leukaemia inhibitory factor (LIF) gene mutations in women with unexplained infertility and recurrent failure of implantation after IVF and embryo transfer. European Journal of Obstetrics, Gynecology, and Reproductive Biology. 2004;**112**(1):69-73

[112] Lessey BA et al. Integrin adhesion molecules in the human endometrium. Correlation with the normal and abnormal menstrual cycle. The Journal of Clinical Investigation. 1992;**90**(1):188-195

[113] Lessey BA et al. Further characterization of endometrial integrins during the menstrual cycle and in pregnancy. Fertility and Sterility. 1994;**62**(3):497-506

[114] Lessey BA et al. Integrins as markers of uterine receptivity in women with primary unexplained infertility. Fertility and Sterility. 1995;**63**(3):535-542

[115] Dorostghoal M et al. Endometrial expression of beta3 integrin, calcitonin and plexin-B1 in the window of implantation in women with unexplained infertility. International Journal of Reproductive BioMedicine (Yazd, Iran). 2017;**15**(1):33-40

[116] von Wolff M et al. Endometrial osteopontin, a ligand of beta3-integrin, is maximally expressed around the time of the "implantation window". Fertility and Sterility. 2001;**76**(4):775-781

[117] Carson DD et al. Changes in gene expression during the early to mid-luteal (receptive phase) transition in human endometrium detected by high-density microarray screening. Molecular Human Reproduction. 2002;**8**(9):871-879

[118] Ponnampalam AP et al. Molecular classification of human endometrial cycle stages by transcriptional profiling. Molecular Human Reproduction. 2004;**10**(12):879-893

[119] Punyadeera C et al. Oestrogen-modulated gene expression in the human endometrium. Cellular and Molecular Life Sciences. 2005;**62**(2):239-250

[120] Talbi S et al. Molecular phenotyping of human endometrium distinguishes menstrual cycle phases and underlying biological processes in normo-ovulatory women. Endocrinology. 2006;**147**(3):1097-1121

[121] Bartel DP. MicroRNAs: Genomics, biogenesis, mechanism, and function. Cell. 2004;**116**(2):281-297

[122] Shukla GC, Singh J, Barik S. MicroRNAs: Processing, maturation, target recognition and regulatory functions. Molecular and Cellular Pharmacology. 2011;**3**(3):83-92

[123] Lim LP et al. Microarray analysis shows that some microRNAs downregulate large numbers of target mRNAs. Nature. 2005;**433**(7027):769-773

[124] Lee RC, Feinbaum RL, Ambros V. The C. elegans heterochronic gene lin-4 encodes small RNAs with antisense complementarity to lin-14. Cell. 1993;**75**(5):843-854

[125] Weber JA et al. The microRNA spectrum in 12 body fluids. Clinical Chemistry. 2010;**56**(11):1733-1741

[126] Pogribny IP. MicroRNAs as biomarkers for clinical studies. Experimental Biology and Medicine (Maywood, NJ). 2018;**243**(3):283-290

[127] Pan Q, Chegini N. MicroRNA signature and regulatory functions in the endometrium during normal and disease states. Seminars in Reproductive Medicine. 2008;**26**(6):479-493

[128] Hull ML, Nisenblat V. Tissue and circulating microRNA influence reproductive function in endometrial disease. Reproductive Biomedicine Online. 2013;**27**(5):515-529

[129] Katzorke N et al. Diagnosis of endometrial-factor infertility: Current approaches and new avenues for research. Geburtshilfe und Frauenheilkunde. 2016;**76**(6):699-703

[130] Vilella F et al. Hsa-miR-30d, secreted by the human endometrium, is taken up by the pre-implantation embryo and might modify its transcriptome. Development. 2015;**142**(18):3210-3221

[131] Cuman C et al. Human blastocyst secreted microRNA regulate endometrial epithelial cell adhesion. eBioMedicine. 2015;**2**(10):1528-1535

[132] Kuokkanen S et al. Genomic profiling of microRNAs and messenger RNAs reveals hormonal regulation in microRNA expression in human endometrium. Biology of Reproduction. 2010;**82**(4):791-801

[133] Altmae S et al. MicroRNAs miR-30b, miR-30d, and miR-494 regulate human endometrial receptivity. Reproductive Sciences. 2013;**20**(3):308-317

[134] Revel A et al. MicroRNAs are associated with human embryo implantation defects. Human Reproduction. 2011;**26**(10):2830-2840

[135] Shi C et al. Endometrial MicroRNA signature during the window of implantation changed in patients with repeated implantation failure. Chinese Medical Journal. 2017;**130**(5):566-573

[136] Estella C et al. miRNA signature and dicer requirement during human endometrial stromal decidualization in vitro. PLoS One. 2012;**7**(7):e41080

[137] Zullo J et al. The cell secretome, a mediator of cell-to-cell communication. Prostaglandins & Other Lipid Mediators. 2015;**120**:17-20

[138] Casado-Vela J et al. Comprehensive proteomic analysis

of human endometrial fluid aspirate. Journal of Proteome Research. 2009;**8**(10):4622-4632

[139] Boomsma CM et al. Endometrial secretion analysis identifies a cytokine profile predictive of pregnancy in IVF. Human Reproduction. 2009;**24**(6):1427-1435

[140] Demiral I et al. Genomic, proteomic and lipidomic evaluation of endometrial receptivity. Turkish Journal of Obstetrics and Gynecology. 2015;**12**(4):237-243

[141] Berlanga O et al. How endometrial secretomics can help in predicting implantation. Placenta. 2011;**32**(Suppl 3):S271-S275

[142] Vilella F, Ramirez LB, Simon C. Lipidomics as an emerging tool to predict endometrial receptivity. Fertility and Sterility. 2013;**99**(4):1100-1106

[143] Vilella F et al. PGE2 and PGF2alpha concentrations in human endometrial fluid as biomarkers for embryonic implantation. The Journal of Clinical Endocrinology and Metabolism. 2013;**98**(10):4123-4132

[144] Peter Durairaj RR et al. Deregulation of the endometrial stromal cell secretome precedes embryo implantation failure. Molecular Human Reproduction. 2017;**23**(8):582

[145] Ansbacher R, Boyson WA, Morris JA. Sterility of the uterine cavity. American Journal of Obstetrics and Gynecology. 1967;**99**(3):394-396

[146] Perez-Munoz ME et al. A critical assessment of the "sterile womb" and "in utero colonization" hypotheses: Implications for research on the pioneer infant microbiome. Microbiome. 2017;**5**(1):48

[147] Moreno I et al. Evidence that the endometrial microbiota has an effect on implantation success or failure. American Journal of Obstetrics and Gynecology. 2016;**215**(6):684-703

[148] Aagaard K et al. The placenta harbors a unique microbiome. Science Translational Medicine. 2014;**6**(237):237ra65

[149] Verstraelen H et al. Characterisation of the human uterine microbiome in non-pregnant women through deep sequencing of the V1-2 region of the 16S rRNA gene. Peer Journal. 2016;**4**:e1602

[150] Franasiak JM et al. Endometrial microbiome at the time of embryo transfer: Next-generation sequencing of the 16S ribosomal subunit. Journal of Assisted Reproduction and Genetics. 2016;**33**(1):129-136

[151] Mitchell CM et al. Colonization of the upper genital tract by vaginal bacterial species in nonpregnant women. American Journal of Obstetrics and Gynecology. 2015;**212**(5):611.e1-611.e9

[152] Moreno I, Franasiak JM. Endometrial microbiota-new player in town. Fertility and Sterility. 2017;**108**(1):32-39

[153] Ravel J et al. Vaginal microbiome of reproductive-age women. Proceedings of the National Academy of Sciences of the United States of America. 2011;**108**(Suppl 1):4680-4687

[154] Sha AG, Liu JL, Jiang XM, Ren JZ, Ma CH, Lei W, Su RW, Yang ZM. Genome-wide identification of micro-ribonucleic acids associated with human endometrial receptivity in natural and stimulated cycles by deep sequencing. Fertil Steril. 2011;**96**(1):150-155 e155

Chapter 3

Aneuploidy Rates Inversely Correlate with Implantation during *In Vitro* Fertilization Procedures: In Favor of PGT

Elizabeth Schaeffer, Leonardo Porchia, Almena López-Luna, Dinorah Hernández-Melchor and Esther López-Bayghen

Abstract

Aneuploidy, the hold of an abnormal number of chromosomes that differs from the normal karyotype, is a recognized leading cause of miscarriage and congenital disabilities. In human gametes and embryos, aneuploidy rates are prevalent, and these rates increase with advanced maternal age; additionally, it has been suggested that hormonal stimulation for achieving *in vitro* fertilization (IVF) protocols further increases aneuploidy rates. Although about 65% of chromosomally abnormal embryos culminate in spontaneous miscarriages, there is still evidence of live births harboring crucial aneuploidies. Furthermore, although some frequent aneuploidies are consistent, others differ between countries, making it harder to focus on a specific set of anomalies but vital to focus regionally on those more prevalent. Preimplantation genetic testing (PGT) is a highly endorsed technique in assisted reproductive treatments to evaluate possible embryo aneuploidies, genetic defects, and congenital disorders. On this subject, this study shows that IVF aneuploidy rates in embryo cohorts of high morphological quality are inversely associated with implantation rates. In its entirety, this study reinforces the utility of PGT for embryo evaluation.

Keywords: aneuploidy, preimplantation genetic testing, embryo implantation, *in vitro* fertilization, karyotype

1. Introduction

Aneuploidy is defined as a chromosome number that is not an exact multiple of the usually haploid number [1]. The terms haploid and diploid that describe single (n) and double (2n) chromosome sets in cells originate from the Greek terms *haplóos* meaning single and *diplóos* meaning double. The term ploidy was subsequently derived to describe the total chromosome content of cells. Consequently, the term euploid refers to a chromosome with an exact multiple of the haploid number [2]. Human body cells (somatic cells) are diploid, carrying two complete sets of chromosomes: one set of 23 chromosomes from their father and one set of 23 chromosomes from their mother; the two sets combined provide a full complement of 46 chromosomes. Human gametes (or sex cells), sperm and oocytes, are haploid and contain only one set of 23 chromosomes.

Aneuploidies can occur either by chromosome gains (trisomies) and losses (monosomies) due to chromosome segregation errors, the so-called "whole chromosomal" aneuploidy or due to rearrangements of chromosomal parts, often accompanied by deletions, amplifications, or translocations of large regions of the genome that is referred to as a "structural" or "segmental" aneuploidy [3]. Whole chromosomal aneuploidies might arise due to random and sporadic chromosome missegregation events that occur with low frequency during any cell division. The missegregation levels range from 1/1000 to 1/10,000 in human cells [4].

Meiosis generates haploid gametes through a specialized cell division process that consists of one round of DNA replication followed by two cell divisions. The first division, or meiosis I (MI), involves the segregation of homologous chromosomes from each other, whereas meiosis II (MII) involves the segregation of the sister chromatids. Missegregation can also occur in germline cells, and the errors that arise in meiosis result in aneuploid embryos [5]. This chapter aims to provide evidence that supports the use of PGT for embryo evaluation and euploid embryo selection due to a positive correlation with fertilization rates.

2. Incidence of aneuploidy

Errors in meiotic chromosome segregation frequently occur during oogenesis (~20%), especially during the first meiotic division; this incidence of meiotic errors in oocytes is more elevated in women with advanced maternal age and may be due to the prolonged time that oocytes spend arrested at meiosis I stage, before ovulation [6]. However, some patterns of nondisjunction appear to be chromosome-specific; almost all cases of trisomy 16 are linked to errors at maternal MI, while MII errors are surprisingly common in trisomy 18. Oppositely, in sperm the incidence of aneuploidy is only 2%. Another considerable percentage of errors (~20%) arise during the first mitosis after fertilization. Among clinically recognized spontaneous abortions (fetal deaths occurring between 6 and 8 weeks and 20 weeks gestation), the incidence increases to ~50% [7]; the most common specific abnormalities are sex-chromosome monosomy (45,X), accounting for nearly 10% of all spontaneous abortions, and trisomies 16, 21, and 22, which together constitute 50% of all trisomies identified in spontaneous abortions. The incidence among stillbirths (fetal

Chr	Case report	References
1	Pure duplication 1q41-qter: further delineation of trisomy 1q syndromes	[9]
	Partial duplication 1q: reports of five cases and review of the literature	[10, 11]
2	Duplication 2q2.1-q3.1	[12, 13];
	Partial trisomy 2q: two cases	[14]
3	Duplication 3p syndrome: reports of three cases and review of the literature	[15–17]
4	Patient with trisomy 4p	[18]
	Partial duplication and duplication region 4q28.3-qter in monozygotic twins with discordant phenotypes	[19, 20]
5	Trisomy 5p: reports of four cases report and review of the literature	[21–23]
6	De novo "pure" partial trisomy (6)(p22.3→pter): case report/review	[24]
	Familial trisomy 6p in mother and daughter	[25]
7	Interstitial de novo tandem duplication of 7 (q31.1-q35)	[26]
	New case of pure partial 7q duplication	[27]

Chr	Case report	References
8	Trisomy 8: report of four cases	[28]
9	Pure 9p trisomy derived from a terminal balanced unreciprocal translocation Trisomy 9: review and report of two new cases	[29] [30]
10	Distal 10q trisomy with copy number gain in chromosome region 10q23.1–10q25.1 Proximal 10q duplication in a child with severe central hypotonia	[31] [32]
11	Partial 11q trisomy syndrome: two cases	[33, 34]
12	Clinical report of a patient with de novo trisomy 12q23.1q24.33	[35]
13	Trisomy 13, Patau's syndrome: reports of three cases	[36–38]
14	Partial proximal trisomy 14	[39]
15	Duplication of distal 15q: reports of 14 cases	[40, 41]
16	Partial trisomy/long arm of chromosome 16: case report/review of literature Complete trisomy 16: a case report	[42] [43]
17	A 790 kb chromosome 17p13.3 microduplication: case report/literature review	[44, 45]
18	Trisomy 18, Edward's syndrome: reports of five cases and discussion	[46–48]
19	Three cases of trisomy 19	[49, 50]
20	20q11.2 duplication syndrome and pure trisomy 20p	[51, 52]
21	Cardiovascular and general health status of adults with trisomy 21	[53]
22	Trisomy 22 syndrome: a report of four cases in newborns and literature review	[54–56]
X	Fragile X syndrome: a case report/review of clinical and molecular diagnoses	[57, 58]
Y	Morphology and pathogenesis of 47,XYY/47,XY patients super male syndrome	[59]

Chr, chromosome number.

Table 1.
Case reports of live births with complete or partial chromosomal abnormalities (mosaic or multiple aberrations are not considered).

deaths occurring between ~20 weeks gestation and term) is ~4% with the types of abnormality being similar to those identified in newborns, and ~0.3% of live-born are aneuploid with the most common abnormalities being trisomies 21, 18, and 13 and sex-chromosome trisomies 47,XXX, 47,XXY, and 47,XYY [5, 8].

2.1 Aneuploidies and live births

Although about 65% of chromosomally abnormal embryos culminate in spontaneous miscarriages, there is still evidence of live births harboring crucial aneuploidies. **Table 1** describes cases that are well documented.

3. Impact of aneuploidy on the efficiency of ART

Assisted reproduction is a solution in many of the growing cases of infertile couples worldwide. A high rate of embryos produced *in vitro* presents chromosomal aneuploidy (~50%), and such aneuploid embryos have reduced the potential for achieving a viable pregnancy. Such abnormalities are recognized as the leading cause of implantation failure and spontaneous miscarriage [60]. Among conceptions that survive to term, aneuploidy is the leading genetic cause of developmental disabilities and mental retardation [5]. **Table 2** describes data from different infertility centers predominantly showing that aneuploidy rates are similar.

The relatively high aneuploidy rate observed in human embryos after an IVF/ICSI cycle has been attributed to the technique itself since this prevalence seems to be lower in natural conceptions [61]. Many hypotheses have been proposed that may explain these findings: (1) controlled ovarian stimulation treatments, (2) factors related to the ICSI technique and (3) lab conditions as embryo culture.

3.1 Ovarian stimulation and the incidence of embryo aneuploidy

To increase the number of oocytes that can be retrieved for IVF, gonadotrophins are commonly used for superovulation in humans. Exogenous administration of gonadotrophins results in higher concentrations of steroids that may affect oocyte and embryo quality. Ovarian stimulation effects have been well characterized mainly in the murine model and have shown that aggressive stimulation leads to a poorer embryo development potential that could increase the chromosomal abnormality rate [79]. In humans, studies are scarce and less conclusive. A recent study in a population of young normovulatory women showed that a high ovarian response after controlled ovarian stimulation with moderate gonadotropin doses did not increase the embryo aneuploidy rate. Indeed, the higher the ovarian response, the more the euploid embryos obtained [80]; the remaining question is whether this can also be extrapolated to infertile patients with good ovarian reserve.

3.2 Intracytoplasmic sperm injection (ICSI) technique and the incidence of embryo aneuploidy

ICSI has become critical for the treatment of severe male infertility. The principal feature of ICSI is the direct injection of spermatozoa into an oocyte, which facilitates the production of fertilized embryos regardless of semen characteristics, such as sperm concentration and motility. However, the chromosomal integrity of ICSI zygotes is degraded compared to zygotes obtained from an *in vitro* fertilization [81, 82]. During the ICSI procedure, a sperm pretreatment is performed to mimic the conditions of natural fertilization and support the progression of fertilization effects. Studies on mouse models revealed that the chromosomal integrity of zygotes derived from ICSI without any pretreatment of spermatozoa was impaired in comparison with zygotes derived from conventional IVF [83]; even the culture sperm conditions may affect the chromosomal stability of the embryo [84]. Chromosomal damage may occur due to the injection of non-capacitated, acrosome-intact spermatozoa, so to reduce the risk of chromosomal aberrations during the ICSI procedure, it is crucial that sperm capacitation and the acrosome reaction be appropriately artificially induced in the proper medium before use [85].

3.3 Embryo culture and the incidence of embryo aneuploidy

Fertilization and embryo development *in vitro* have the potential to introduce (often inadvertently) stress which cannot only impair embryo development in the laboratory but also have downstream effects after transfer.

In vivo, the developing preimplantation embryo is exposed to gradients of nutrients, hormones, cytokines, and growth factors as it progresses through the fallopian tube to the uterus. Within the lumen of the female tract, the embryo resides in a few 100 nanoliters of a complex viscous fluid characterized by high levels of mucins, albumin, and glycosaminoglycans and by reduced levels of oxygen

N samples (country)	Day of biopsy	Aneuploidy rate	Trisomy %	Monosomy %	Most affected chromosomes	Less affected chromosomes	Ref.
87 (India)	3	54	14.9*	42.5*	22, 18	No data	[62]
150 (Japan)	5	40.6	18*	21.3*	15, 22, 21, 16, 18	4, 12	[63]
52 (UK)	5	40.4	51.3#	48.7#	22, 16, 15, 18, 21, X	1, 2, 5, 10, 17, 19	[64]
12 (UK)	3	75	22*	11*	20, 21, 22	6	[65]
5879 (USA)	3	70.6	No data	No data	No data	No data	[66]
	5	47.8					
759 (UK)	3	64.6	40#	60#	16, 22, 21, 4, 5	4, 6	[67]
274 (US)	3	72.3	39.8*	44.5*	22, 16, 7	6, 9, 19	[68]
192 (Italy)	5	55.2	37.9*	42.7*	No data	No data	[69]
240 (Mexico, Center A)	5	36.3	5.34*	~5*	16, 22, XXX, 9	3, 7, 8, 10, 12, 18, 20	[70]
	6	61.1	5.55*	~10*			
210 (Mexico, Center B)	5	48.9	15.96*	~4*	15, 16, 21, 4	1, 2, 3, 5, 8, 9, 10, 11, 12, 14, 17, 20, 22, X, Y	
	6	43.1	14.65*	~5*			
404 (Mexico)	No data	60.89	No data	No data	4, 15, 22, 16	No data	[71]
15,169 (USA)	No data	No data	No data	No data	13, 15, 16, 18, 19, 21, 22	1, 12, 3	[72]
2204 (UK)	0	74	56#	44#	16, 21, 22, 15, 19	No data	[73]
	3	83	49#	51#	22, 16, 19, 21, 13		
	5	58	47#	53#	22, 16, 15, 21, 19		
21 sets (Italy)	0/1	97.4	Chromosomal loss three times more frequent than gain		22, 15, 16, 17	No data	[74]
	3	47.6+					
	5 or 6	80++					
195 (USA)	0/1	65.5	39.86#	60.14#	22, 13, 15, 16, 19, 21	6, 5, Y, 3	[75]
	5/6	45.2	52#	48#	22, X, 16, 18, 21		
1025 (Mexico)	3/5	43.9	59.3#	40.7#	16, 21, 22, 19, 15, 20	8, 4, 3, 2, 7, 1	Current study

**Percentage of the total number of samples.*
#Percentage of the total number of aneuploid samples.
+Rate from the previous stage of development, PBs to blastomere.
++Rate from the previous stage of development blastomere to TE, PBs = polar bodies, TE = trophectoderm.
For the current study, infertile patients who underwent ART at the Ingenes Institute were included. The patients were clinically evaluated according to a standardized protocol that includes family and personal clinical history. The protocol was approved by the Ethics Committee of the Ingenes Institute, and a signed informed consent was obtained from all patients. IVF, embryo biopsy, and mCGH were performed according to the standard protocols of the Institute Ingenes as previously described [76, 77]. Only optimal morphological embryos were considered for this study. Selection and embryo transfer were done on Day 3 or Day 5 of development according to the embryo morphological assessment, using the criteria established by the Istanbul consensus Workshop on Embryo Assessment [78].

Table 2.
Aneuploidy rates of different IVF clinics around the world; when mentioned, the most commonly affected chromosomes are listed.

(typically 2–8%). The embryo is in constant motion, moved by gentle ciliated and muscular action of the female tract [86]. This scenario is in stark contrast to the laboratory environment, where typical gametes and embryos are exposed to relatively large volumes of culture medium, remain static during culture while resting on a polystyrene substrate, and create unstirred layers where the end products of metabolism concentrate and nutrients become limited [87].

Embryos are sensitive to both chemical and physical signals within their microenvironment. Factors within the laboratory as oxygen level, ammonium released from amino acids into the culture, poor laboratory air quality, temperature and pH, oil overlay, embryo culture volume/density, the static nature of culture, light, or even mechanical factors as pipetting, can negatively impact gametes and embryos and generate stress. When more than one stress factor is present in the laboratory, more negative synergies can result, and these factors play a significant role in influencing the development and events post transfer [88]. For example, recent studies have reported that a decrease in temperature has the potential to affect the stability of the oocyte's meiotic spindle, reducing fertilization rates, delaying embryo development, and decreasing clinical pregnancy rates [89]. However, more studies are needed to demonstrate the impact of embryo culture on aneuploidy rates.

4. Aneuploidy detection: techniques for PGT

PGT is the genetic diagnosis analysis performed to identify euploid embryos before uterine transfer [90]. PGT determines the numeric chromosomal constitution of a cellular biopsy sample obtained from a cultured embryo to determine its competence [91, 92].

PGT was first described in 1990 by Handyside et al. [93] when the sex of the six- to eight-cell stage embryos from two couples with a known risk of transmitting X-linked diseases was assessed by DNA amplification of a Y chromosome-specific repeat sequence. The earliest PGT studies in the 2000s were based on the fluorescence in situ hybridization (FISH) technique where 3–12 chromosomes can be analyzed on the cleavage stage or polar body biopsies [90]. Those studies had disappointing results in clinical practice since it had no beneficial effect on live birth rate after IVF [94]. The major drawback of FISH-based PGT is the limited number of chromosomes that can be analyzed considering that aneuploidy can affect any of the 22 autosomes and both sex chromosomes [95]; consequently, there have been dramatic improvements in PGT technology to make it valuable for clinical practice.

Nowadays, several methodologies for 24-chromosome analysis are available for clinical use that aim to increase implantation rates and decrease miscarriage rates associated with IVF [90]: microarray comparative genomic hybridization (mCGH), single-nucleotide polymorphism (SNP) microarray, real-time polymerase chain reaction (qPCR), and next-generation sequencing (NGS) [96, 97]. This review will focus on the relevant aspects of the PGT techniques used in our laboratory.

4.1 Microarray comparative genomic hybridization (mCGH)

mCGH is a ratio labeling protocol to compare the DNA product of a clinical sample to a healthy control. For PGT, biopsied embryonic cells must be lysed to extract the sample's DNA, which will be further amplified by a protocol that provides whole genome coverage [90, 95, 98]. The resulting DNA products are co-hybridized with a standard DNA control sample (46,XY and 46,XX) with a series of site-specific fluorophores on a microarray chip with approximately 4000 markers spaced throughout the genome [90]. Then, a confocal laser platform detects the relative color intensity, and a bioinformatics compares the intensity of each fluorophore in the sample versus the control to identify any bias and determine the ploidy status of the sample [90, 95, 98].

The mCGH analysis reports the ratio of sample DNA to a reference DNA, as a chromosomic profile where the molecular karyotype is represented. Usually, the sample DNA is labeled with a green fluorescent dye, while the reference DNA sample is tagged red [99]. Thus, diploid embryos will have a relatively equal ratio of green-to-red fluorescence in every pair of chromosomes, represented as a continuous horizontal plot line. Monosomy will be represented as a clear downward deviation in the plotted line, indicating a relative lack of green-to-red signal intensity; in contrary, a trisomy will be displayed as an upward deviation in the plotted line due to a relative increase in the green-to-red signal intensity.

The specificity rate of mCGH-based PGT is about 99% [90]. The test results can be available within 12–15 h, considering that the entire analysis can be performed during this short time frame [90, 91]. Additionally, brand-specific features are offered by each manufacturer: Agilent's GenetiSure Pre-Screen Microarray offers a detection rate of 100% for aberrations >10 Mb and 89% for >5.3 Mb [100]; KaryoLite BoBs Kit from Perkin Elmer uses an alternative BACs-on-Beads technology and results are interpreted by the BoBsoft™ analysis software [99]; and RHS's EmbryoCellect Kit is the only mCGH-based PGT validated for mosaicism detection [101]. Recently, Illumina's 24sure PGS Microarray had been discontinued, and the NGS-based VeriSeq PGS is now offered as an alternative solution [102].

mCGH entails some disadvantages: first, the embryo sample requires a previous whole genome amplification (WGA) process to support single-cell diagnostics by mCGH [95], raising the possibility of introducing errors during the amplification [91]; second, mCGH is a semiquantitative technique that only reports the ratio of sample DNA to a reference DNA; it is to say that only imbalances in DNA content can be identified. Therefore, mCGH is unable to detect uniparental disomic or triploid embryos as it cannot discriminate between 46,XX from 69,XXX, and 46, XY from 69,XXY [90, 91, 95]. Last, the mCGH used for PGS cannot identify structural chromosome aberrations or diagnose mosaicism in a trophectoderm sample [90].

4.2 Next-generation sequencing (NGS)

NGS refers to the emerging technology of non-Sanger-based DNA sequencing that allows the sequence in parallel millions of DNA strands with high-throughput yield. In the field of ART, this powerful tool is being applied for PGT to replace cytogenetic microarrays [98, 102].

Different platforms are commercially available for NGS with different technological approaches. Illumina's MiSeq NGS platform applies a sequencing-by-synthesis method, where DNA is attached and amplified in situ to be subsequently used as a template for synthetic sequencing with fluorescent-labeled reversible-terminator nucleotides [103]. Ion Torrent NGS technology, commercialized by ThermoFisher Scientific, is based on collecting data by sensing the hydrogen ions that are released as by-products when nucleotides are incorporated by a template-directed DNA polymerase synthesis on an ion chip [104].

Despite the dissimilarities between platforms, the common basis of chromosome copy number analysis by NGS is the fragmentation of the amplified DNA sample into small segments of 100–200 base pairs that are further sequenced in parallel until the number of reads covering a determined position in the genome is attained, in general, a 30× coverage (sequencing each base pair 30 times) ensures sufficient accuracy. The sequence data obtained are then compared with a reference genome and counted by bioinformatics software. The copy number of a specific chromosome should be proportional to the number of counted sequences; therefore, an increase or reduction in the number of reads will, respectively, represent a trisomy or monosomy [97, 99].

NGS allows to simultaneously perform both qualitative and quantitative analyses of multiple embryos with high-resolution data for chromosomal analysis [96, 97]. The higher sensitivity and precision offered by NGS [96, 105, 106] makes possible to exclude embryos with mosaicism [105, 106] and partial aneuploidies or triploidies [106], improving pregnancy outcomes due to its enhanced capability for detecting those challenging abnormalities.

PGT by NGS can predict not only chromosome copy number for the diagnosis of whole chromosome aneuploidy with 99.98% assignment consistency [97] but also single-gene disorders [107], abnormalities of the mitochondrial genome [108], and segmental chromosome imbalances [97, 99]. Balanced chromosomal rearrangements cannot be detected by NGS [97].

The increasing demand and accelerated development are continuously reducing the cost of NGS technology [109]. Also, potential cost-benefit ratios can be achieved when the full sequencing capacity of the apparatus is exploited [96, 97, 99]. Furthermore, molecular tools, like barcoding, are being implemented to allow multiplex high-throughput sequencing [110]; this promising strategy will reduce the diagnosis' cost per patient by performing simultaneous analysis of multiple embryos from different patients [97].

5. Aneuploidy and women age

In our study, by analyzing the mCGH data, the total number of aneuploidies was found to be 734, and from these, 641 (87.3%) were derived from patients and 93 (12.7%) from donors. Overall, this study displayed similar rates of monosomies, trisomies, double aneuploidies, and multiple aneuploidies. The total number of monosomies (191) was similar to the number of trisomies (194), accounting for 26 and 26.4% of the total aneuploidies, correspondingly. Furthermore, the total number of double (165) and multiple (184) aneuploidies was also very similar, accounting for 22.5 and 25.1% of the total aneuploidies, correspondingly. Nevertheless, it is worth noticing that when considering only the donor group, monosomies seem to be more prevalent: 38.7% of the total donors' aneuploidies were monosomies vs. 24.7% of trisomies, 16.1% of double aneuploidies, and 20.4% of multiple aneuploidies; what is more, the percentage of monosomies in the donor group is higher than that of the monosomies of the patient group (38.7 vs. 24.3%). The most common monosomies affected chromosomes 15, 16, and 22, whereas the most common trisomy affected chromosomes 16, 19, and 21 (**Table 3**).

It has been shown that the lowest risk for embryonic aneuploidy is between ages 26 and 30, with aneuploidy rates steadily increasing with maternal age after 26 years of age [111] and leaping significantly from the age of 39 [112]. For this reason, women of advanced maternal age are encouraged to favor oocyte donation to yield high-quality viable embryos.

Interestingly, some studies have identified that women of younger ages possess an increased prevalence of aneuploidy, with >40% of aneuploidy in women of 23 years and under [111] and 58% of aneuploidy in women of <31 years of age. In the current study, both the donor (≤29 years) and the patient group of ≤29 years

	All	Patients	Donors
Total	734	641	93
Monosomy	191 (26.0%)	155 (24.3%)	36 (38.7%)
−Chr 15	13	13	0
−Chr 16	21	18	3
−Chr 22	22	22	0
−Chr X	10	8	2
−Chr Y	40	26	14
Trisomy	194 (26.4%)	171 (26.7%)	23 (24.7%)
+Chr 16	26	24	2
+Chr 18	9	7	2
+Chr 19	23	21	2
+Chr 20	17	13	4
+Chr 21	20	20	0
+Chr 22	19	19	0
+Chr X	10	8	2
+Chr Y	1	1	0
Dual	165 (22.5%)	150 (23.4%)	15 (16.1%)
Multiple	184 (25.1%)[a]	165 (25.7%)[b]	19 (20.4%)[c]

The current study included 441 patients, resulting in 474 cycles. A total of 1629 embryos were analyzed; from those, 54 were excluded due to failed WGA, leaving 1575 embryos for analysis, 1258 from patients, and 317 from donors. Biopsies were performed at the blastomere (Day 3, patients = 238 and donors = 50) and blastocyst stages (Day 5, patients = 1020 and donors = 267). Finally, 734 embryos (46.6%) were found to be aneuploid (patients = 641 and donors = 93). The total number of monosomies and trisomies is provided along with their respective percentages; furthermore, the number of the most common aneuploidies of the mCGH data is listed.
[a]Nine embryos had completely abnormal mCGH profiles.
[b]Even embryos had completely abnormal mCGH profiles.
[c]Two embryos had completely abnormal mCGH profiles.

Table 3.
Most frequent types of aneuploidies in the mCGH data of the current study.

displayed high aneuploid rates, 28.5 and 27.4%, respectively (**Table 4**). Given the high rates of aneuploidy in younger women, attention should be paid in detecting aneuploidy in embryos from women of young maternal age, especially since this group of patients is not routinely encouraged to perform a PGT. Still, whether there is a difference between the distribution of aneuploidies between donors and patients remains uncertain.

When stratifying our analysis in age groups (a, ≤29; b, 30–34; c, 35–37; d, 38–40; e, 41–43; and f, ≥44 years of age), a visible continuous increase in aneuploidy rate can be observed as maternal age increases (**Table 4**); furthermore, this increase in aneuploidy goes hand in hand with a continuous decrease in implantation, as it can be observed in the decrease of positive beta-human chorionic gonadotropin (β-hCG) values as age increases (β-hCG values ≥10 mUI/ml from Day 14 after transference were considered positive).

Category	≤29 a	30–34 b	35–37 c	38–40 d	41–43 e	≥44 f	Donors g
Number of samples (n, cycles)	27	66	72	105	90	31	83
Age (years)	$27.0 \pm 2.4^{b,c,d,e,f,g}$	$32.5 \pm 1.3^{a,c,d,e,f,g}$	$36.1 \pm 0.7^{a,b,d,e,f,g}$	$39.0 \pm 0.8^{a,b,c,e,f,g}$	$41.8 \pm 0.8^{a,b,c,d,f,g}$	$45.0 \pm 1.6^{a,b,c,d,e,g}$	$22.8 \pm 3.0^{a,b,c,d,e,f}$
Body mass index (kg/m^2)	23.6 ± 3.8^{g}	24.6 ± 3.8^{g}	24.3 ± 3.9^{g}	24.4 ± 3.7^{g}	24.8 ± 3.8^{g}	24.6 ± 2.8^{g}	$21.8 \pm 2.5^{a,b,c,d,e,f}$
Ova collected (n)	$16.8 \pm 7.9^{e,f}$	$17.3 \pm 9.8^{d,e,f}$	$16.1 \pm 9.0^{d,e,f}$	$12.0 \pm 6.3^{b,c,g}$	$11.0 \pm 5.9^{a,b,c,g}$	$8.9 \pm 6.0^{a,b,c,g}$	$15.8 \pm 8.1^{d,e,f}$
Ova fertilized (n)	$14.3 \pm 6.9^{e,f}$	$15.2 \pm 8.6^{d,e,f}$	$13.6 \pm 7.5^{d,e,f}$	$10.3 \pm 5.2^{b,c,g}$	$9.8 \pm 5.5^{a,b,c,g}$	$7.6 \pm 4.8^{a,b,c,g}$	$13.9 \pm 6.2^{d,e,f}$
Embryos (n)	$11.2 \pm 6.3^{e,f}$	$11.7 \pm 7.9^{d,e,f}$	$10.6 \pm 6.0^{d,e,f}$	$7.7 \pm 4.2^{b,c,g}$	$7.5 \pm 4.6^{a,b,c,g}$	$6.2 \pm 3.9^{a,b,c,g}$	$10.4 \pm 4.6^{d,e,f}$
Fertilization rate (%)	77.9 ± 17.0	76.3 ± 15.1	79.7 ± 14.2	76.6 ± 16.6	77.0 ± 18.9	83.6 ± 15.2	76.1 ± 16.0
Aneuploidy rate (%)	$27.4 \pm 33.3^{d,e,f}$	$36.0 \pm 29.2^{d,e,f}$	$35.7 \pm 30.3^{d,e,f}$	$57.7 \pm 34.9^{a,b,c,f,g}$	$66.6 \pm 37.7^{a,b,c,g}$	$87.4 \pm 19.5^{a,b,c,d,g}$	$28.5 \pm 28.7^{d,e,f}$
Pregnancy rate (%)	$60.87^{b,c,d,e,f,g}$	$50.99^{a,c,d,e,f,g}$	$49.09^{a,b,d,e,f,g}$	$44.00^{a,b,c,e,f,g}$	$55.17^{a,b,c,d,f,g}$	$42.86^{a,b,c,d,e,g}$	$50.79^{a,b,c,d,e,f}$

Values are shown as mean ± standard error. Significance was determined by one-way ANOVA followed by a Bonferroni or Dunnett's T3 post hoc *test. Superscripts indicate a significant difference ($p < 0.05$, two-tailed):*
[a] *Versus 29 years old group.*
[b] *Versus 30–34 years old group.*
[c] *Versus 35–37 years old group.*
[d] *Versus 38-40 years old group.*
[e] *Versus 41–43 years old group.*
[f] *Versus 44 years old group.*
[g] *Versus donors.*

Table 4.
Comparative of aneuploidy and pregnancy rates between age groups.

DOI: http://dx.doi.org/10.5772/intechopen.81884

6. Remarks

One of the most critical reasons for unsuccessful IVF procedures is implantation failure due to aneuploid embryos. Aneuploidies are the primary cause of perinatal death and genetic abnormalities; consequently, the detection of chromosomal disorders constitutes the most frequent indication for PGT. Here, we report on the aneuploidy rates found in IVF procedures in Mexico. Even though there are studies that assert that PGT does not improve pregnancy rates, we show that aneuploidy rates inversely correlate with implantation and that levels of aneuploidy among high morphological quality embryos are still an important issue to be faced in everyday ART practice, and this evidence works in favor of continuing to use PGT analysis.

Acknowledgements

This work was supported by the Consejo Nacional de Ciencia y Tecnología (Conacyt-Programa de Estímulos a la Innovación 2018-253224).

Abbreviations

ANOVA	analysis of variance
ART	assisted reproductive techniques
Chr	chromosome
FISH	fluorescence in situ hybridization
ICSI	intracytoplasmic sperm injection
IVF	*in vitro* fertilization
mCGH	microarray comparative genomic hybridization
MI	meiosis I
MII	meiosis II
NGS	next-generation sequencing
OR	odds ratio
PGS	preimplantation genetic screening
PGT	preimplantation genetic testing
qPCR	real-time polymerase chain reaction
SNP	single-nucleotide polymorphism
WGA	whole genome amplification
β-hCG	beta-human chorionic gonadotropin

Author details

Elizabeth Schaeffer[1], Leonardo Porchia[1,3], Almena López-Luna[1,2], Dinorah Hernández-Melchor[1,2] and Esther López-Bayghen[1,2]*

1 Laboratorio de Investigación y Diagnóstico Molecular, Instituto de Infertilidad y Genética SC, INGENES, CDMX, México

2 Departamento de Toxicología, Cinvestav-IPN, CDMX, México

3 Tecnologías DAAT SA de CV, Xalapa, Veracruz, México

*Address all correspondence to: ebayghen@cinvestav.mx

References

[1] Torres EM, Williams BR, Amon A. Aneuploidy: Cells losing their balance. Genetics. 2008;**179**(2):737-746

[2] Orr B, Godek KM, Compton D. Aneuploidy. Current Biology. 2015; **25**(13):R538-R542

[3] Storchova Z. The causes and consequences of aneuploidy in eukaryotic cells. In: Aneuploidy in Health and Disease. Croatia: InTech; 2012. pp. 3-22

[4] Hassold T, Hall H, Hunt P. The origin of human aneuploidy: Where we have been, where we are going. Human Molecular Genetics. 2007;**16**:R203-R208. Spec no. 2

[5] Hassold T, Hunt P. To err (meiotically) is human: The genesis of human aneuploidy. Nature Reviews. Genetics. 2001;**2**(4):280-291

[6] Battaglia DE et al. Influence of maternal age on meiotic spindle assembly in oocytes from naturally cycling women. Human Reproduction. 1996;**11**(10):2217-2222

[7] Kim JW et al. Chromosomal abnormalities in spontaneous abortion after assisted reproductive treatment. BMC Medical Genetics. 2010;**11**:153

[8] Hassold T et al. Human aneuploidy: Incidence, origin, and etiology. Environmental and Molecular Mutagenesis. 1996;**28**(3):167-175

[9] Kulikowski LD et al. Pure duplication 1q41-qter: Further delineation of trisomy 1q syndromes. American Journal of Medical Genetics. Part A. 2008;**146A**(20):2663-2667

[10] Campos TCaF, Puntero B. Trisomía parcial 1q por translocación materna. Anales Españoles de Pediatría. 2000;**52**: 178-184

[11] Rasmussen SA et al. Partial duplication 1q: Report of four patients and review of the literature. American Journal of Medical Genetics. 1990;**36**(2): 137-143

[12] Acosta Sabatés MM, Hernández García I, García Martínez DA, Lavaut Sánchez K. Duplication 2 q 2.1- q 3.1: A case report. Revista Cubana de Pediatría. 2008;**80**:1

[13] Sebold CD et al. Partial trisomy 2q: Report of a patient with dup (2) (q33.1q35). American Journal of Medical Genetics. Part A. 2005;**134A**(1):80-83

[14] Schumacher RE, Rocchini AP, Wilson GN. Partial trisomy 2q. Clinical Genetics. 1983;**23**(3):191-194

[15] Dworschak GC et al. Comprehensive review of the duplication 3q syndrome and report of a patient with Currarino syndrome and de novo duplication 3q26.32-q27.2. Clinical Genetics. 2017;**91**(5):661-671

[16] Natera-de Benito D et al. A patient with a duplication of chromosome 3p (p24.1p26.2): A comparison with other partial 3p trisomies. American Journal of Medical Genetics. Part A. 2014;**164A**(2): 548-550

[17] Charrow J, Cohen MM, Meeker D. Duplication 3p syndrome: Report of a new case and review of the literature. American Journal of Medical Genetics. 1981;**8**(4):431-436

[18] Varley J, Wehner T, Sisodiya S. Diaphragm myoclonus followed by generalised atonia in a patient with trisomy 4p: Unusual semiology in an unusual condition. Epileptic Disorders. 2015;**17**(4):473-477

[19] Collia A et al. Partial duplication of chromosome 4 in a patient with bilateral ocular coloboma. Archivos Argentinos de Pediatría. 2012;**110**(4):e59-e62

[20] Celle L et al. Duplication of chromosome region 4q28.3-qter in monozygotic twins with discordant phenotypes. American Journal of Medical Genetics. 2000;**94**(2):125-140

[21] Velagaleti GV, Morgan DL, Tonk VS. Trisomy 5p. A case report and review. Annales de Génétique. 2000;**43** (3–4):143-145

[22] Fujita M et al. A new case of "complete" trisomy 5p with isochromosome 5p associated with a de novo translocation t(5;8)(q11;p23). Clinical Genetics. 1994;**45**(6):305-307

[23] Orye E, Benoit Y, van Mele B. Complete trisomy 5p owing to de novo translocation t(5;22)(q11;p11) with isochromosome 5p associated with a familial pericentric inversion of chromosome 2, inv 2(p21q11). Journal of Medical Genetics. 1983;**20**(5):394-396

[24] Sivasankaran A et al. De-novo 'pure' partial trisomy (6)(p22.3→pter): A case report and review of the literature. Clinical Dysmorphology. 2017;**26**(1): 26-32

[25] Savarese M et al. Familial trisomy 6p in mother and daughter. American Journal of Medical Genetics. Part A. 2013;**161A**(7):1675-1681

[26] Zelante L et al. Interstitial "de novo" tandem duplication of 7(q31.1-q35): First reported case. Annales de Génétique. 2003;**46**(1):49-52

[27] Alfonsi M et al. A new case of pure partial 7q duplication. Cytogenetic and Genome Research. 2012;**136**(1):1-5

[28] Mellado C, Moreno R, López F, Sanz P, Castillo S, Villaseca C, et al. Trisomia 8: Reporte de cuatro casos. Revista Chilena de Pediatría. 1997;**53**(2):93-98

[29] Brambila-Tapia AJ et al. Pure 9p trisomy derived from a terminal balanced unreciprocal translocation. Genetic Counseling. 2014;**25**(3):289-297

[30] Arnold GL et al. Trisomy 9: Review and report of two new cases. American Journal of Medical Genetics. 1995;**56**(3): 252-257

[31] Wong SL et al. Distal 10q trisomy with copy number gain in chromosome region 10q23.1-10q25.1: The Wnt signaling pathway is the most pertinent to the gene content in the region of copy number gain: A case report. BMC Research Notes. 2015;**8**:250

[32] Manolakos E et al. Proximal 10q duplication in a child with severe central hypotonia characterized by array-comparative genomic hybridization: A case report and review of the literature. Experimental and Therapeutic Medicine. 2014;**7**(4):953-957

[33] Utine GE et al. Partial trisomy 11q syndrome (11q23.1→11qter) due to de novo t (11q; 13q) detected by multicolor fluorescence in situ hybridisation. Genetic Counseling. 2005;**16**(3):291-295

[34] Pihko H, Therman E, Uchida IA. Partial 11q trisomy syndrome. Human Genetics. 1981;**58**(2):129-134

[35] Geckinli BB et al. Clinical report of a patient with de novo trisomy 12q23.1q24.33. Genetic Counseling. 2015;**26**(4):393-400

[36] Oka N et al. Norwood procedure performed on a patient with trisomy 13. International Heart Journal. 2016;**57**(1): 121-122

[37] Tunca Y, Kadandale JS, Pivnick EK. Long-term survival in Patau syndrome.

Clinical Dysmorphology. 2001;**10**(2): 149-150

[38] Lu HT, Han XH. One case report of Patau syndrome. Zhonghua Er Ke Za Zhi. 2011;**49**(7):555-556

[39] Dutta UR, Pidugu VK, Dalal A. Partial proximal trisomy 14: Identification and molecular characterization in a girl with global developmental delay. Genetic Counseling. 2013;**24**(2):207-216

[40] Lacro RV et al. Duplication of distal 15q: Report of five new cases from two different translocation kindreds. American Journal of Medical Genetics. 1987;**26**(3):719-728

[41] Schnatterly P et al. Distal 15q trisomy: Phenotypic comparison of nine cases in an extended family. American Journal of Human Genetics. 1984;**36**(2): 444-451

[42] Laus AC et al. Karyotype/phenotype correlation in partial trisomies of the long arm of chromosome 16: Case report and review of literature. American Journal of Medical Genetics. Part A. 2012;**158A**(4):821-827

[43] Aviña Fierro JA, Blum ER, Aviña DAH. Trisomía 16 completa. Reporte de un caso clínico. Revista Mexicana de Pediatría. 2005;**72**(5):237-239

[44] Ho AC et al. A newborn with a 790 kb chromosome 17p13.3 microduplication presenting with aortic stenosis, microcephaly and dysmorphic facial features—Is cardiac assessment necessary for all patients with 17p13.3 microduplication? European Journal of Medical Genetics. 2012;**55**(12):758-762

[45] Belligni EF et al. 790 kb microduplication in chromosome band 17p13.1 associated with intellectual disability, afebrile seizures, dysmorphic features, diabetes, and hypothyroidism. European Journal of Medical Genetics. 2012;**55**(3):222-224

[46] Saldarriaga W, Rengifo-Miranda H, Ramirez-Cheyne J. Trisomy 18 syndrome: A case report. Revista Chilena de Pediatría. 2016;**87**(2):129-136

[47] Zellweger H, Beck K, Hawtrey CE. Trisomy 18. Report of a case and discussion of the syndrome. Archives of Internal Medicine. 1964;**113**:598-605

[48] Bharucha BA et al. Trisomy 18: Edward's syndrome (a case report of 3 cases). Journal of Postgraduate Medicine. 1983;**29**(2):129-132

[49] Jung SI et al. Two cases of trisomy 19 as a sole chromosomal abnormality in myeloid disorders. The Korean Journal of Laboratory Medicine. 2008;**28**(3): 174-178

[50] Humphries JE, Wheby MS. Trisomy 19 in a patient with myelodysplastic syndrome and thrombocytosis. Cancer Genetics and Cytogenetics. 1990;**44**(2): 187-191

[51] Avila M et al. Delineation of a new chromosome 20q11.2 duplication syndrome including the ASXL1 gene. American Journal of Medical Genetics. Part A. 2013;**161A**(7):1594-1598

[52] Sidwell RU et al. Pure trisomy 20p resulting from isochromosome formation and whole arm translocation. Journal of Medical Genetics. 2000; **37**(6):454-458

[53] Hayes SA et al. Cardiovascular and general health status of adults with trisomy 21. International Journal of Cardiology. 2017;**241**:173-176

[54] He X, Yao D, Zhao ZY. Trisomy 22 syndreom: A report of 2 cases. Zhongguo Dang Dai Er Ke Za Zhi. 2015; **17**(5):524-525

[55] Heinrich T et al. Live-born trisomy 22: Patient report and review. Molecular Syndromology. 2013;**3**(6):262-269

[56] Petersen MB, Hansen M, Djernes BW. Full trisomy 22 in a newborn infant. Annales de Génétique. 1987; **30**(2):101-104

[57] Ciaccio C et al. Fragile X syndrome: A review of clinical and molecular diagnoses. Italian Journal of Pediatrics. 2017;**43**(1):39

[58] Vazquez Gonzalez B. Turner syndrome; case report. Ginecología y Obstetricia de México. 1958;**13**(2): 103-110

[59] Song Y et al. Morphology and pathogenesis of 47, XYY/47, XY, +mar identified in patients with super male syndrome. Zhonghua Yi Xue Yi Chuan Xue Za Zhi. 2015;**32**(3):378-380

[60] Lathi RB, Westphal LM, Milki AA. Aneuploidy in the miscarriages of infertile women and the potential benefit of preimplanation genetic diagnosis. Fertility and Sterility. 2008; **89**(2):353-357

[61] Simon C et al. Increased chromosome abnormalities in human preimplantation embryos after in-vitro fertilization in patients with recurrent miscarriage. Reproduction, Fertility, and Development. 1998;**10**(1):87-92

[62] Majumdar G et al. Preimplantation genetic screening for all 24 chromosomes by microarray comparative genomic hybridization significantly increases implantation rates and clinical pregnancy rates in patients undergoing *in vitro* fertilization with poor prognosis. Journal of Human Reproductive Sciences. 2016;**9**(2): 94-100

[63] Chung MK et al. Comprehensive chromosome analysis of blastocysts before implantation using array CGH. Molecular Cytogenetics. 2013;**6**(1):22

[64] Fragouli E et al. Cytogenetic analysis of human blastocysts with the use of FISH, CGH and aCGH: Scientific data and technical evaluation. Human Reproduction. 2011;**26**(2):480-490

[65] Wells D, Delhanty JD. Comprehensive chromosomal analysis of human preimplantation embryos using whole genome amplification and single cell comparative genomic hybridization. Molecular Human Reproduction. 2000;**6**(11):1055-1062

[66] Harton GL et al. Diminished effect of maternal age on implantation after preimplantation genetic diagnosis with array comparative genomic hybridization. Fertility and Sterility. 2013;**100**(6):1695-1703

[67] Gutierrez-Mateo C et al. Validation of microarray comparative genomic hybridization for comprehensive chromosome analysis of embryos. Fertility and Sterility. 2011;**95**(3): 953-958

[68] Rabinowitz M et al. Origins and rates of aneuploidy in human blastomeres. Fertility and Sterility. 2012; **97**(2):395-401

[69] Fiorentino F et al. Application of next-generation sequencing technology for comprehensive aneuploidy screening of blastocysts in clinical preimplantation genetic screening cycles. Human Reproduction. 2014; **29**(12):2802-2813

[70] Sánchez-Usabiaga RA, González Becerra J, Vera-Aguado MG, Ramírez EG, Durand-Montaño C. Incidence and parental origin of aneuploidies in blastocysts: Comparison between two centers of assisted reproduction in Mexico. Ginecología y Obstetricia de México. 2017;**85**(5):289-297

[71] López-Rioja MJ, Aguinaga-Ríos M, Sánchez-González CM, Recio-López Y, Zavala-González PN, García-Sánchez R, et al. Preimplantation genetic testing for aneuploidies PGT-A: Results of the transition between different technologies. Ginecología y Obstetricia de México. 2018;**86**(2):96-107

[72] Franasiak JM et al. Aneuploidy across individual chromosomes at the embryonic level in trophectoderm biopsies: Changes with patient age and chromosome structure. Journal of Assisted Reproduction and Genetics. 2014;**31**(11):1501-1509

[73] Fragouli E et al. The origin and impact of embryonic aneuploidy. Human Genetics. 2013;**132**(9): 1001-1013

[74] Capalbo A et al. Sequential comprehensive chromosome analysis on polar bodies, blastomeres and trophoblast: Insights into female meiotic errors and chromosomal segregation in the preimplantation window of embryo development. Human Reproduction. 2013;**28**(2):509-518

[75] Fragouli E et al. Comprehensive chromosome screening of polar bodies and blastocysts from couples experiencing repeated implantation failure. Fertility and Sterility. 2010; **94**(3):875-887

[76] Schaeffer E et al. Whole genome amplification of day 3 or day 5 human embryos biopsies provides a suitable DNA template for PCR-based techniques for genotyping, a complement of preimplantation genetic testing. BioMed Research International. 2017;**2017**:1209158

[77] Camargo-Diaz F et al. Colony stimulating factor-1 and leukemia inhibitor factor expression from current-cycle cannula isolated endometrial cells are associated with increased endometrial receptivity and pregnancy. BMC Womens Health. 2017; **17**(1):63

[78] Medicine, A.S.I.R. and E.S.I.G. Embryology. Istanbul consensus workshop on embryo assessment: Proceedings of an expert meeting. Reproductive Biomedicine Online. 2011; **22**(6):632-646

[79] Elbling L, Colot M. Abnormal development and transport and increased sister-chromatid exchange in preimplantation embryos following superovulation in mice. Mutation Research. 1985;**147**(4):189-195

[80] Labarta E et al. A higher ovarian response after stimulation for IVF is related to a higher number of euploid embryos. BioMed Research International. 2017;**2017**:5637923

[81] In't Veld PA et al. Intracytoplasmic sperm injection (ICSI) and chromosomally abnormal spermatozoa. Human Reproduction. 1997;**12**(4): 752-754

[82] Macas E, Imthurn B, Keller PJ. Increased incidence of numerical chromosome abnormalities in spermatozoa injected into human oocytes by ICSI. Human Reproduction. 2001;**16**(1):115-120

[83] Tateno H. Possible causal factors of structural chromosome aberrations in intracytoplasmic sperm injection of the mouse. Reproductive Medicine and Biology. 2009;**8**(3):89-95

[84] Tateno H, Kamiguchi Y. Evaluation of chromosomal risk following intracytoplasmic sperm injection in the mouse. Biology of Reproduction. 2007; **77**(2):336-342

[85] Watanabe H. Risk of chromosomal aberration in spermatozoa during intracytoplasmic sperm injection.

The Journal of Reproduction and Development. 2018 Oct 12;**64**(5):371-376. DOI: 10.1262/jrd.2018-040. Epub 2018 Jul 7

[86] Gardner DK et al. Environment of the preimplantation human embryo in vivo: Metabolite analysis of oviduct and uterine fluids and metabolism of cumulus cells. Fertility and Sterility. 1996;**65**(2):349-353

[87] Gardner DK, Lane M. Culture of viable mammalian embryos. In: Cibelli J, Wilmut I, Jaenisch R, Gurdon J, Lanza R, West M, Campbell K, editors. Principles of Cloning. 2nd ed. Elsevier; 2014. pp. 63-84. DOI: https://doi.org/10.1016/C2010-0-66663-6

[88] Wale PL, Gardner DK. The effects of chemical and physical factors on mammalian embryo culture and their importance for the practice of assisted human reproduction. Human Reproduction Update. 2016;**22**(1):2-22

[89] Wang WH et al. Limited recovery of meiotic spindles in living human oocytes after cooling-rewarming observed using polarized light microscopy. Human Reproduction. 2001;**16**(11):2374-2378

[90] Brezina PR, Anchan R, Kearns WG. Preimplantation genetic testing for aneuploidy: What technology should you use and what are the differences? Journal of Assisted Reproduction and Genetics. 2016;**33**(7):823-832

[91] Brezina PR, Kutteh WH. Clinical applications of preimplantation genetic testing. BMJ. 2015;**350**:g7611

[92] Brezina PR, Brezina DS, Kearns WG. Preimplantation genetic testing. BMJ. 2012;**345**:e5908

[93] Handyside AH et al. Pregnancies from biopsied human preimplantation embryos sexed by Y-specific DNA amplification. Nature. 1990;**344**(6268):768-770

[94] Mastenbroek S et al. Preimplantation genetic screening: A systematic review and meta-analysis of RCTs. Human Reproduction Update. 2011;**17**(4):454-466

[95] Kaser DJ, Ginsburg ES. Embryo biopsy for aneuploidy detection in the general infertility population. Seminars in Reproductive Medicine. 2014;**32**(2):100-106

[96] Imudia AN, Plosker S. The past, present, and future of preimplantation genetic testing. Clinics in Laboratory Medicine. 2016;**36**(2):385-399

[97] Fiorentino F et al. Development and validation of a next-generation sequencing-based protocol for 24-chromosome aneuploidy screening of embryos. Fertility and Sterility. 2014;**101**(5):1375-1382

[98] Kurahashi H et al. Preimplantation genetic diagnosis/screening by comprehensive molecular testing. Reproductive Medicine and Biology. 2016;**15**(1):13-19

[99] Sermon K. Novel technologies emerging for preimplantation genetic diagnosis and preimplantation genetic testing for aneuploidy. Expert Review of Molecular Diagnostics. 2017;**17**(1):71-82

[100] Agilent. GenetiSure Pre-Screen Kit Details & Specifications. Available from: https://www.genomics.agilent.com/article.jsp?pageId=8100002

[101] RHS. EmbryoCellect. Available from: http://www.rhsc.com.au/our-products/embryocellect/

[102] Illumina. 24sure PGS Microarray. Available from: https://support.illumina.com/array/array_kits/24sure-pgs-microarray-kit.html

[103] Bentley DR et al. Accurate whole human genome sequencing using reversible terminator chemistry. Nature. 2008;**456**(7218):53-59

[104] Rothberg JM et al. An integrated semiconductor device enabling non-optical genome sequencing. Nature. 2011;**475**(7356):348-352

[105] Fragouli E. Next generation sequencing for preimplantation genetic testing for aneuploidy: Friend or foe? Fertility and Sterility. 2018;**109**(4): 606-607

[106] Friedenthal J et al. Next generation sequencing for preimplantation genetic screening improves pregnancy outcomes compared with array comparative genomic hybridization in single thawed euploid embryo transfer cycles. Fertility and Sterility. 2018; **109**(4):627-632

[107] Treff NR et al. Evaluation of targeted next-generation sequencing-based preimplantation genetic diagnosis of monogenic disease. Fertility and Sterility. 2013;**99**(5):1377-1384 e6

[108] Fragouli E et al. Altered levels of mitochondrial DNA are associated with female age, aneuploidy, and provide an independent measure of embryonic implantation potential. PLoS Genetics. 2015;**11**(6):e1005241

[109] Yin X et al. Massively parallel sequencing for chromosomal abnormality testing in trophectoderm cells of human blastocysts. Biology of Reproduction. 2013;**88**(3):69

[110] Knapp M, Stiller M, Meyer M. Generating barcoded libraries for multiplex high-throughput sequencing. Methods in Molecular Biology. 2012; **840**:155-170

[111] Franasiak JM et al. The nature of aneuploidy with increasing age of the female partner: A review of 15,169 consecutive trophectoderm biopsies evaluated with comprehensive chromosomal screening. Fertility and Sterility. 2014;**101**(3):656-663 e1

[112] Lukaszuk K et al. Next generation sequencing for preimplantation genetic testing of blastocysts aneuploidies in women of different ages. Annals of Agricultural and Environmental Medicine. 2016;**23**(1):163-166

Chapter 4

Screening (Bi Test, Triple Test, Panorama Test) and Amniocentesis for Early Diagnosis of Congenital Malformations

Gladys Cristina Al Jashi and Isam Al Jashi

Abstract

The genetic consult is very important in the diagnosis of early fetal malformations and its complications at birth and after it. Our research is based on a 3-year research on 6097 pregnant women who underwent screening *Bi-Test or Triple Test*. We discovered 408 pregnant women who were found positive and needed amniocentesis for a diagnostic of certitude. Out of them, 14 had a positive result from which 10 were found with Down syndrome and 4 with Edwards syndrome. In Romania, amniocentesis has become the most used method of prenatal diagnosis for pregnant women at 35 or above with a family history of hereditary congenital anomalies. However, the latest screening test from maternal blood, the Panorama test, can discover many malformations (for chromosomes 21, 18, and 13 and the abnormality of the sex chromosome). The accuracy for false positive is 2% and false negative 98%. In that light, the purpose of our study is to decrease the use of amniocentesis and to introduce the latest tests (Panoramic) for the early diagnosis of fetal malformation, the use of maternal blood, and the avoidance of using invasive medical procedures.

Keywords: screening *Bi-Test*, *Triple Test*, *Panorama Test*, amniocentesis for early detection of fetal malformations

1. Introduction

Talking with the patient about his or her family medical history, such as if the couple had a person with Down syndrome in their family or any other malformation, is important. That can be considered as a 'genetic consultation' in order to find out about any possible genetic problems before screening or any other tests. The specialist tries, in the first place, to make the patient understand the nature of his disease he is confronting, what's the most possible evolution of it, and last but not least what are the possibilities of treating it. A major objective of the genetic consult is to make the patient understand what are the factors that lead to the development of the disease, what are the mechanisms that transmit the disease, and how high is the risk of appearance to other family members. The doctor establishes, after running different tests, the correct diagnosis of the disease, its evolution, its prognosis, as well as the possibilities of treatment for a higher quality of life.

The genetic consult is very important in the diagnosis of early fetal malformations and their complications at birth and after it. The consult is necessary for pregnant women with high risk in developing a malformation (e.g., if the woman has in the family history a newborn with malformations, if she gave birth before to a baby with malformations, if she lives in a toxic environment).

In Romania, it is mandatory that every pregnant woman should do the BI-TEST or TRIPLE TEST, depending on the pregnancy weeks.

The cytogenetic prenatal diagnosis refers to early detection of intrauterine fetal abnormalities and genetic malformations, very often determined by genetics, which complicate and threaten the life of 3–5% of newborn. These abnormalities explain the 20% deaths in the newborn period and the much higher percentage of deaths in childhood and puberty [1]. From the fetal abnormalities, 95% are unexpected and appear at pregnancies that are not considered at risk for these situations. They can be light or severe.

The severe abnormalities are not compatible with life or can cause a long-term handicap [2].

There are multiple purposes for the detection of intrauterine prenatal illness such as:

- Establishing the high-risk diagnosis and possibly the termination of the pregnancy (abortion), the pregnancy representing a high risk of death both for the mother and fetus
- Identifying a situation that can influence the date, the place, or the childbirth method
- Identifying fetuses that will benefit of early pediatric intervention
- Identifying situations that can affect future pregnancies

Opinions are different when it comes to knowing the diagnosis; many families decide to keep the pregnancy even if there are severe malformations, thinking that they can fight this problem by taking care of the baby and not terminating the pregnancy. But most of the families decide to terminate the pregnancies and seek treatments for future pregnancies.

In conclusion, the most important thing in the prenatal diagnosis is obtaining very early in the pregnancy of correct informations about the fetus and making a correct and easy decision in conformity with family principles [3].

The indications for prenatal diagnosis are:

- Pregnancy—women at age of 35 or above
- Husband aged 45 or above
- Past pregnancy with a fetus with chromosomal abnormality
- One of the parents has a chromosomal abnormality including being a carrier of a balanced translocation or other structural rearrangements
- Down syndrome
- Pregnancies with risk of Mendelian or polygenic diseases (e.g., Tay-Sachs disease)

- Abnormal fetus diagnosed by the ultrasound procedure within the first weeks of pregnancy

- Maternal anxiety

- Determination of fetal risk in pregnancies with high risk for hereditary x-linked diseases, for that there is not yet a specific prenatal diagnosis (it is obtained better by determining the Y-chromosomal material with the new ADN techniques)

- History of two or more abortions without the possibility of effectuation of the cytogenic analysis to both parents or a noninvasive prenatal diagnosis [4]

The screening programs focus on the most common chromosomal abnormality—the Down syndrome (trisomy 21)—and they not only detect very often other fetal abnormalities.

2. The BI-TEST

The screening of the first trimester is a prenatal text that offers informations about early pregnancy and the risk of the baby to develop some specific chromosomal abnormalities—the Down syndrome (trisomy 21) and the Edwards syndrome (trisomy 18).

The Down syndrome determines a lifetime affectation of the mental and social development and also some physical modifications. As for Edwards syndrome, it is a much more severe condition, and often it is fatal until the early age of 1 year [5].

The screening tests from the first trimester of pregnancy do not evaluate the risk of defects of the neural tube, such as spina bifida.

We do the BI-TEST (the screening of the first trimester in pregnancy) because it can be done much earlier than other tests of screening; the results can be available in the first weeks of pregnancy, which offers the parents much more time in taking the best decision regarding additional tests, medical treatment, and the course of pregnancy [6].

Normally, the screening of the first trimester (the BI-TEST) is made between weeks 11 and 14 of pregnancy. The results of the test are formulated as positive or negative and also like a probability, e.g., 1:5000 risk to have a baby with Down syndrome. In general, the test is considered positive if the risk is 1:300 or above. The BI-TEST implies two steps: a blood test that measures the level of two hormones that are specific in pregnancy (pregnancy-associated plasma protein A (PAPP-A) and human chorionic gonadotropin (hCG)) and an ultrasound examination which measures the size of the transparent space behind the neck of the fetus (the nuchal translucency) [7]. Correlating the values obtained from the blood tests with the ultrasound examination and some specific parameters of the mother (age, height, weight, blood pressure, smoking, diabetes, assisted reproductive techniques), the specialist can calculate the risk of the mother of having a baby with chromosomal abnormalities [6].

Frequently, a positive result will enforce the performance of secondary diagnosis tests:

- The chorial biopsy—it is done normally in the first trimester of pregnancy, and it is an invasive test which implies cropping a piece of the placenta, with a low risk of abortion. It is used to detect chromosomal abnormalities, like the Down syndrome.

- Amniocentesis—it can be used to identify chromosomal abnormalities but also defects of the neural tube, like spina bifida. During the amniocentesis, which frequently is done in the second trimester of pregnancy, a small amount of amniotic liquid is estranged. It also had a low risk of abortion [6].

3. The TRIPLE TEST

A continuance of the BI-TEST from the first trimester and the screening TRIPLE TEST (made in the second trimester of pregnancy) is more and more often requested by us. With the help of it, we can evaluate the risks of the baby developing a neural tube disorder (spina bifida) or chromosomal abnormalities such as the Down syndrome (trisomy 21) or the Edwards syndrome (trisomy 18) [6].

Between the 15 and 20 weeks of pregnancy (with optimal results in the weeks 16–18), pregnant women can be tested with a simple blood test known as the TRIPLE TEST (or triple marker test or second trimester screening). It is made in combination with an ultrasound examination [3].

For the triple marker testing, a blood test is required that will measure the levels of:

- hCG (human chorionic gonadotropin) secreted by the placenta.
- Unconjugated estriol, a product of the placenta and the fetus.
- Alpha-fetoprotein (AFP gene) produced by the fetus's liver.
- Sometimes it also tested the inhibin A (produced by the placenta), because it considerably grows the chances of detection of the Down syndrome, which is also known as the quadruple test.

The level of each substance present in the blood helps the doctor identify if the fetus is exposed to the risk of developing congenital disorders and chromosomal abnormalities. Exactly as in the case of the BI-TEST, multiples of the median (MoM) is measured depending on the clinical data of the mother (age, height, weight, race, smoker, age of the fetus) [4]. The results of the tests are available in 2 weeks from the blood tests. Usually, if the results of the TRIPLE TEST show a high level of alpha-fetoprotein, it indicates a neural tube defect. The doctor will recommend an ultrasound examination of the fetal cranium and spinal cord. The mother is also purposed to do an amniocentesis, which involves extracting a small amount of amniotic liquid from the uterus for further investigation.

On the other hand, low levels of alpha-fetoprotein (AFP) and unconjugated estriol indicate a high risk of Down syndrome. In this situation, it is recommended that an amniocentesis should be done.

4. Amniocentesis

Amniocentesis is a medical procedure used in prenatal diagnosis of chromosomal abnormalities, in which a small amount of amniotic fluid which contains fetal tissues is sampled from the amniotic sac and which is afterward tested.

It is a two-person procedure which requires a lot of experience and care from the side of the doctor. One of the doctors assures the ultrasound examination, and the other doctor places the puncture needle. This way, he can manipulate the needle with all the attention and care to attain the desired target. There has to be an incredible collaboration between both doctors, because they have to anticipate each other's moves.

Before the procedure, it is mandatory that the sanguine groups ABO and RH be determined and also to perform an ultrasound examination, which will determine the number of fetuses, the position and viability, the localization of the placenta, the amount of amniotic fluid, and the largest amniotic sac that will be punctured and will assure that the fetus is anatomically normal. Also the gestational age of the fetus by the fetal biparietal diameter, the length of the thigh bone, and the abdominal circumference is determined.

Choosing the place of puncture is made after the initial evaluation. When doing the amniocentesis, it is important to choose the place of the puncture needle in the amniotic cavity wisely, namely by carefully avoiding the placenta, the umbilical cord insertion place and the fetus. Colored Doppler ultrasound can facilitate the choosing of the place for the transplacental puncture, to avoid fetal hemorrhage.

Diagnosis of chromosomal abnormalities and diagnosis of the neural tube defects are the most common indications for amniocentesis in the beginning of the second trimester of pregnancy. As for the evaluation of the fetal pulmonary maturation, it is the most common indication for amniocentesis in the third trimester [8].

At the end of the first trimester and the beginning of the second trimester, the indications for amniocentesis are:

- Cytogenetic diagnosis
- Diagnosis of neural tube defects
- Diagnosis of metabolic disorders

At the end of the second trimester and the beginning of the third trimester, the indications for amniocentesis are:

- Evaluation of the severity of Rh immunization
- Evaluation of fetal pulmonary maturation
- Diagnosis of intra-amniotic infections
- Confirmation of the ruptured membranes

Last, but not the least, the therapeutic purpose of the amniocentesis has indications in the polyhydramnios drainage and the medical treatment of fetal diseases.

Ultrasound examination (localization of the placenta, fetus, and amniotic liquid bag) and the insertion of the needle can take up to 20 min. Extraction of the amniotic liquid can take up to 5 min.

Some women don't feel pain (painless procedure), while other pregnant women feel cramps as the needle punctures the uterus or a pressure sensation during the procedure of extracting amniotic liquid.

Local anesthetic is not normally used. Associated factors to the high level of pain during the amniocentesis are considered to be maternal anxiety, presence of menstrual cramps in the past, an anterior amniocentesis, and the insertion of the needle in the inferior part of the uterus, where the pain is less felt by the patient [9], although there aren't any data to support this practice [10].

5. Panoramic Test

The Panorama Test is a prenatal noninvasive test, which analyzes the fetal DNA from the maternal blood and establishes the risk of common fetal aneuploidies in pregnant women, most interestingly the 21, 18, and 13 chromosomes but also the analyzes the sexual chromosomes X and Y.

Aneuploidies are chromosomal abnormalities characterized by the alteration of the number of chromosomes, for example, more or less chromosomes instead of the exact number of chromosomes; the trisomy, when there is an extra chromosome; or monosomy, when there is missing a chromosome.

Trisomy 21 is caused by the presence of an extra chromosome to the 21st pair, and it's known as the Down syndrome. It is the most frequent cause of mental retardation. The patients with Down syndrome present cardiac disease or other diseases which require surgical interventions or specific medical treatment. Other symptoms are, for example, ophthalmological problems [11].

Trisomy 18 is caused by the presence of an extra chromosome to the 18th pair, and it's known as the Edwards syndrome. It is a serious cause of mental retardation. The majority of the children with trisomy 18 present severe cerebral malformations, severe heart malformations, or severe malformations of other organs. Often, the growth retardation is found, which can lead to the termination of pregnancy (abortion). Most of the children who are born with the growth retardation die within their first year of life. Patients who survive present severe intellectual disability and have problems of growth and development [12].

Trisomy 13 is caused by the presence of an extra chromosome to the 13th pair, and it's known as the Patau syndrome. It is a cause of severe mental retardation. The majority of the patients present serious cerebral and other organs abnormalities. In many situations the pregnancy doesn't get to its due date, and the newborn dies very fast. Half of the born babies die in their first month of life, and 90% from those who survive in the first year of life present heart malformations, renal malformations, and renal malformations.

Aneuploidies of the sexual chromosomes are caused by the presence or the absence of one of the sexual chromosomes. The test can detect the risk for XXX, XYY, XXYY, and XXY (the Klinefelter Syndrome) and for the X monosomy (the Turner syndrome). A significant variability exists in regarding the severity of these genetic conditions, but the majority of the affected individuals are characterized by slight modifications physically and mentally.

During pregnancy some fragments of the fetal DNA run through the maternal blood. The fetal DNA is detectable since the 5th week of pregnancy. The concentration increases successively in the next weeks of pregnancy and disappears right after childbirth. The circulating fetal DNA quantity is found from the 9th to 10th weeks of pregnancy, and it's sufficient to guarantee the relevance and the sensibility of the Panorama Test.

The test is done by collecting a blood sample from the pregnant woman from at least the 10th week of unique and twin pregnancies. In the laboratory a complex analysis is made, where the fetal DNA from the maternal blood is isolated. This procedure identifies the eventual chromosomal abnormalities, like aneuploidy, and cutting-edge sequencing technology is used, along with advanced bioinformatics analysis.

The results obtained in the Panorama Test are:

- "High risk" implies that there is a probability that the fetus can present an aneuploidy of the 21st, 18th, and 13th pairs of chromosomes or an aneuploidy of

X and Y sexual chromosomes. This result indicates that there is a high chance that the fetus can present one of the abovementioned abnormalities, but it is not sure the fetus will have that condition. The endorsed advice is an invasive prenatal diagnosis test, like chorionic villus biopsy or amniocentesis.

- "Low risk" implies that the test detected a risk <1/10,000 (0.01%) for the fetus to be affected by a chromosomal aneuploidy.

- In some cases, when the quantity of fetal DNA is not enough, the test can show.

- Incomplete results or no results. In this situation, a new blood sample will be collected for the repetition of the test. In case the sexual analysis is demanded, this result can also be provided.

The accuracy of the Panorama Test

	% diagnosis (sensibility)	% false positive (specificity)
Trisomy 21 (the Down syndrome)	99 (>99%)	<0.1% (>99%)
Trisomy 18 (the Edwards syndrome)	99 (>99%)	<0.1% (>99%)
Trisomy 13 (the Patau syndrome)	08/10/16	<0.1% (>99%)
Determination of the fetal sex	99 (>99%)	<0.1% (>99%)

The test can detect the risk of a sexual chromosomal aneuploidy with a rate of detection which varies depending on the type of detected aneuploidy: 13th, 18th, and 21st.

The test was validated for women with unique and twin pregnancies, with at least 10 weeks of pregnancy. This test evaluates just the chromosomal aneuploidies of the 13th, 18th, 21st, X, and Y. In the case of pregnancy with more fetuses, informations about the sexual chromosomes will not be provided. This procedure does not replace the invasive prenatal diagnosis (choriocentesis or amniocentesis).

Also, the analysis cannot highlight the equilibrated chromosomal rearrangements, partial abnormalities of the analyzed chromosomes, structural chromosome abnormalities, fetal or placental chromosomal mosaicism in a small percentage (the presence of two cellular lines with a different set of chromosomes, with a cellular line weakly represented), point mutations, methylation defects, and polyploidy. In particular, the analysis does not reveal the presence of hereditary genetic disorders with the Mendelian transmission. The test can give a false-positive result in the case of chromosomal mosaicism, but this thing can limit itself to the placenta. The sex of the fetus is mentioned as masculine or feminine, based on the presence or absence of the Y chromosome, not on the presence or absence of the SRY gene.

Pregnancies with abnormal ultrasound results should be studied with other types of prenatal investigations, such as fetal karyotype from the chorionic villi or amniotic liquid. Although this test is very precise, the results are not a diagnosis, and they should be evaluated in the clinical context of the patient and the family medical history. The test is limited to offer just informations about the risks of aneuploidies for the analyzed chromosomes.

A "low-risk" result can significantly reduce the possibility for the fetus to present an aneuploidy of the examined chromosomes, but it cannot guarantee that these chromosomes are normal or that the fetus is healthy.

This test can be done for 13th, 18th, and 21st aneuploidies also in case of twin pregnancies and of pregnancies obtained by fertilization in vitro.

This study seeks the degree of accuracy of the prenatal screening which indicates a high risk. The results are established after the confirmation or information of the diagnosis by the invasive diagnosis method for genetical abnormalities, the amniocentesis procedure. The interventions have been made by the obstetrician, helped by different doctors, based on the "2 operatives" technique. The puncture and the extraction of the amniotic liquid will be done by the obstetrician, and the doctor helping him would monitor by ultrasound the whole intervention.

Our study is based on a 3-year research (2012–2015). We evaluated 6097 pregnant women, from which 408 had an indication for effectuation of amniocentesis with the purpose of establishing a correct diagnosis. The dynamics of the number of pregnancies in the analyzed period presents significant decreases from 15%–14.65% in 2014 to 11.15% in 2015.

From the total amount of 408, 114 were made in the year 2012, 102 in the year 2013, 97 in the year 2014, and 95 in the year 2015.

Year	Number of pregnancies	Number of amniocentesis
2012	1641	114
2013	1596	102
2014	1402	97
2015	1458	95
TOTAL	**6097**	**408**

Comparing the dynamics of the number of amniocentesis procedures done in 2015 with the amniocentesis procedures done in 2012, they have a pronounced decreasing rhythm, so that in 2013 the level decreased by 10.53%, in 2014 the level decreased by 14.91%, and in 2015 the level decreased by 16.67% due do the apparition of the Panorama Test sampled from the maternal blood.

Year	Number of pregnancies	Number of amniocentesis procedures	Dynamics than 2012	
			Number of pregnancies	Number of amniocentesis procedures
2012	1641	114	–	–
2013	1596	102	97.26	89.47
2014	1402	97	85.44	85.09
2015	1458	95	88.85	83.33
Total	**6097**	**408**	–	–

By the fact that, in medical practice, age above 35 years old in pregnant women represents a risk factor; the patients have been divided in two groups such as patients aged until 35 years old and patients above 35 years old.

Group of age	2012	2013	2014	2015	TOTAL
Under 35 years old	74	66	57	54	251
Above 35 years old	40	36	40	41	157
Total number of patients	**114**	**102**	**97**	**95**	**408**

In the year of our study, 408 amniocentesis procedures have been made. Two hundred and fifty-one procedures were done on patients with age under 35 years old, which represents a percentage of 61.5%. One hundred and fifty-seven procedures were made on patients with age above 35 years old, with a percentage of 38.5%.

The distribution of the patients based on the pregnancy period and the number of amniocentesis procedures done in percentage:

Weeks of pregnancy	2012	2013	2014	2015	Total	%
Week 15	15	15	11	14	58	14.22
Week 16	36	34	32	39	141	34.56
Week 17	31	28	26	24	109	26.72
Week 18	14	13	IO	8	45	11.03
Week 19	9	5	10	6	30	7.35
Week 20	7	5	3	2	17	4.17
Week 21	2	2	2	2	8	1.96
Total	**114**	**102**	**97**	**95**	**408**	**100%**

Even though the number of amniocentesis procedures has slightly decreased, the number of BI-TESTS has significantly increased since year 2012. In year 2013, the increasing number has been a percentage of +115.4%, in 2013 it has been +315.45%, and in the year 2015, it has been 269.2%. In the last 2 years, the percentage for positive tests for Down syndrome in amniocentesis has been 50%. In all our period of research, 35.05% of the patients have effectuated the double-positive test for the Down syndrome.

Year	Number of amniocentesis procedures	Triple Test positive for Down syndrome	Triple Test positive for Edwards syndrome	Percentage of positive Triple Test (minus the number of amniocentesis procedures) for:	
				Down syndrome	Edwards syndrome
2012	114	93	5	81	4.39
2013	102	64	3	62.75	2.94
2014	97	36	1	37.11	1.03
2015	95	33	1	34.74	1.05
Total	**408**	**226**	**10**	**55.39**	**2.45**

In the period of our research, the percentage for Down syndrome and Edwards syndrome of the TRIPLE TEST decreased in relation to the number of amniocentesis procedures. In this period the result positive for these tests is considered:

- In 2012 the test for Down syndrome represented 81.58%, and 4.39% for Edwards syndrome.
- In 2013 the test for Down syndrome represented 62.75% and 2.94% for Edwards syndrome.

- In 2014 the test for Down syndrome represented 37.11% and 1.03% for the Edwards syndrome.

Although in 2012 the highest number of amniocentesis procedures (114) was recorded, a single case was recorded in which amniocentesis had as an indication the maternal age, which represents a percentage of 0.88%. In 2013, a slight increase with a percentage of 6.86% is noted (in 2014, 4.12%, and in 2015, 3.16%). In all, there were 15 procedures made after the indication of the maternal age above 35 years old, with a percentage of 3.68%.

Year	Number of amniocentesis procedures	Age of the mother >35	Percentage %
2012	114	1	0.88
2013	102	7	6.86
2014	97	4	4.12
2015	95	3	3.16
Total	**408**	**15**	**3.68**

The distribution of the amniocentesis procedures based on priors:

- Down syndrome in the family
- Prior pregnancy with Down syndrome
- Prior pregnancy with Turner syndrome

Year	Number of amniocentesis procedures	Down syndrome in the family	Prior pregnancy with Down syndrome	Prior pregnancy with Turner syndrome
2012	114	1	1	0
2013	102	0	0	0
2014	97	1	0	1
2015	**95**	**0**	**0**	**0**

It is observed that in 2012, from the amount of 114 amniocentesis procedures, 2 of them had as indication the pathological chromosomal history. One of the patients presents in her family history a member of the family who has Down syndrome. Another patient had a prior pregnancy with Down syndrome. These indications represent 1.75% of the indications for year 2012. In 2014 two more pregnant women have been diagnosed. One of them was coming from a family with Down syndrome, and the other one had a prior pregnancy with Turner syndrome. These indications represent a percentage of 2.61% for the year 2014.

From a total of 408 pregnancies, the indications for pathological chromosomal history represent a percentage of 0.97%.

Number of amniocentesis procedures	408	100%
BI-TEST	131	32.10%
TRIPLE TEST	258	63.23%
Indication for maternal age 35 years or above	15	3.67%

Down syndrome in the family history	2	0.49%
Prior pregnancy with Down syndrome	1	0.24%
Prior pregnancy with Turner syndrome	1	0.24%

From the amount of 408 pregnancies which underwent the amniocentesis procedure, 149 of them were on their first pregnancy, with a percentage of 36.52%. One hundred and ninety-eight of the pregnant women were on their second pregnancy (48.53%), and 61 were on their third or more pregnancy (14.95%).

Distribution of the patients depending on environmental residence

Environment	**2012**	**2013**	**2014**	**2015**	**Total**	**%**
Rural	16	34	25	20	95	23.28
Urban	98	68	72	75	313	76.72
Total	**114**	**102**	**97**	**95**	**408**	**100**

It is concluded that only 23.28% of the pregnant women came from a rural environment; the rest of 76.72% came from an urban environment.

Distribution of patients depending on chronic diseases

Chronic diseases	**Number of patients**	**%**
Diabetes mellitus	11	2.70
Hepatitis A and hepatitis B	5	1.23
High blood pressure	9	2.21
Obesity	10	2.45
Kidney diseases	3	0.74

Distribution of patients depending on vicious behavior

	2012	**2013**	**2014**	**2015**	**Total**	**%**
Smoker	38	49	36	37	**160**	**39.22**
Non-smoker	76	53	61	58	**248**	**60.78**
Total	**114**	**102**	**97**	**95**	**408**	**100**

It can be observed that, despite the fact that smoking affects the pregnancy and the fetus, a quite high percentage of pregnant women smoke (39.22%).

Distribution of the patients depending on the result of the amniocentesis

Result	**Number of patients**	**%**
Normal results	394	96.57
Down syndrome	10	2.45
Edwards syndrome	4	0.98
Total	**408**	**100**

The percentage of Down syndrome reported to the total number of pregnancies (6097 patients) represents 1.6% above the European statistics. The percentage of Edwards syndrome reported to the total number of pregnancies represents 0.656% above the European statistics.

Maternal ages for women diagnosed with pregnancies with Down syndrome were 28, 31, 32, 38, and 40 years old.

Age of the mother diagnosed with pregnancies with Down syndrome	Number of cases
28	2
31	1
32	2
38	3
40	2

All the patients with babies diagnosed with Down syndrome or Edwards syndrome decided to terminate the pregnancy.

The 14 cases positive for Down syndrome and Edwards syndrome (4 cases positive for Edwards syndrome and 10 cases positive for Down syndrome) were actually found because of the indications of the TRIPLE TEST, 4 more cases positive for Down syndrome were found because of the indications for BI-TEST, and just 1 case of Down syndrome was found because of the indications for maternal age.

Amniocentesis has become the most used prenatal diagnosis method for women aged 35 and above or even for younger women with a family history specific for hereditary congenital abnormalities.

In our studied group, 157 pregnant women asked for a prenatal diagnosis for Down syndrome. They were aged 35 or above 35 and demanded the association of some pathological biochemical markers (double test, TRIPLE TEST).

The frequency of the structural chromosomal abnormalities (translocations, deletions, inversions) did not show a similar numeric relationship regarding the maternal age.

According to the literature, the screening tests have a margin of error in the diagnosis of fetal malformations, so the diagnosis of certainty will result after the amniocentesis.

The majority of the specialists consider that amniocentesis is an invasive procedure, so it is recommended that the number of these procedures should be decreased. In this regard, our study did very well: the result of the utilization of the other methods (BI-TEST, TRIPLE TEST) was positive. The BI-TEST can detect trisomy 21 (the Down syndrome) and trisomy 18 (the Edwards syndrome).

There are tests like Panorama Test which are cutting-edge technology, and they reduce the use of amniocentesis. But the cost of this test is high and not covered by the National Health, which is the reason why it is not accessible for all patients.

Our study has been made on a large segment of the population which had access to all the screening tests (BI-TEST, TRIPLE TEST), but not all of the patients had this possibility.

The TRIPLE TEST in pregnancy can help in tracking down the neural tube defects and chromosomal abnormalities such as trisomy 18 and trisomy 21. So, if a high level of alpha-fetoprotein (AFP gene) is found in the blood, it is likely that the fetus presents a neural tube defect (more exactly the brain and the spinal cord), defects such as spina bifida or anencephaly. If the results of the TRIPLE TEST show a low level of alpha-fetoprotein (AFP gene) and estriol, it can indicate trisomy 21 (the Down syndrome), trisomy 18 (Edwards syndrome), or other chromosomal abnormalities.

Before we proceeded with the amniocentesis procedure, we acquired from every patient and her partner a written consent after informing them about what the procedure implies.

In obstetrical practice the risk of carrying out an amniocentesis is evaluated permanently, compared to the individual risk of numerical chromosomal abnormalities or structural chromosomal abnormalities. We recommend the amniocentesis as an invasive diagnosis prenatal method when the risk of a detectable abnormality is higher than the risk of the procedure itself.

It is important that the patient should be informed about the pam felt during the amniocentesis.

Numerical chromosomal abnormalities (aneuploidies)—in the case of amniocentesis procedures done to pregnant women with age 35 or above from our consignment—have a frequency of 3.18% (5 cases out of 157).

The majority of the patients questioned said that the procedure was a painful method. Pain is a multidimensional sensation and complex, which varies in quality, intensity, duration, localization, and inconvenience from an individual to another. The intensity and inconvenience of the pain is not directly connected to the nature and the extension of the tissue lesion. Cultural, genetic, familial, growth, psychological, and social variables have a strong impact in the perception of the pain and in its expression from the patient.

The objective quantification of the pain is difficult. Before any invasive procedure effectuated throughout pregnancy, the biggest concern of the patient is the risk and the intensity of the pain during procedures.

Ultrasound conclusions in cases of chromosomal abnormalities

Trisomy	Ultrasound conclusions
Trisomy 21 (Down syndrome)	• Duodenal atresia, tracheoesophageal fistula, and hydramnios are present usually if there are gastrointestinal damages • Heart abnormalities—defects of the atrial septum • Hypoplasia of the middle phalanx of the 5th finger of the hand • Conclusions from the second trimester: increased cervical skinfold (>6 mm), ratio of the actual femur (length), and the expected femur = 0.91
Trisomy 18 (Edwards syndrome)	• Growth retardation in uterus • Hydramnios • Hand wrapped with superimposed fingers (the index covers the third finger, and the fifth finger covers the fourth finger) • *Ankle equinus* with prominent calcaneus and clubfoot. • Heart abnormalities: ventricular septal defect • Omphalocele, diaphragmatic hernia • Cyst of the choroid plexus
Trisomy 13 (Patau syndrome)	• Holoprosencephaly (defects of the facial break line) • Cleft palate • Heart abnormalities such as ventricular septal defect • Polydactylism • Omphalocele • Polycystic kidneys

6. Conclusions

In conclusion, all the pregnant women who underwent the amniocentesis procedure and had a negative result had an on-term pregnancy and delivered healthy newborn babies.

As for the pregnant women who underwent the amniocentesis procedure and had a positive result, all of them decided to terminate their pregnancy due to existent fetal malformations.

The purpose of our study was to decrease the use of amniocentesis procedures and to introduce a cutting-edge technology, which is the Panorama Test for early diagnosis of fetal malformations. We tried to implement a noninvasive method (Panorama Test made from the mother's blood sample) over an invasive method which is the amniocentesis.

Our purpose has been achieved by the decrease of amniocentesis and by increasing the use of cutting-edge technology without affecting the final results in the percentage of the fetal malformations statistics worldwide.

Author details

Gladys Cristina Al Jashi[1*] and Isam Al Jashi[2]

1 Primary Physician in Obstetrics and Gynecology at Sfantul Pantelimon Emergency Hospital, Bucharest, Romania

2 Faculty of Medicine, Titu Maiorescu University, Primary Physician in Obstetrics and Gynecology at Regina Maria Private Hospital, Bucharest, Romania

*Address all correspondence to: dr.jashi@yahoo.com

References

[1] D'Alton ME, DeCherney AH. Prenatal diagnosis. New England Journal of Medicine. 1993;**328**:114-120

[2] O'Reilly DB, Bottomley C, Rymer J. Essentials of Obstetrics and Gynaecology. Elsevier Health Sciences; 2005. pp. 219-235

[3] Wieacker P, Steinhard J. The prenatal diagnosis of genetic diseases. Deutches Ärzteblatt International. 2010;**107**(48): 875-862. DOI: 10.3238/arztebl.2010.0857

[4] Rary JM, Park IJ, Heller RH, Jones HW Jr, Baramki TA. Prenatal cytogenetic analysis of women with high risk for genetic disorders. Journal of Heredity. 1974;**65**(4):209-212

[5] Wang J, Cui YX. Advances in rapid prenatal detection of fetal chromosome abnormalities. Zhonghua Nan Ke Xue. 2010;**16**(4):359-363 Review. Chinese

[6] Available from: https://www.reginamaria.ro/articole-medicale/testele-screening-de-trimestru-1-bitestul-sau-testul-combinat-0

[7] Nanu D et al. Esențialul în Obstetrică. Medicală Amaltea; 2015

[8] Feinstein SJ, Vintzileos AM, Lodeiro JG, Campbell WA, Weinbaum PJ, Nochimson DJ. Amniocentesis with premature rupture of membrane. Obstetrics and Gynecology. 1986

[9] Leukskul N, Tannirandorn Y. The location of needle insertion effect on maternal pain in amniocentesis. Journal of the Medical Association of Thailand. 2006

[10] Verny RT. Pre-And Peri-Natal Psychology: An Introduction. New York: Human Sciences Press;

[11] Available from: https://en.wikipedia.org/wiki/Down_syndrome

[12] Leigh Simpson J. Genetic Counseling and Prenatal Diagnosis. pp. 215-238, chapter 9 in Gabe SG

[13] Niebil JR, Simpson JL. Obstetrics: Normal and Problem Pregnancies. 3rd ed. Churchill Livingstone; 1996

Section 3

Cancer Genetics

Chapter 5

Genetic Contributors to Hereditary Cancer Predispositions: Do We Have Enough Information?

Tom Nolis and Rodney J. Scott

Abstract

Genomics medicine and molecular genetics are experiencing a surge of interest as well as a push for a more prominent role in mainstream medicine. This, coupled with the advancement of next-generation sequencing, along with a national, if not global, steering of funding to support the advancement and development of genetics is suggesting that we are entering a new age of medicine. As this push begins to gain some momentum, the impact of genomics medicine on clinical utility and the influence of supporting data on genes that make their way from research to diagnostic medicine are worth reviewing.

Keywords: genetics, genomics, utility, cancer, *BRCA1*, *BRCA2*, *PALB2*, population, family, epidemiology, multigene, panels

1. Introduction

One of the most fundamental clinical validity and clinical utility questions currently at the forefront of molecular genetic testing is as important today as it was nearly 30 years ago, "is the genetic variation detectable in the genes of interest actually associated with a clear and quantifiable risk for disease?" or in other words, "are the variants that we are finding in these genes relevant for the disease of interest?"

The question above is a very important one, and to be able to answer, it we must take a step back and explore clinical utility and epidemiology in genetics more thoroughly. The answer to this question requires an inevitable focus on cancer genetics as breast and ovarian cancer are excellent examples of both past and present accomplishments in genomics medicine. These disease entities are topical and have enough data to appropriately highlight the genetic journey previously taken into genomics medicine, and they are also able to shed light on how new genetic players are entering the diagnostic scene (e.g., *PALB2*).

Finally, the following text will briefly compare and contrast the significant influence of family-based and population studies on genetic data. This chapter will close on a more general note by reviewing the current cancer assessment guidelines and how these reflect the current clinical utility of genomics medicine.

1.1 Key points

- Clinical utility in genetics is largely a continual revamp of the ACCE framework.
- Genetic epidemiology has followed a natural flow paralleling the advancement of technology and detection.
- Family-based studies and population data are still at the forefront of both research genetics and diagnostic genomics.
- *BRCA1* and *BRCA2* are excellent models for useful genetics input.
    - This has led to a complete change in the identification and treatment of a very tangible cancer entity.
- Less penetrant genes in breast and ovarian cancer, such as *PALB2*, along with multigene panels have their place in genomics medicine, but, generally, quality supportive data are lacking.

2. Clinical utility in genetic testing

Clinical utility is a broad concept and one that is deeply fundamental in the world of medicine. At its most basic level, it is the essence of what propels medical advancement. It answers the question of "what should I do next?" when this question is asked by physicians, but it can also be a much larger almost all encompassing concept; as if to ask, "what is worth doing next?" in any field of medicine. Clinical utility, by many sources, can be simply defined as "the balance of benefits to risks" or more broadly refer to any use of test results to inform clinical decision-making [1]. A genetic test can be defined as a "test that involves the analysis of chromosomes, deoxyribonucleic acid (DNA), ribonucleic acid (RNA), genes, or gene products (e.g., enzymes and other proteins) to detect heritable or somatic variations related to disease or health. Whether a laboratory method is considered a genetic test also depends on the intended use, claim or purpose of a test" [2].

In some fields, such as in colorectal surgery, it is easy to determine clinical utility: if the patient's morbidity or mortality is improved by removing a tumor, then you remove it; otherwise you do not. In the realm of genetic testing, the concept of clinical utility can be difficult to precisely define—or even quantify. Indeed, this was reinforced when experts in evidence-based medicine and genetics for the Centers for Disease Control and Prevention (CDC) in 2005 failed to come to a consensus on the term "clinical utility" despite initially claiming to be confident in its meaning beforehand [1]. As a consequence, the term had to be elaborated on by the Analytic validity, Clinical validity, Clinical utility, and Ethical, legal and social implications (ACCE) project that was carried out by the Foundation for Blood Research with support from the CDC [1]. In screening or in diagnostic testing, clinical utility broadly refers to the ability of a test to prevent or ameliorate adverse health outcomes such as mortality, morbidity, or disability through the adoption of efficacious treatments conditioned on test results (**Figure 1**) [1]. The perceived value of genotypic information, at this point in time, includes a more thorough understanding of a patient's diagnosis, prognosis, risk, and disease or treatment susceptibility; the caveat is that this knowledge may not influence clinical management at all. Clinical utility in the ACCE framework was expanded to include

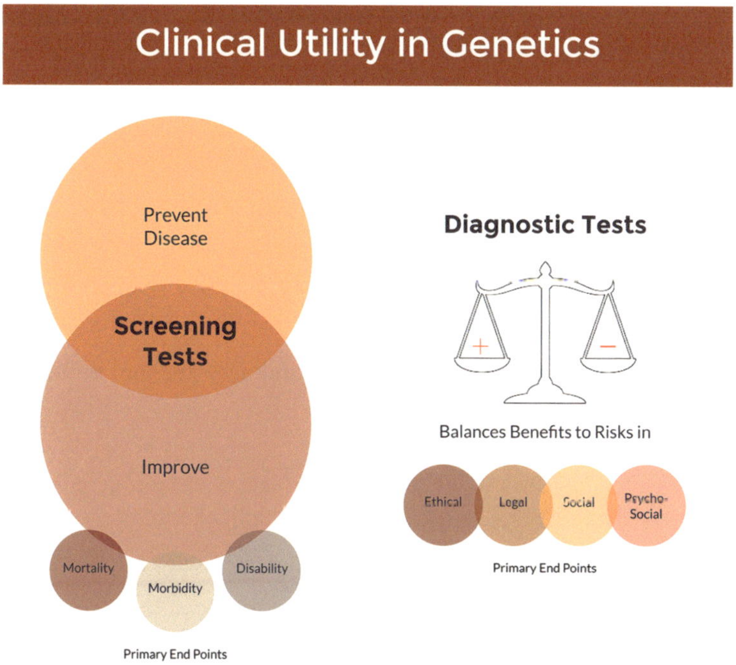

Figure 1.
Clinical utility in genetics. The clinical utility in screening tests is to provide preventative care and to improve primary end points such as overall morbidity; in a diagnostic setting, the value of genetic testing lies in the balance of benefits to risks in more psychological domains.

contextual or implementation issues (e.g., availability of resources) and that clinical utility can also include psychological benefits.

2.1 Evaluating genetic tests

Evaluating genetic tests is often methodologically difficult largely due to small patient populations and the resulting dearth of high-quality studies. Adapted from Morrison et al. 2012, the following subsections are generally agreed upon as common characteristics for reviewing genetic tests [3, 4].

2.1.1 Overview of disease and underlying genetics

- Information on the disease prevalence, treatments, and outcomes of the disease as well as overall cost
- Description of the genetic causes, including inheritance patterns
- Classifying the mutational spectrum, along with the prevalence and penetrance
- Determination if there is any "gold standard" tests to compare to

2.1.2 Target population and intended use

- Prevalence for target population involves age, ethnicity, and eligibility for testing

- Purpose of the test: is it diagnostic, for treatment, prognosis, management, carrier testing, prenatal testing, or other?

2.1.3 Laboratory information

- Validation details
    - Test new or already in use?
        - If already in use:
            - Number and rate of positive and negative mutations
            - Turn-around-time for results
            - Similar tests available
            - Current testing activity and expected with appropriate funding figures
            - Whether there are other laboratories that could offer the test
            - Infrastructure requirements
- Quality assurance, maintenance, and improvement programs—both internally and externally

2.1.4 Economic considerations

- Cost estimates for the test including equipment, personnel, and consumables/reagents
- Costs of disease burden with or without treatment
- Costs saved by employing test

2.1.5 Analytic validity

- Precision, reliability, accuracy, sensitivity, and specificity of the genetic test, and how these compare with other employed screening and diagnostic tests

2.1.6 Clinical validity

- Specificity and sensitivity, positive predictive value, negative predictive value, likelihood ratios, and how these compare with other employed screening and diagnostic tests

2.1.7 Clinical utility

- Benefits and risks of the test
- Treatment of the patient

- Patient's health outcome
- Effect on patient and patient's family members
- Alternative diagnostic tests compare and contrast to genetic testing

2.1.8 Ethical, legal, and social

- Details on ethical, legal, and societal issues include support and follow-up

2.2 ACCE model and beyond

The CDC furthered the ACCE framework by establishing the Evaluation of Genomic Applications in Practice and Prevention (EGAPP) Initiative. The EGAPP supports the development and implementation of an evidence-based process for evaluating genetic tests and other genomic applications for clinical and public health practice. An independent, nonfederal EGAPP Working Group (EWG), consisting of a multidisciplinary expert panel selects topics of interest, reviews evidence, and recommends courses of action [2]. Key objectives of the EWG are to develop an openly accountable process, reduce conflicts of interest, and provide a connection between well-established evidence and the EWG recommendations [2, 5].

The ACCE model has stood the test of time and has proven to be a basis for the technical appraisal of genetic tests across the globe (**Figure 2**). The UK Genetic Testing Network and the Andalusian Agency for Health Technology Assessment expanded the ACCE model to guide the introduction of new genetic tests into their

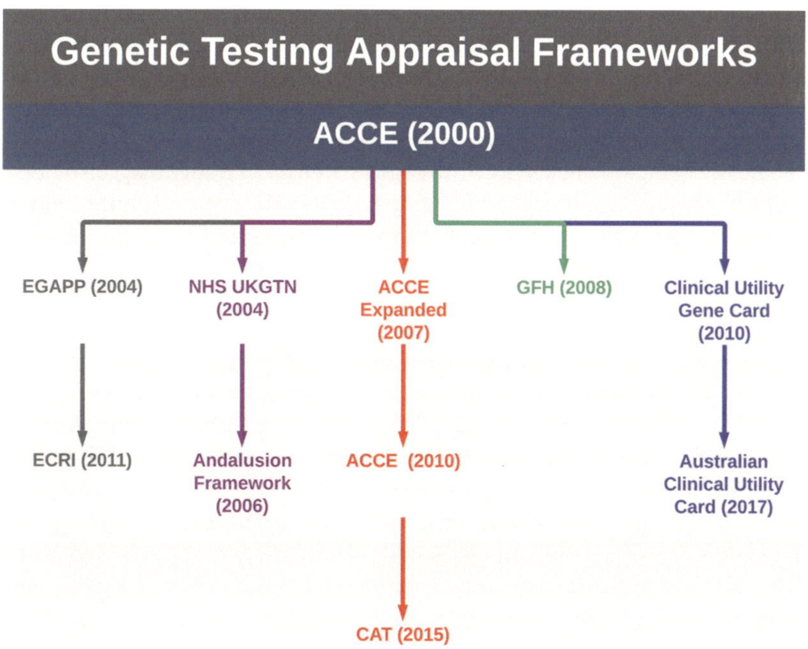

Figure 2.
Genetic testing appraisal frameworks. Testing appraisal frameworks around the globe have largely stemmed from the analytic validity, clinical validity, clinical utility, and ethical, legal and social implications (ACCE) framework announced in 2000. EGAPP, evaluation of genomic applications in practice and prevention; ECRI, Emergency Care Research Institute; NHS UKGTN, National Health Service UK Genetic Testing Network; CAT, companion test assessment; GFH, German Society of Human Genetics.

public health system, creating the 2004 NHS UKGTN Gene Dossier and the 2006 Andalusian Framework. In 2007, the ACCE model was again modified by adding health quality measures to the evaluation process. The process was made more streamlined, shortening the systematic review process for emerging genetic tests. In 2010, the ACCE model was adapted to particular types of genetic testing through the Complex Disease Framework and the ACHDNC Newborn Screening Framework of the Advisory Committee on Heritable Disorders in Newborns and Children. The ACCE model also spawned two European frameworks: the 2008 GFH Indication Criteria of the German Society of Human Genetics and the 2010 Clinical Utility Gene Card of EuroGentest. The 2015 Companion test Assessment Tool (CAT) build upon the ACCE model to determine which tests needed further evaluation. In 2011, the ECRI Institute utilized the EGAPP process to develop a set of analytical frameworks for various testing scenarios and other stakeholder aspects. Lastly, the EuroGentest inspired the 2017 Australian Clinical Utility Card [5].

We cannot ignore the influence of the stakeholders who will invariably have wildly different opinions as to which outcomes are considered relevant. Thus, the types of outcomes that must be considered in evaluating the utility of a genetic test will also depend on the purpose of the test and the congregation of those who make the final decision. For example, a state-funded public health program will likely focus on the overall impact on morbidity and mortality versus privately funded sectors concentrating on the greater net profit. Coverage decisions by third-party payers may be based in large part on perceptions that test results are useful for timely or accurate diagnoses and clinical management. For a test that is offered to families in a clinical setting on a voluntary basis, the value of information for making career, residential, and reproductive decisions take on greater relevance.

Before a genetic test can be accepted into clinical practice, data must be collected to demonstrate the benefits and risks that accumulate from both positive and negative results [1]. Referring back to the colorectal surgery example above, the term "clinical utility" fits in very well; however, when applied to genetics or genomics medicine, the term may be too limited. The clinical end points are extremely important, but the utility or actionable context must remain broad in genetics to embody an overall net benefit. Genetic testing is particularly useful in the psychosocial, ethical, legal, and social (ELS) realms as well as in many diagnostic cases (**Figure 1**). By including clinical and ELS together in the concept, improvements in health outcomes such as enhancements in morbidity, mortality, and in disability become strong primary end points when assessing the utility of genetic testing.

3. Genetic epidemiology: population and family-based studies

The process of genetic epidemiology has flowed sequentially from observing phenotypic differences between populations to demonstrating that diseases can run in families, to examining feasible genetic susceptibility models, to tracking co-segregating genetic markers with disease in families, to narrowing the region of candidate genes, to association analyses with candidate genes, to cloning and mutation identification, to functional and structural characterization of a gene, and, ultimately, to extending the phenotype characterization even further based on the genotype identified (**Figure 3**) [6].

Population and family-based studies are at the heart of genetic epidemiology. Population-based association studies are generally regarded as more statistically powerful than family-based studies, as they tend to have more subjects and are overall easier to execute [6]. Family-based designs are influential methods that use the proband as well as their relatives to assess the genetic and molecular

Figure 3.
Genetic epidemiological historical progression.

epidemiology of disease (**Figure 4**). The various types of studies available include familial aggregation, twins, segregation, linkage, and association.

Linkage and association studies directly evaluate genetic markers and require the collection of DNA from the study subjects—as opposed to twin studies, segregation studies, and familial aggregations studies. Family-based studies have been the long-standing primary approach to detect disease-causing genes. Segregation and linkage studies are highly valuable methods for assisting in

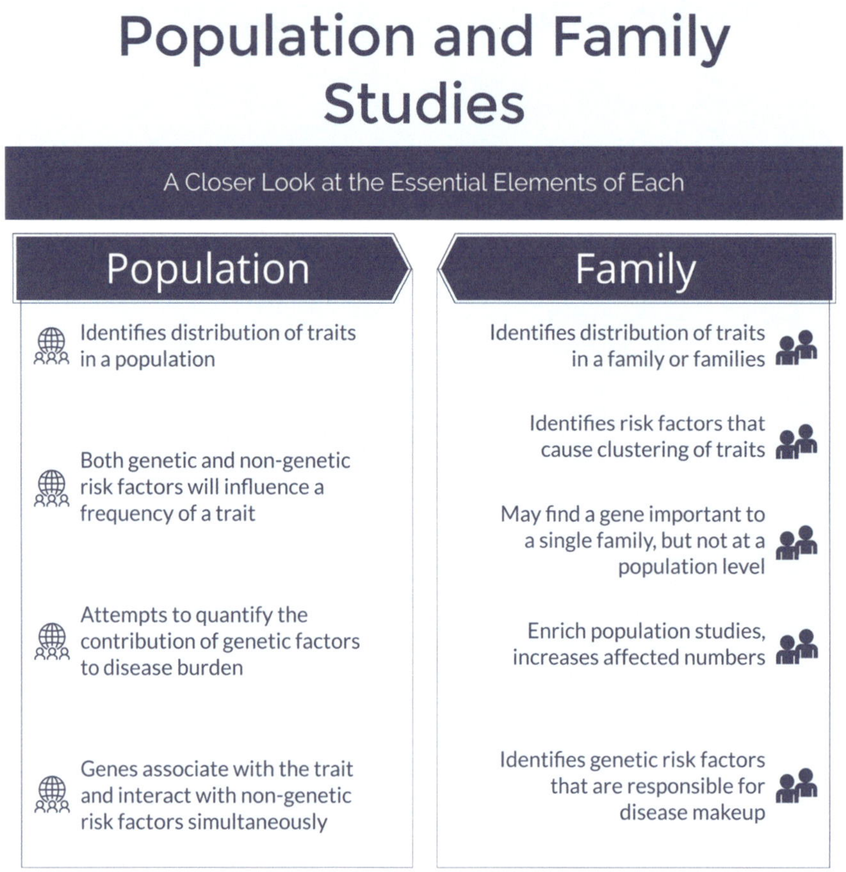

Figure 4.
A closer look at family-based and population studies.

cloning highly penetrant rare disease-causing genes. Family-based association studies strengths lie in their ability to control for confounding bias due to population stratification—albeit they do so at a loss of power [1, 5, 6].

Population stratification is when contrived associations can be detected if cases and controls come from different source populations that have systematic differences in ancestral allele frequencies. A great advantage of employing family-based study designs is that population stratification can be circumvented. By studying parents and their offspring/siblings, cases and controls within each family are virtually guaranteed to arise from the same sample source or population. Due to the increasing efficacy of identification of association with disease, the importance of family-based studies has seemingly subsided; however, it is worth pressing that family-based studies are arguably more important than they have ever been as they are still the only way to truly link a causative variant to disease [1, 5, 6].

Family-based studies can help determine whether a disease or trait is genetically influential by studying familial aggregation. Results can be furthered with a segregation analysis to identify the mode of inheritance. The results from the segregation analysis can add power to a linkage analysis, which searches across the entire genome in an attempt to locate regions containing causal genes [1, 5, 6].

Segregation analysis is a method of establishing the genetic inheritance of disease and is performed exclusively with family data. This approach assists in determining if a disease segregates with a variant in a single gene; in addition to this, the mode of inheritance can also be ascertained. Very large pedigrees and families with a plethora of affected individuals are exceptionally informative for identifying specific genes [1, 5, 6].

Once a family is collected, and studied, in a segregation analysis, that family is generally made available for further analyses (e.g., linkage). Linkage occurs when two loci/alleles on the same chromosome are inherited together. Since recombination during meiosis can occur virtually anywhere in the genome it stands to reason that the closer two loci are to each other the less likely they will be separated after a recombining event (i.e., the more likely they are to stay together after recombination), that is, it says they are "linked." Linkage analyses utilize this phenomenon by investigating co-segregating genetic markers along with a disease trait seen within the family (or families)—the trait can be either qualitative or quantitative. If the markers and traits are observed to co-segregate within families, it can be logically inferred that the disease-causing variants are within close proximity to the markers [1, 5, 6].

This point reveals that linkage is "intrafamilial," whereby the co-segregating marker allele may very well be different in different families. Families are generally recruited into linkage studies on the basis of having at least one identified affected individual, and the families are either quite large or have affected siblings. Generally, the markers are spaced evenly over the entire genome, and linkage can be performed by utilizing these markers.

4. Breast and ovarian cancer: the journey of finding *BRCA1* and *BRCA2*

Establishing penetrance is an arduous undertaking even for some of the most well-studied genes (e.g., with *BRCA1* and *BRCA2*) simply because no two genetic studies have yielded the same findings. Many of the data and results discrepancies can be linked to differences in the populations studied and to the methodologies employed. It is vanishingly rare, if not impossible, for these studies to have similar methodologies to allow for perfect reproduction of results. The range of penetrance found for the BRCA variants and genes has guided the clinical recommendations for breast and ovarian cancer surveillance and prevention and has provided a sort of genetic "gold standard" by which all other genes are now compared and contrasted against [7, 8].

A strong family history of breast cancer is associated with an early age of onset, the addition of ovarian cancer, bilateral tumors, and a rarely affected male. *BRCA1* (and similarly for *BRCA2*) has been identified through the study of women who have presented with a strong family history of breast and ovarian cancer (**Figure 5**). In 1988, more than 1500 families with multiple cases of breast and ovarian cancer were studied; the data generated was subjected to a segregation analysis. The results illuminated that roughly 5% of cases, particularly those with early-onset disease, could be heritable in a Mendelian fashion. Selected families, those with near-Mendelian pedigree patterns, were aggregated for linkage analysis. A major susceptibility locus, *BRCA1* (OMIM 113705), was mapped to 17q21 in 1990. 17q-linked families that were above 45 years of age at diagnosis were given negative LOD scores.

In 1994, a linkage analysis performed on 15 large families with breast cancer that were also determined to not be linked to 17q helped identify the *BRCA2* locus located on 13q12 (OMIM 600185). Later that year *BRCA1* was cloned, and in 1995 *BRCA2* was cloned. *BRCA1* variants accounted for a very large proportion of

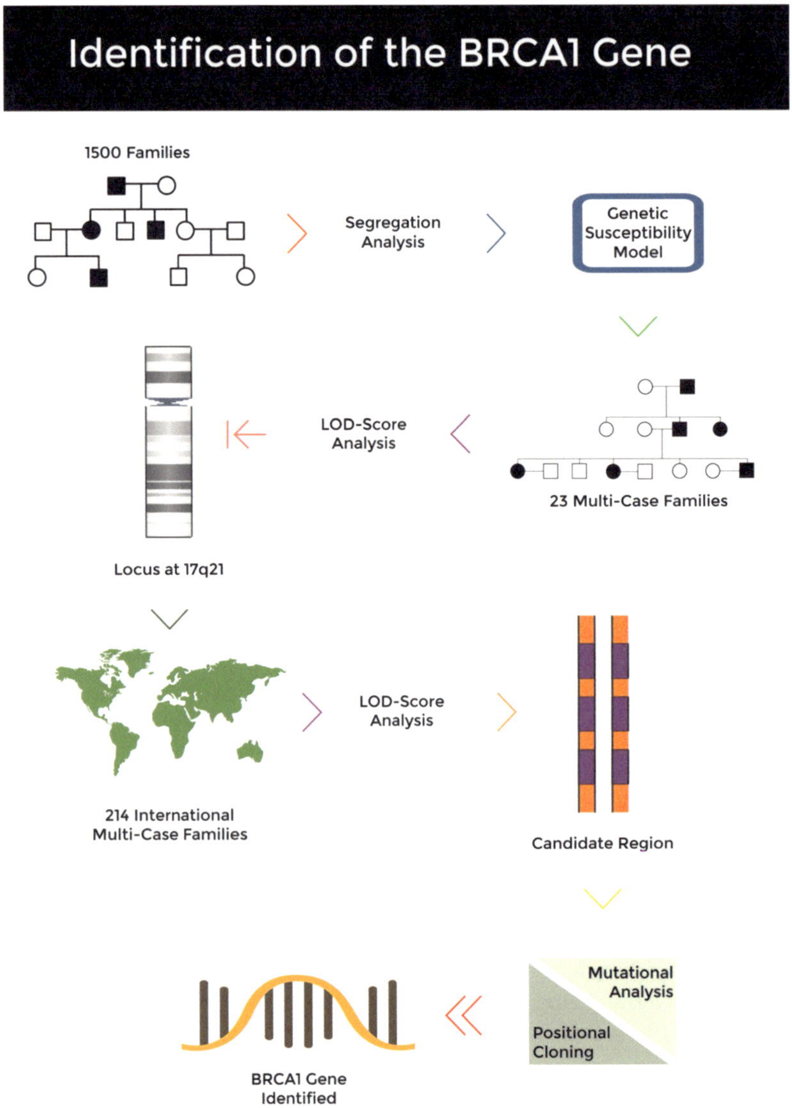

Figure 5.
The progression of discovering the BRCA1 gene.

families with both breast and ovarian cancers, while male breast cancer was predominately observed in *BRCA2* variant families [9].

BRCA1 and *BRCA2* were identified by focusing on a small number of specific families. Other candidate genes were established from focused studies.

- In 2004, the variant c.1100de1C in the *CHEK2* gene was observed in 201 cases (1.9%) and 64 controls (0.7%) in 10,860 breast cancer cases and 9065 controls from 10 case-control studies in 5 countries (estimated odds ratio (OR) 2.34; 95% CI 1.72–3.20; P = 0.0000001) [10].

- In 2006, the *ATM* gene was screened in individuals from specific families, and 12 mutations were found in affected individuals and in 2 controls (P = 0.0047) from 443 familial breast cancer pedigrees and 521 controls (estimated relative risk of 2.37 (95% CI 1.51–3.78, P = 0.0003) [11].

- *BRIP1* was initially described as a breast cancer predisposition gene in 2006. The analysis of 1212 women with familial breast cancer along with 2081 controls yielded mutations in 9 cases and in 2 controls (estimated relative risk of 2.0; 95% CI = 1.2–3.2; P = 0.012) [12].

- *PALB2* mutations were found in 10 cases of 923 individuals with familial breast cancer, and no mutations were found in 1084 controls (P = 0.0004) (estimated odd ratio of 2.3; 95% CI = 1.4–3.9; P = 0.0025) [13].

These factors are the most relevant in families where the disease and the variant are actually segregating together, but at a population level, their overall implication is surprisingly small. For instance, *CHEK2*:c.1100de1C, at a population level, is seen in only 1.9% of cases [14].

5. Less penetrant genes: *PALB2*, *BRIP1*, and *RECQL*

Traditionally, epidemiological studies tend to emphasize the inclination that the greatest benefits to the population are found in interventions that decrease risk factors for the bulk of the population, not in targeting a small number of individuals at the extreme ends of the risk spectrum. As we can see from the sample multigene panel for hereditary breast cancer the genes selected for testing range from quite influential (e.g., *BRCA1* and *BRCA2*) to marginally influential (e.g., *STK11*) (**Table 1**). However, it is quite interesting to explore the data for lower-risk genes such as *PALB2* and *BRIP1* and also for genes that are suspected of having involvement in breast or ovarian cancer, but the data is not quite yet congruent with the theory (e.g., *RECQL*).

Less penetrant genetic loci are mainly represented by the single-nucleotide polymorphisms (SNPs) or by the variants of uncertain significance discovered through genome-wide association studies. Variants associated with a minor increased risk, unlike high-risk mutations, can only account for a small portion of cancers seen in family histories of carriers. Thus, the cumulative risk for carriers with a positive family history will largely depend on the risk levels attributed to both their family history for that cancer and the risk induced by the variant itself. For example, if a carrier of a variant associated with an increased odds ratio of cancer also has a family history sufficient to quadruple her risk, her cumulative risk will be about that of a woman with a 4 x OR increased risk [7].

Mutations in *ATM*, *PALB2*, and *CHEK2*:c.1100delC, in conjunction with a strong family history, are very likely to be associated with a high absolute risk of breast cancer [17]. It is important to note that the family history creates a context whereby it changes the penetrance of these mutations. This interpretation is clinically important and would justify testing for these mutations in multi-case breast cancer families such as those seen by typical cancer family genetics services.

Genetic risk factors, which are familial by their very definition, will be more frequent in women with positive family histories involving multiple breast cancers through their direct association with breast cancer as a disease entity and with their direct association with the familial aspect of breast cancer. This can be illustrated in a simple example where we consider a rare mutation whose presence doubles the risk of breast cancer relative to the general population: it will be roughly four times more common in women who are affected and who have an affected first-degree relative, which will square the ratio [17].

Sample multigene hereditary breast cancer panel	
Gene	**Notes**
ATM c.7271 T > G	—
other *ATM* mutations	Lack of data regarding penetrance and surveillance except for c.7171 T > G mutation. If mutation has been identified, a careful assessment of residual risk for relatives who are noncarriers is needed
BARD1	—
BRCA1	—
BRCA2	—
BRIP1	—
CDH1	Increased risk of lobular breast cancer. Gastric cancer risk is unknown when mutations are identified in absence of a positive family history of gastric cancer
CHEK2 c.1100delC	Mutations are rare, but high penetrance in some families
other *CHEK2* mutations	Lack of penetrance data except for in specific mutations
NBN	—
NF1	Can identify mutation carriers by clinical phenotype and then perform gene specific test
PALB2	Mutations are rare, but high penetrance in some families
other *PALB2* mutations	—
PTEN	Can usually be identified by clinical phenotype
RAD50	—
STK11	Can usually be identified by clinical phenotype. Low penetrance in breast cancer
TP53	Breast cancer risk management inferred from other genes. High penetrance for breast cancer, but mutations are rare
Adapted from: Invitae and eviQ [15, 16].	

Table 1.
Sample multigene panel offered for hereditary breast cancer.

5.1 PALB2

The Finnish founder mutation was found in 0.9% (18 of 1918) of cases without selecting for or emphasizing family history. Likewise, a French-Canadian founder mutation was found in 0.6% (2 of 356) of cases also without selecting for or emphasizing family history selected. The numbers of the *PALB2* founder mutation carriers were too small to make precise risk inferences, but modified segregation analyses of data from the families of case-carriers were used to estimate risk for carrier families, which has demonstrated the importance of these founder mutations in the risk of developing breast cancer [13, 14, 18, 19].

Blanco et al. found that the frequency of *PALB2* mutations was ~1.5% after investigating the incidence of mutations in *PALB2* patients with breast cancer that were also negative for any variants in *BRCA1* and *BRCA2* +/− a family history of pancreatic cancer; previous studies had the mutation rate for similar cohorts range from 0 to 4.8%. Dansonka-Mieszkowska et al. conducted similar research among Polish women; their study showed a minor but significant *PALB2* mutation presence at ~0.6% [13, 14, 18, 19].

Likewise, Bogdanov et al. conducted a study on the occurrence of *PALB2* mutations among Russian and German women and reported a mutation rate of ~2%. In 2014, Antoniou et al. tested 362 women from 154 families and found that the risk of breast cancer in women 40 years of age or younger with *PALB2* mutation was ~9x greater, ~8x greater in patients aged 40–60 years, and ~5x greater in patients 60 years and older when compared to the general population. The absolute risk of breast cancer in women with *PALB2* mutations under 70 years of age ranges from 33 to 58% for women without and with a positive family history of breast cancer, respectively. Interestingly, Hartley et al. conducted a study that confirmed that as the number of cases of breast cancer in a family increases, the likelihood of *PALB2* mutations increases as well. Women with 3+ positive cases of familial breast cancer have a 2.6% greater likelihood of having a *PALB2* mutation than those without multiple cases [13, 14, 18, 19].

Southley et al. studied Australian women to determine the occurrence rate and penetrance of *PALB2* mutations. The study found that the women with breast cancer and a positive family history of having *PALB2* mutations had *PALB2* mutations present in 1% of patients. The study identified that the women with breast cancer and no family history of *PALB2* mutations had *PALB2* mutations present in only ~0.4% of patients. There were no *PALB2* mutations detected in the control/healthy population of women. Heikkinen et al. studied southern Finnish women and found *PALB2* mutations in 2% of patients with a positive familial history of breast cancer and also in 0.6% of women with a sporadic breast cancer presentation [13, 14, 18, 19].

In 2016, a seminal paper by Thompson et al. examined 2000 predominantly breast cancer-affected women with a strong family history that were also *BRCA1* and *BRCA2* variant-negative and compared them to 1997 controls. They observed that a significant proportion of mutations were only in *PALB2* (26 cases vs. 4 controls) and in TP53 (5 cases vs. none in controls), whereas no mutations were identified in *STK11* [14]. *PALB2* is a great example of how penetrance estimates can depend on the population, family history, and age at onset as well as other considerations [17].

5.2 BRIP1

BRIP1 was initially described as a breast cancer predisposition gene in 2006. The analysis of 1212 women with familial breast cancer along with 2081 controls yielded mutations in 9 cases and in 2 controls (estimated relative risk of 2.0; 95% CI = 1.2–3.2; P = 0.012). However, recently *BRIP1*'s association with breast cancer has grown suspect, while its association with ovarian cancer has risen sharply. *BRIP1* mutations confer a high ovarian risk in familial index patients (OR = 20.97, 95% CI = 12.02–36.57; P < 0.0001) and in the subgroup of patients with late onset ovarian cancer (OR = 29.91, 95% CI = 14.99–59.66; P < 0.0001) [12].

5.3 RECQL

In a screen of 144 Polish and 51 French-Canadian women with early-onset familial breast cancer, 2.6% possessed truncating mutations in *RECQL*. Validation studies that reviewed the *RECQL* variant c.1667_1667C3delAGTA in over 13,000 breast cancer patients with 4702 Polish controls showed the *RECQL* mutation appeared in 0.23% of cases and in 0.04% in controls. Likewise, the *RECQL* variant c.634C > T (p. Arg215*) seen in the French-Canadian population was further screened in 538 patients and 7136 newborn controls and was detected in 5 patients and in one control—a nearly 50x increase in frequency in affected versus unaffected individuals. Studies of patients in northern China revealed a pathogenic *RECQL* mutation in 2.0% of the 448 familial breast cancer patients compared to the 0.06% seen in 1588 control subjects [20]. By whole exome sequencing 0 early-onset familial breast cancer patients without *BRCA1*/2 mutations and by screening the *RECQL* gene in an additional 439 unrelated familial breast cancer patients, 9 index cases were found to carry a pathogenic mutation in the *RECQL* gene among the 448 BRCA-negative familial breast cancer patients. It was determined that the pathogenic mutation rate of the *RECQL* gene in familial breast cancer in *BRCA1-*/*BRCA2*-negative breast cancer patients was 2.0%. Further to these results, no loss of heterozygosity was found in the *RECQL* mutation carriers, suggesting that *RECQL*-associated tumorigenesis is likely through classical haploinsufficiency [21].

In 2018, Li et al. sequenced all the exons of RECQL and at least 10 bp of the exon–intron flanking regions in 9112 subjects from Australia. The case subjects were females diagnosed with breast or ovarian cancer from 4536 families with a negative result after BRCA1 and BRCA2 mutation testing. The controls were 4576 women who were above 40 years of age and were cancer-free as of May 2016. Thirteen loss-of-function mutations in the cases and 25 in the controls were identified (0.29 versus 0.55%, odds ratio 0.52, 95% confidence interval 0.25–1.06, P = 0.072 by two-tailed Fisher's exact test) [22]. Missense variants observed between cases and controls were not statistically significant (54 cases, 1.19%, versus 37 controls, 0.81%; P = 0.073) [22]. It is generally accepted that a predisposition gene is considered actionable only if the 90% confidence limit of the estimated relative risk is greater than four [22]; therefore, RECQL, can be excluded based on these findings.

5.4 Lack of data and the impact of finding a variant of unknown significance

Despite the intense and widespread shared enthusiasm to reduce the risk of predictable cancers, the adoption of the more recent breast cancer predisposition genes, such as *PALB2*, has been sluggish. This has been largely due to the communal appreciation that following a variant of unknown significance result, the complexity of interpretation may lead to a subsequent clinical utility hindrance. An additional 3.9% of patients tested by multiple gene panels had pathogenic mutations identified in other breast cancer predisposition genes, namely, in *PALB2*, *CHEK2*, and *ATM*; however, many of these multigene panel tests also identified many variants of uncertain significance, where the classification is either uncertain or simply just not possible—and this information cannot safely be utilized clinically [19].

Indeed, this consideration is shared in recent literature and has been extended by Thompson et al. where the authors examined various genes that are common on hereditary breast cancer panels. They observed that the frequency of mutations in most breast cancer panel genes is quite low, or even, in most cases, similar to the frequency of mutations observed in cancer-free population controls. They concluded that panels have the potential to provide clinical misinformation and harm at the individual level if the data is not interpreted with extreme caution [14]. This lack of evidence for new genetic players is not limited to breast and ovarian cancer, but it also plagues many other genetic diseases and also afflicts a great number of other multigene panels [19].

Due to this lack of supportive data, international large-scale studies into the genes included in these multigene panels are absolutely critical to increase the utility of the information yielded and also to ensure that the new genetic information presented is both safe and useful in a clinical setting.

6. Conclusions

Keeping in-line with global regulation entities such as CAT, EGAPP, NHS UKGTN, and the ACCE backbone discussed above, the recent initiative eviQ provides a variety of guidelines for cancer genetics investigations. EviQ is part of the Cancer Institute NSW (New South Wales) and provides evidence-based information to support health professionals in the delivery of cancer treatments available at the time treatment decisions are being made. We can see from the eviQ's general practitioner referral guidelines for cancer genetics assessment that generally, even in oncology, genetics plays a large part in the diagnostics of disease [23]. We can also note that there must be a strong clinical suspicion of the diagnosis involving a genetic element (e.g., quite young, strong family history, hailing from a region with restricted gene flow, etc.) to warrant testing and for it to be useful.

The clinical utility of genetics is highly variable and dependent on the gene or disorder involved, but genomics medicine appears to be very good at revealing a diagnosis, and, at times, it can help explain why the phenotype is the way it is. In oncology, genetic testing plays an integral role in disease management by influencing treatment options or by being a major inclusion component for clinical trials. Where genetic testing is taking off and, perhaps, where its true potential lies is in its ability to offer predictive and preventative medicine, particularly for families, as opposed to adhering to a purely reactive approach that is typically employed in mainstream medicine.

Genomics is by no means irrelevant; it is revealing much about human disease and pathophysiology, but barring the current leaps and bounds observed in oncology, genomics medicine is still in an informational gathering phase—albeit, it is doing so at an alarming and unparalleled rate. It will take a great deal more data, analysis, and time before it will be considered true mainstream medicine, but there is no doubt that genomics is the future of medicine.

Conflict of interest

The authors declare no conflict of interest.

Author details

Tom Nolis and Rodney J. Scott*
NSW Health Pathology North, Hunter Medical Research Institute, Newcastle, NSW, Australia

*Address all correspondence to: rodney.scott@newcastle.edu.au

References

[1] Grosse SD, Khoury MJ. What is the clinical utility of genetic testing? Genetics in Medicine. 2006;**8**(7): 448-450

[2] Teutsch SM, Bradley LA, Palomaki GE, Haddow JE, Piper M, Calonge N, et al. The evaluation of genomic applications in practice and prevention (EGAPP) initiative: Methods of the EGAPP working group. Genetics in Medicine. 2009;**11**(1):3-14

[3] Morrison A, Boudreau R. Evaluation Frameworks for Genetic Tests [Environmental Scan issue 37]. Ottawa: Canadian Agency for Drugs and Technologies in Health; 2012

[4] Murad MH, Montori VM, Ioannidis JPA, Jaeschke R, Devereaux PJ, Prasad K, et al. How to read a systematic review and meta-analysis and apply the results to patient care: Users' guides to the medical literature. JAMA - The Journal of the American Medical Association. 2014;**312**(2):171-179

[5] Pitini E, De Vito C, Marzuillo C, D'Andrea E, Rosso A, Federici A, et al. How is genetic testing evaluated? A systematic review of the literature. European Journal of Human Genetics. 2018;**26**(5):605-615. Available from: http://dx.doi.org/10.1038/s41431-018-0095-5

[6] Fardo DW, Charnigo R. Families or unrelated: The evolving debate in genetic association studies. Journal of Biometrics and Biostatistics. 2012;**3**(4): 10-11. Available from: http://www.omicsonline.org/2155-6180/2155-6180-3-e108.digital/2155-6180-3-e108.html

[7] Lindor NM, Hopper J, Dowty J. Estimating cumulative risks for breast cancer for carriers of variants in uncommon genes. Familial Cancer. 2016;**15**(3):367-370

[8] Guidugli L, Shimelis H, Masica DL, Pankratz VS, Lipton GB, Singh N, et al. Assessment of the clinical relevance of BRCA2 missense variants by functional and computational approaches. American Journal of Human Genetics. 2018:233-248

[9] Miki Y, Swensen J, Shattuck-eidens D, Futreal PA, Harshman K, Tavtigian S, et al. Strong candidate for the breast and ovarian cancer susceptibility gene BRCA1. Science (80-). 1994;**266** (October):66-71

[10] Causeway W. CHEK2*1100delC and susceptibility to breast cancer: A collaborative analysis involving 10,860 breast cancer cases and 9,065 controls from 10 studies. American Journal of Human Genetics. 2004;**74**(6):1175-1182. Available from: http://linkinghub.elsevier.com/retrieve/pii/S0002929707627049

[11] Renwick A, Thompson D, Seal S, Kelly P, Chagtai T, Ahmed M, et al. ATM mutations that cause ataxia-telangiectasia are breast cancer susceptibility alleles. Nature Genetics. 2006;**38**:873. Available from: http://dx.doi.org/10.1038/ng1837

[12] Weber-Lassalle N, Hauke J, Ramser J, Richters L, Groß E, Blümcke B, et al. BRIP1 loss-of-function mutations confer high risk for familial ovarian cancer, but not familial breast cancer. Breast Cancer Research. 2018;**20**(1):1-6

[13] Rahman N, Seal S, Thompson D, Kelly P, Renwick A, Elliott A, et al. PALB2, which encodes a BRCA2-interacting protein, is a breast cancer susceptibility gene. Nature Genetics. 2007 Feb;**39**(2):165-167. Available from: http://www.nature.com/articles/ng1959

[14] Thompson ER, Rowley SM, Li N, McInerny S, Devereux L, Wong-Brown

MW, et al. Panel testing for familial breast cancer: Calibrating the tension between research and clinical care. Journal of Clinical Oncology. 2016; **34**(13):1455-1459

[15] Invitae. Invitae Breast Cancer Panel [Internet]. 2018 Available from: https://www.invitae.com/en/physician/tests/01202/ [Accessed: 2018-07-16]

[16] Clinical Resource: Panel Testing for Familial Breast Cancer 2018 V.1, eviQ Cancer Treatments Online, Cancer Institute NSW, viewed 16th July 2018, https://www.eviq.org.au/cancer-genetics/genetic-testing-using-cancer-gene-panels/3413-panel-testing-for-famil.

[17] Byrnes GB, Southey MC, Hopper JL. Are the so-called low penetrance breast cancer genes, ATM, BRIP1, PALB2 and CHEK2, high risk for women with strong family histories? Breast Cancer Research. 2008;**10**(3):208. Available from: http://breast-cancer-research.biomedcentral.com/articles/10.1186/bcr2009

[18] D MWA, E MJ. The risk of breast cancer due to PALB2 gene mutations. Advances in Clinical and Experimental Medicine. 2017;**26**(2):339-342

[19] Winship I, Southey MC. Gene panel testing for hereditary breast cancer. The Medical Journal of Australia. 2016; **204**(5):188-190.e1

[20] Banerjee T, Brosh RMJ. RECQL: A new breast cancer susceptibility gene. Cell Cycle. 2015;**14**(22):3540-3543

[21] Sun J, Wang Y, Xia Y, Xu Y, Ouyang T, Li J, et al. Mutations in RECQL gene are associated with predisposition to breast cancer. PLoS Genetics. 2015; **11**(5):4-15

[22] Li N, Rowley SM, Goode DL, Amarasinghe KC, McInerny S, Devereux L, et al. Mutations in RECQL are not associated with breast cancer risk in an Australian population. Nature Genetics. 2018;**89**:1. Available from: http://www.nature.com/articles/s41588-018-0206-9

[23] Clinical Resource: General Practitioner Referral Guidelines for Cancer Genetics Assessment 2017 V.4, eviQ Cancer Treatments Online, Cancer Institute NSW, viewed 16th July 2018, https://www.eviq.org.au/footer-pages/faqs/protocols-and-documents

Computational Health Informatics for Biomedical Applications

Edited by

Aryan Chaudhary
Sardar M. N. Islam (Naz)

AAP APPLE ACADEMIC PRESS

First edition published 2023

Apple Academic Press Inc.
1265 Goldenrod Circle, NE,
Palm Bay, FL 32905 USA

760 Laurentian Drive, Unit 19,
Burlington, ON L7N 0A4, CANADA

CRC Press
6000 Broken Sound Parkway NW,
Suite 300, Boca Raton, FL 33487-2742 USA

4 Park Square, Milton Park,
Abingdon, Oxon, OX14 4RN UK

Library and Archives Canada Cataloguing in Publication

Title: Computational health informatics for biomedical applications / edited by Aryan Chaudhary, PhD, Sardar M. N. Islam (Naz), PhD.
Names: Chaudhary, Aryan, editor. | Islam, Sardar M. N., 1950- editor.
Description: First edition. | Includes bibliographical references and index.
Identifiers: Canadiana (print) 2022048578X | Canadiana (ebook) 20220485887 | ISBN 9781774912539 (hardcover) | ISBN 9781774912546 (softcover) | ISBN 9781003331681 (ebook)
Subjects: LCSH: Medical informatics. | LCSH: Medicine—Data processing.
Classification: LCC R858 .C66 2023 | DDC 610.285—dc23

Library of Congress Cataloging-in-Publication Data

CIP data on file with US Library of Congress

ISBN: 978-1-77491-253-9 (hbk)
ISBN: 978-1-77491-254-6 (pbk)
ISBN: 978-1-00333-168-1 (ebk)

Dedication

This book is dedicated to Lord Ram and
My Beloved Parents (Navneet and Parul).
—Aryan Chaudhary

About the Editors

Aryan Chaudhary
Research Head, Nijji HealthCare Pvt Ltd., Kolkata, West Bengal, India

Aryan Chaudhary is the Research Head and Lead Member of the research project launched by Nijji Healthcare Pvt Ltd. He focuses on implementing technologies such as artificial intelligence, deep learning, IoT (Internet of Things), cognitive technology, and the blockchain to better the healthcare sector. He has published academic papers on public health and digital health in international journals and has participated as a keynote speaker at many international and national conferences. His research includes the integration of IoT and sensor technology for gathering vital signs through one-time/ambulatory monitoring and then the functionality of artificial intelligence and machine learning leading to big data analytics for faster intervention, tracking of prognosis, and research on these vast data for effective clinical research for future development of treatment, drug, pathological tests, and supply systems. He is editor of many books on biomedical science and the Chief Editor of a CRC book series. He also serves as a guest editor of many special issues in reputed journals. He has been awarded with being named Most Inspiring Young Leader in Healthtech Space 2022 by Business Connect and the best project leader at Global Education and Corporate Leadership. He is the senior member of many international associations in science.

Sardar M. N. Islam (Naz), PhD
Professor, Institute for Sustainable Industries & Liveable Cities; Lead, Decision Sciences and Modeling Program, Victoria University, Australia

Sardar M. N. Islam (Naz), PhD, is currently a Professor at the Institute for Sustainable Industries and Liveable Cities and Lead of the Decision Sciences and Modeling Program at Victoria University, Australia. He is also a Distinguished Visiting Professor of Artificial Intelligence at Sriwijaya University (UnSri), Indonesia, and a Distinguished Visiting Professor (2019–2021) at American University of Ras Al Khaimah (AURAK), United Arab of Emirates. His academic work has gained international acclaim resulting in many honors and awards, visiting and adjunct professorial appointments in different countries, appointments in editorial roles of journals, as a keynote speaker at international conferences in several countries. He has published over 31 scholarly authored academic books in different disciplines; each of these books makes significant scientific contributions to the literature. He has also published about 250 articles, including some in the top leading international journal articles in his specialized research areas.

Contents

Contributors

Mohit Angurala
Khalsa College of Engineering and Technology, Amritsar, Punjab India,
E-mail: mohitpit@gmail.com

Manju Bala
Khalsa College of Engineering and Technology, Amritsar, Punjab, India

Rasika Bhise
Vishwakarma Institute of Information Technology, Pune, Maharashtra, India

Raman Chadha
UIE, Department of Computer Science and Engineering, Chandigarh University, Gharuan, Punjab, India, E-mail: dr.ramanchadha@gmail.com

Aryan Chaudhary
Nijji Health Care Pvt Ltd., Kolkata, West Bengal, India;
Email: research@nijjihealthcare.com / aryanchaudhary500@gmail.com

Pravin Futane
Vishwakarma Institute of Information Technology, Pune, Maharashtra, India

Ranjith B. Gowda
Department of Electronics and Communication Engineering, Dayananda Sagar University, Bangalore, Karnataka, India

Sardar M. N. Islam (Naz)
Institute for Sustainable Industries & Liveable Cities; Decision Sciences and Modeling Program, Victoria University, Australia; Email: Sardar.Islam@vu.edu.au / profdrsardarmni@gmail.com

Jisu Elsa Jacob
Department of Electronics and Communication Engineering, Sree Chitra Thirunal College of Engineering, Thiruvananthapuram, Kerala, India, E-mail: jisuelsa@sctce.ac.in

Chinju John
Department of Computer Science and Engineering, Indian Institute of Information Technology, Kottayam, Kerala, India, E-mail: chinjuj.phd201001@iiitkottayam.ac.in

Venusamy Kanagraj
Engineering Department, University of Technology and Applied Sciences-Al Mussanah, Sultanate of Oman

Keshav Kaushik
School of Computer Science, University of Petroleum and Energy Studies, Dehradun, Uttarakhand, India, E-mail: officialkeshavkaushik@gmail.com

Tejas Khopade
Vishwakarma Institute of Information Technology, Pune, Maharashtra, India

R. Maheswari
School of Computer Science and Engineering, Vellore Institute of Technology, Chennai, Tamil Nadu, India, E-mail: maheswari.r@vit.ac.in

Priyanka Mandal
Vishwakarma Institute of Information Technology, Pune, Maharashtra, India

N. Manjunath
Department of Mechanical Engineering, College of Engineering Trivandrum, Kerala, India, E-mail: iammanjunathn@gmail.com

R. Faerie Mattins
School of Computer Science and Engineering, Vellore Institute of Technology, Chennai, Tamil Nadu, India

Akarsh K. Nair
Department of Computer Science and Engineering, Indian Institute of Information Technology, Kottayam, Kerala, India

V. Pattabiraman
Vellore Institute of Technology, Chennai, Tamil Nadu, India

Suvarna Pawar
School of Computing, MIT ADT University, Pune, E-mail: pawar.suvarna@gmail.com

M. S. Guru Prasad
Graphic Era (Deemed to be University), Dehradun, Uttarakhand, India

Tejas Rasane
Vishwakarma Institute of Information Technology, Pune, Maharashtra, India

Jayakrushna Sahoo
Department of Computer Science and Engineering, Indian Institute of Information Technology, Kottayam, Kerala, India

Preeta Sharan
Department of Electronics and Communication Engineering, The Oxford College of Engineering, Bangalore, Karnataka, India, E-mail: sharanpreeta@gmail.com

Kirandeep Singh
Chandigarh University, Punjab, India

Prabhdeep Singh
Graphic Era (Deemed to be University), Dehradun, Uttarakhand, India, E-mail: ssingh.prabhdeep@gmail.com

S. Srivarshan
School of Computer Science and Engineering, Vellore Institute of Technology, Chennai, Tamil Nadu, India

Nilesh Uke
Trinity Academy of Engineering, Pune, Maharashtra, India

Anup M. Upadhyaya
Department of Mechanical Engineering the Oxford College of Engineering, Bangalore, Karnataka, India

Pavitra Vasudevan
School of Computer Science and Engineering, Vellore Institute of Technology, Chennai, Tamil Nadu, India

Rohit Kumar Verma
Assistant Professor, Department of Computer Science, Himachal Pradesh University Regional Center, Dharamshala, Kangra, Himachal Pradesh, India

Abbreviations

1D	one-dimensional
ADAMM	automated device for asthma monitoring and management
ADME	absorption, distribution, metabolism, excretion
AES	advanced encryption standard
AI	artificial intelligence
AMD	advanced micro device
AM-FM	amplitude- and frequency-modulated
ANN	artificial neural network
ApEn	approximate entropy
API	application program interface
ARPT	active reader passive tag
ASD	autistic spectrum disorder
AUC	appropriate use criteria
BCI	brain-computer interfaces
BHT	bio-heat transfer
BLE	bluetooth low energy
BP	blood pressure
BPD	bipolar disorder
CA	number of vessels
CAD	computer-aided diagnosis
CADD	computer-aided drug discovery
CAE	computer-aided engineering
CD	correlation dimension
CEA	carcinoembryonic antigen
CEM	cumulative effective minutes
CHD	coronary heart disease
CHOL	serum cholesterol
CHPP	continuous hyperthermic peritoneal perfusion
CMS	central monitoring system
CNN	convolutional neural network
COVID	coronavirus disease
CP	chest pain
CPS	central pain syndrome

CRM	customer relationship management
CSS	cascading style sheets
CSV	comma separated values
CT	computed tomography
CT	continuous-time
CVD	chemical vapor deposition
CVRP	captivated vehicle routing problem
DBR	distributed Bragg reflector
DL	deep learning
DNA	deoxyribonucleic acid
DNN	deep neural networks
DoS	denial of service
DR	diabetic retinopathy
DWT	discrete wavelet transform
ECG	electrocardiogram
EDPS	early detection and prevention system
EEG	electroencephalogram
EHR	electronic health record
ELISA	enzyme-linked immunosorbent assay
EMR	electronic medical records
EMW	electromagnetic wave
ERD	event-related desynchronization
EXANG	exercise-induced angina
FBS	fasting blood sugar
FD	fractal dimension
FDA	Food and Drug Administration
FFT	fast Fourier transform
FNN	false nearest neighbor
FP	fiber optic
GPS	global positioning system
HDL	high-density lipoprotein
HIE	health information exchange
HIFU	high-intensity focused ultrasound
HIMS	health information management systems
HIMSS	Healthcare Information and Management Systems Society
HIS	hyperspectral imaging
HPC	high power command
HR	heart rate
HSPs	heat shock proteins

HT	hyperthermia therapy
HTML	hypertext markup language
HTVS	high throughput virtual screening
ICT	information and communication technology
ICU	intensive care unit
IMFs	intrinsic mode functions
IoMT	internet of medical things
IoT	internet of things
IoV	internet of vehicles
IrDA	infrared data association
KNN	K-nearest neighbor
LAT	latitude
LBDD	ligand-based drug discovery
LBVS	ligand-based virtual screening
LDL	low-density lipoprotein
LIGA	lithography galvanoformung abformung
LLE	largest Lyapunov exponent
LOC	lab-on-chip
LONG	longitude
LOS	length of stay
LR	linear regression
L-RRN	layer-recurrent neural networks
LSTM	long short-term memory
LTE	long-term evolution
MAFIA	maximal frequent item set algorithm
MDS	medical drone system
ME-Drone	medication extender drone
MEG	magnetoencephalography
MEMS	micro-electro-mechanical system
MG	mammography
MG/DL	milligrams per deciliter
MI	mutual information
MIF	micro-immunofluorescence
ML	machine learning
MMSE	mini-mental state examination
MRI	magnetic resource imaging
MWA	microwave ablation
NADE	neural autoregressive distribution estimation
NB	Naïve Bayesian

NFC	near-field communication
NGS	next-generation sequencing
NHS	National Health Service
NLP	natural language processing
NMR	nuclear magnetic resonance
NPV	negative prescient worth
NPV	net present value
OECD	Organization for Economic Cooperation and Development
OLDPEAK	ST depression
ORIM	requirements-oriented modeling
OTP	one-time password
PACS	picture archiving and communication system
PC	photonic crystals
PDE	partial differential equation
PET	positron discharge tomography
PII	personally identifiable information
PIMS	photonic integrated microcantilever sensor
PPV	positive predictive value
PSD	power spectral density
PSOWNN	particle swarm optimized wavelet neural network
PUFs	physical unclonable functions
Q	quality factor
QAM	quadrature amplitude modulation
QR code	quick response code
RAT	radio access technology
RESTECG	rest electrocardiograph
RF	radio frequency
RF	rheumatoid factor
RFA	radiofrequency ablation
RFID	radiofrequency identification
RI	refractive index
RIM	reference information model
RL	deep reinforcement learning
RMD	random matrix discriminant
RMS	root means square
RNN	recurrent neural network
ROC	receiver operating characteristic
RPM	remote patient monitoring

RT-PCR	reverse transcription–polymerase chain reaction
S	sensitivity
SAML	security assertion markup language
SampEn	sample entropy
SAR	structure-activity relationship
SARS-CoV	2-severe acute respiratory syndrome coronavirus
SBDD	structure-based drug discovery
SDN	software-defined network
SN	sensor node
SPA	Sanus per aqua
SPECT	single-photon emanation registered to picture
SPR	surface plasmon resonance
SSVEP	steady-state visual evoked potential
SVC	support vector machine classification
SVM	support vector machine
SVR	support vector machine regression
THAL	thalassemia
THALCH	maximum heart rate
TMM	transfer matrix method
TRESTBPS	resting blood pressure
TSP	traveling salesman problem
TV	television
UAV	unmanned aerial vehicles
UI/UX	user interface/user experience
US	ultrasound
VAMP	variational approach for Markov processes
VMs	virtual machines
VoIP	voice over IP
VR	virtual reality
VTC	video teleconferencing
VTK	visualization tool kit
WBH	whole-body hyperthermia
WDM	wavelength division multiplexing
WHO	World Health Organization
WPAN	wireless personal area network
WSN	wireless sensor network
XML	extensible markup language

Symbols

%	percentage
∇	gradient operator
A	Arrhenius constant
B	magnetic flux density
c	velocity of light
C(r)	correlation sum
Cbl	specific heat
CEM43°C	cumulative number of equivalent minutes at 43°C
D	electric flux density and
db4	Daubechies wavelet
E	electric field vector
E	heat of inactivation in kcal/mole
FWHM	full-width half-maximum
H	magnetic flux vector
J	current density
K	Bloch wave number
k_1, k_2	wavenumber
l_H	geometrical length of material 1
l_L	geometrical length of material 2
Lm (k)	length of the curve
m	embedding dimension
M	magnetic polarization
n_H	refractive index of material 1
n_L	refractive index of material 2
P	electric polarization
Q_{met}	metabolic heat generated
R	reflection
R	the Molar gas constant in Kcal/mole-K
RIU	refractive Index unit
T	absolute temperature in K
t	time interval in minutes
T	transmission
Ta	arterial blood temperature

X	horizontal axis
Y	vertical axis
α	alpha sub-band
β	beta sub-band
Δ	delta sub-band
Δn	change in refractive index
$\Delta\lambda$	change in resonant wavelength
ε	electric permittivity
θ	incident angle of light
θ	theta sub-band
λ	wavelength of incident light
μ	magnetic permittivity
ρ	charge density
ρbl	blood density
τ	time delay
ω	blood perfusion rate

Foreword

It is my pleasure to write the foreword for this book on computational health. Being in this domain for a considerable amount of time, I feel highly connected to the topic and look forward to the contribution of this book to existing literature.

Although the domain itself is very nascent, previous approaches have established promising potentials in healthcare computation. The potential is that the WHO itself, in its June 2021 global report, has directed the attention of various governments and stakeholders to the ethics and governance of artificial intelligence (AI) in healthcare. Over the last couple of years, computational prowess has aided in combating the COVID-19 pandemic to date. It, itself, is an example of growing computational support and interest in healthcare. Healthcare has witnessed phenomenal approaches from varied dimensions apart from AI, such as computation related to electro-mechanical observations in health, the internet of medical things, drone-enabled healthcare, health signal processing, and extensive health data analysis, to name a few. The broad domain of learning algorithms has contributed enormous proposals for healthcare informatics, taking aid from machine learning, deep learning, transfer learning, and recent advances in reinforcement learning. Precision healthcare computation involves autonomous health device management, which is also considered to be one of the highly sought-after domains. With the rise of connected and intelligent health devices, monitoring security at the device and framework levels extends up to the edge and the cloud is imperative. In these regards, the present book looks promising in terms of adding value to the body of knowledge.

Contributions made to this book cover a wide area of computational healthcare. It ranges from analysis of health data, connecting, and computation based on the health sensors, use of learning algorithms for healthcare, biomedical signal processing, security concerns in connected devices and healthcare computation, and recent and emerging trends in healthcare, including the future dimensions. This book envelopes almost all of the prevalent and attention-specific areas in computational health informatics to the best of my belief. The use of computation in healthcare typically concentrates on two wide branches: the analysis of clinical data, and the other focuses on drug, dosage monitoring, and analysis. Also, it can be argued that, to a certain extent, drug composition calculations and genetic computation are also a part of the computational healthcare domain. The book also has chapters that are dedicated to the possibilities of AI in the health segment, emerging technologies for drug development, and the future dimensions for research on the topic.

Given the recent influence of AI in almost all research areas, healthcare is posed to extract the most benefit from the rise of technology. Books like these are poised to spread the latest research and developments and generate interest in incoming researchers to explore the promising domain of computational healthcare. Many universities around the globe have already started courses on computational healthcare in collaboration with leading technological giants. The interest of many global firms in this domain has been observed over the last few years. Health assistive technology has helped society in almost all age groups.

So, in my view, books like this are certainly going to generate interest in readers and researchers at par, together with spreading the recent knowledge in the domain. This book will definitely add value to the existing body of knowledge.

—**Diganta Sengupta, PhD**
Meghnad Saha Institute of Technology,
Kolkata, West Bengal, India

Acknowledgments

"Presentation, inspiration, and motivation have always played an important role in the success of any venture."

Firstly, I thank the Almighty for his inspiration and benevolence for giving me the opportunity to design this book.

I want to acknowledge the help of all the people involved in this project and, more specifically, the authors and reviewers that took part in the review process. Without their support, this book would not have become a reality. I want to thank each one of the authors for their contributions.

I pay my sincere gratitude to Tanya, who always told me to aim high and says that do not to aim low; otherwise, I will miss the mark. I am immensely obligated to my two teachers, R. K. Khangar and Arjun Singh, for their elevating inspiration, encouraging guidance, and kind supervision throughout the journey.

I wish to thank the officials at Apple Academic Press for their invaluable efforts, great support, and valuable advice for this project towards the successful publication of this book.

I am deeply grateful to my parents for their support, appreciation, encouragement, and a keen interest in my academic achievements.

—**Aryan Chaudhary**
Research Head, Nijji Healthcare Pvt. Ltd., India

Preface

The explosion of technology in healthcare has rapidly changed the health sector. Over the past few years, technologies like artificial intelligence (AI), machine learning (ML), neural networks, along with integration of the internet of medical things (IoMT), have evolved to tackle the need for a remote healthcare system, offering a less-contact diagnostic and colossal scope for augmenting healthcare system in a self-sustainable way. The waves of technologies have opened an opportunity to transform every industry. Various methodologies of computational health informatics techniques are evolving in the health sector, resulting in the model's efficiency and precision. The aim of this book is to prepare the research scholar, healthcare professional, information technology professional, and biomedical engineer to understand the computational techniques used in health informatics and related health science concepts.

The book describes computational tactics, including AI, ML, signal analysis, computer-aided design, robotics, and automation, biomedical imaging, telemedicine, and a lot more.

The book contains 14 chapters, which provide a solid framework to generate a modern class of medical gearheads who will understand and apply the computational mechanisms to expedite patient-amiable automation to improve healthcare.

Chapter 1 examines all of the essential points of view and compares the importance and necessity of each in terms of healthcare IoT services, applications, and security. Advancements in modern technology introduce many new methods for diagnosing major life-threatening diseases. In its preliminary stage, a meticulous detection of the disease will decrease the risk influence at a higher stage. Chapter 2 includes a 1D photonic crystal-based DBR used to detect foreign bodies like cancer cells or the infected stages of blood cells for quick medical diagnosis. Biosensor chips are considered an easy and simple detection and diagnosis system for different diseases.

Chapter 3 consists of the model of different types of photonic MEMS sensors, which are designed and discussed for different biomedical applications. The sensing mechanisms, fabrication process, and sensing principle of each kind of sensor are discussed. Chapter 4 summarizes the emphasis on various chaotic and non-linear features for EEG to diagnose multiple neurological diseases. Various artificial intelligence techniques have already been explored and utilized in the healthcare sector, including machine learning, deep learning, etc. Chapter 5 provides a pragmatic analysis of the application of Ai and Dl in healthcare. Machine learning (ML) might be used to track the disease, predict its progress, and design tactics and legislation to control it.

Predictive analysis has become a critical component for future prediction as the science of machine learning has progressed. Heart disease is often recognized as a cardiovascular disease that denotes a range of conditions that stress the heart and has become the prominent cause of death worldwide in recent decades. Chapter 6 is enhanced with an enriched content of various machine learning methods, grouping, and association rules, vector machine assistance, and evolutionary algorithms.

Chapter 7 starts with predicting the coronavirus outbreak and detection through chest X-rays with a machine learning approach. In the work, two supervised learning models experimented with anticipating the future using the COVID-19 time-series dataset during the second pandemic wave from 20 January 2020 to 20 March 2020. The evolution and implementation of information and communication technologies into healthcare delivery hold enormous promise for patients, providers, and payers in future healthcare systems.

Chapter 8 contains a detailed study on numerical simulations of the different procedures in the field of 'thermal medicine' that are analyzed and compared by utilizing computer-aided designing. Powerful artificial intelligence algorithms can reveal clinically significant information hidden in vast amounts of data.

Chapter 9 highlights the role of artificial intelligence in healthcare; it also enlightens the readers about the research scope of deep learning in healthcare. Drones can deliver payloads faster and more efficiently because they are small and lightweight. For this reason, drones in health centers have endless applications. Chapter 10 focuses on simulating a ME-drone, using the tool CoppeliaSim for drug distribution within hospital premises.

Chapter 11 addresses the advancement of information and communication technologies that are used in the development of the health sector. Biomedical signal processing is utilized to separate signs of revenue and perform all things considered examination. Chapter 12 provides the signal processing advancement used in healthcare.

The conventional data storage and management approach for healthcare data leads it to be highly disseminated, resulting in lesser quality and value of data. Thus, the need for aggregating data was always high as it is an idealistic alternative for ensuring data usability in the future. Chapter 13 concludes that applications of AI in the healthcare sector have a never-ending scope.

Chapter 14 is an integrated view of the state of the art of digitalization in the healthcare informatics literature, is presented to highlight the important aspects and commercial uses of the latest technologies in the healthcare system.

CHAPTER 1

IoT-Based Healthcare Systems and Their Security Concerns

MOHIT ANGURALA[1], MANJU BALA[1], and PRABHDEEP SINGH[2]

[1]*Khalsa College of Engineering and Technology, Amritsar, Punjab, India*

[2]*Graphic Era (Deemed to be University), Dehradun, Uttarakhand, India India, E-mail: rtyagi@ddn.upes.ac.in (R. Tyagi)*

ABSTRACT

Many IoT-based healthcare systems are being used in the healthcare business to monitor or diagnose a patient's condition. A wide range of technological advances has led to a rise in health concerns among people, including the use of mobile phones has an impact on the eyes and brain, X-ray equipment raises the risk of cancer among patients. Many industries and researchers have come up with low-cost, powerful smart gadgets as a result of the rise in the population and the spread of disease. IoT-based healthcare solutions are becoming more popular as a result of the rapid advancement in wireless communication technology. Biosignal sensors and devices' dependability and resilience are critical to the implementation of an IoT-based healthcare system at the highest level. These biosignals are generated as a result of changes in the membrane potentials of cells and include a wealth of physiological information. To find abnormalities in humans, the deviations are compared to normal patterns. These biosignals require unique bio-compatible measuring techniques that have certain qualities, such as low noise exposure. Additional steps are needed to ensure worldwide access to information,

Computational Health Informatics for Biomedical Applications. Aryan Chaudhary and Sardar M. N. Islam (Naz) (Eds.)

including digitization, filtering, and processing of data signals. This chapter examines all of the important points of view and compares the importance and necessity of each in terms of healthcare IoT services, applications, and security. IoT-based healthcare systems are introduced in this chapter, along with security and other issues that must be addressed to ensure optimal resource usage.

1.1 INTRODUCTION

The Internet of Medical Things (IOMTs) is the amalgamation of clinical gadgets with the Internet of Things (IoT). In today's era, recent medical care frameworks in the future will depend upon IOMT where each clinical gadget will be associated and observed over the internet through medical services experts. This offers a quicker and lower cost of medical services as it advances. Figure 1.1 illustrates an architecture of IOMT in which the details about the patient vitals are gathered or collected by the means of sensors and are then forwarded to the IOMT applications via the internet. The data is then streamed to the medical professional person (expert) and clinical staff and finally, a reaction is sent to the patients.

Ever since the COVID-19 outbreak, medical professionals (doctors) are presently giving telehealth services to patients remotely via IoMT. This is happening because of the social distance rule, which has to be followed by every individual living in the society. Great well-being and prosperity are some of the significant objectives of sustainable development goal. At present, IoMT can satisfy the objective of good well-being and prosperity. This chapter is tied in with giving an outline of IoMT, IoMT, and its wireless technologies, Smart E-healthcare, and a lot more. By 2025, the IoMT is anticipated to have an economic effect of $6 trillion, making it one of the most disruptive technologies in history. These gadgets will allow tens of millions of people to communicate with one another. About 60% of Indian hospitals have already used the IoMT to assist patient care activities. Cloud-based patient records may be accessed from any computer, tablet, or smartphone with internet connectivity, within or outside the hospital. It's no surprise that IoMT has become so highly sought after. As part of the healthcare business, hospitals depend on IoMT to provide quick service. The introduction of IoMT has boosted the productivity of the workforce. There has been

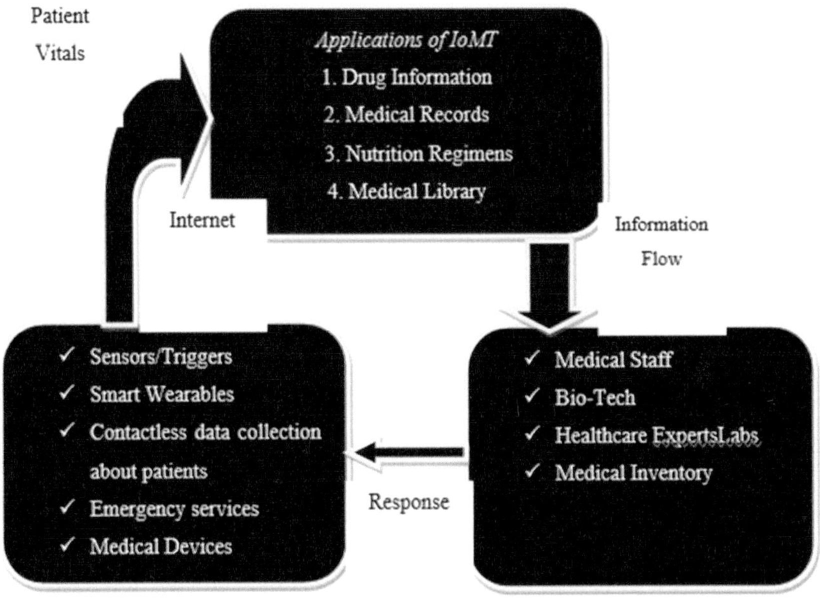

FIGURE 1.1 Internet of medical things.

a decrease in the cost of medical treatment as well. IoMT may also be used to keep tabs on patients daily, depending on the situation. This information is quickly shared with the appropriate individuals so that patients' treatment plans may be tailored to their specific needs. Hospitals keep this information on file for future use in the event of the same patient or comparable circumstances. The data created in the healthcare business, on the other hand, is very sensitive. A breach of privacy is a violation of this data's confidentiality, and the industry must do all in its power to keep it safe. Unfortunately, there isn't much that IoMT does to secure data. As a consequence, sensitive information is exposed. As a result, healthcare organizations that use IoMT are particularly concerned about data security. Among the seven healthcare industries in which IoMT has provided timely answers, there are seven. These regions are seeing an increase in the number of clinical cases. Increasing clinical efficiency, increasing consumer or home monitoring after patient discharge, fitness wearables that quickly provide patients with specific numbers indicating their health status, increasing infant monitoring, increasing biometric sensors and wearables, introducing sleep monitors, and introducing brain sensors in the field of neurotechnology are just a few of these areas. The advantages of IoMT are many. First and foremost, it provides round-the-clock monitoring for patients. This has the potential to expedite patient diagnoses while also raising diagnostic efficacy. Patients may also be monitored from afar through remote access. As a result, the expenditures of medical treatment have been greatly decreased. The management of the disease is made easier. For each patient, the doctors design a treatment plan that is tailored to their specific requirements. There is room for improvement in drug management as well.

1.2 APPLICATIONS OF IOMT

Healthcare IoT applications The Measurement of Blood Pressure (BP) One of the most critical measures of a person's health is their BP. Monitors that are both safe and easy to use are becoming ubiquitous. IoT devices or sensors are used to link healthcare equipment/systems, making it easier for physicians and caregivers to communicate with patients. IoT sensors are linked to an electronic BP monitor to obtain real-time data on patient BP levels. The system of rehabilitation People with impairments may benefit from a rehabilitation system by having their functional skills enhanced and restored, as well as their overall quality of life improved, while also helping to alleviate issues associated with an aging population and a scarcity of healthcare professionals. There is a smart rehabilitation approach that is community-based and successful. An ontology-based automated design technique linked to an IoT-based smart rehabilitation system may make it easy to communicate and allocate medical resources according to patient needs.

1.2.1 SATURATION OF OXYGEN IN THE BLOOD

The pulse oximeter is a non-invasive device that continually checks the patient's blood oxygen saturation. As wireless networks and medical sensors improve in terms of power consumption and loss, the market for these devices is increasing. In many medical applications, pulse oximeters are used to measure blood oxygen levels and heart rate (HR). The patient's HR and oxygen levels may be monitored and sensed by the IoT sensor attached to the patient's body.

1.2.2 MANAGEMENT OF WHEELCHAIRS

Wheelchairs are used by those who are unable to walk due to disease or other physical limitations and must rely on outside assistance. It is possible to employ a wireless body area network (WBAN) as a person-centric sensing (sensor) device for wheelchair users. It will be possible to detect whether a person is slipping out of a wheelchair using a pressure cushion (a resistive pressure sensor). There is a second accelerator sensor on a smart wheelchair to detect a wheelchair fall. From the hospital, the doctor or caregiver may keep an eye on the patient's data at all times.

1.2.3 SOLUTIONS FOR HEALTHCARE

As a result of this, healthcare providers may profit greatly from the use of mobile devices and healthcare applications (HCPs). Health records, information, time, contact and consultation with physicians, continuous patient monitoring, and effective clinical decision-making are just a few of the numerous medical healthcare apps that are now readily available and accessible. Improved patient outcomes will be aided by increasing access to healthcare via smartphone applications and sensors. Modifications such as sensors, signal converters, and communication modems may turn existing medical equipment into IoMT devices that collect real-time data for patient monitoring. Wearable devices, home medical devices, point-of-care kits, and mobile healthcare apps are all examples of IoMT devices. They can connect with medical specialists in faraway places over the internet. Additionally, they have been used to prevent and treat illness, promote fitness, and provide remote assistance in times of crisis. A list of IoMT application areas is discussed in further sections.

1.2.4 MANAGEMENT OF LONG-TERM ILLNESS

Promising options for managing chronic morbid diseases such as hypertension, heart failure, and diabetes are available with IoMT-enabled devices. For example, BP, blood sugar levels, body weight, and electrolyte concentrations may all be monitored with these devices. It is possible to anticipate the progression of a disease using the real-time vital data gathered by these devices and analyzed at a higher level. To learn more about a given disease's prevalence in a particular community, researchers might take advantage of centralized data collecting.

1.2.5 RESPITE CARE IN A REMOTE LOCATION (TELE HEALTH)

The doctor's office maintains a central database where information from network devices may be accessed. Healthcare automation is enabled by the collection and processing of patient-specific data, which compares new data to previous records and determines the patient's future course of treatment. Using IoMT machines for routing, monitoring, and field administration reduces the amount of money service providers have to spend on deploying and maintaining follow-up resources and infrastructure. Members' drop-out rates have decreased, and healthcare resources are working more efficiently as a result of remote monitoring. For cardiac monitoring, Body Guardian Remote Monitoring System is a marketed system that separates the patient's identity information and observation data to preserve confidentiality. This is further supported by the use of encryption technologies for transmitting and storing crucial information.

1.2.6 HEALTH PROMOTION AND DISEASE PREVENTION (LIFESTYLE ASSESSMENT)

Monitoring systems for food, exercise, and quality of life have been made possible by IoMT-enabled devices. Continuous data on patient activity and associated essential changes may be tracked using cutting-edge technology such as wearables, implanted chips, and embedded systems in biomedical equipment. Users of smart devices may link numerous important events with local health issues thanks to the advanced sensors, converters, and firmware in these devices. Additionally, these devices' remote networking capabilities allow for professional help at any distant place in the event of an emergency.

1.2.7 INTERVENTION FROM A DISTANCE

Physicians may use real-time data from sensors to give medications and analyze their reactions in the event of an emergency. Hospitalization costs are reduced by providing high-tech medical support on time.

1.2.8 MORE EFFECTIVE DRUG ADMINISTRATION

RFID tags based on the IoT are used to address supply and distribution issues. Radio-frequency identification and supply chain management have been recommended by the Food and Drug Administration (FDA). Manufacturers can assure supply chain quality with the inclusion of tags on pharmaceutical packaging. IoT "smart" pills produced by WuXi PharmaTech and TruTag Technologies may also be used to assist monitor medicine dosages and the patient's pharmacodynamics. Pharmaceutical firms may benefit from these types of solutions by reducing the risks and costs associated with the distribution and administration of their products. Applications in the following areas:

- Access to electronic health records (PHR/EHR) without losing medical information.
- Online protein analysis and correctness of composition.
- Training courses and coaching representation to paramedical personnel chronic illness management has the most potential for IoT-based devices while taking into account the aforementioned use cases.

1.3 LITERATURE SURVEY

Wu et al. [1] discussed the functions or roles which information and communication technologies play in pursuing sustainable development goals. The authors in this work identify various gaps in the research and found that research works to date have mostly focused on the technical aspects therefore, the way to revolutionize information and communication technologies to assist nations in achieving sustainable development goals to become important. Naresh et al. [2] provided an IoT survey in the healthcare field for smart healthcare along with a few applications of IoT in the medical field. Wang et al. [3] explored the way of applying blockchain to the IoT for Industries and proposed a few recommendations to direct blockchain researchers for the future. Taylor et al. [4] recognizes traditional works which make use of blockchain for cyber security use. They also presented a proper examination of the applications of blockchain securities. Dilawar et al. [5] discussed the blockchain technology and then presented an architecture for IoMT-based securities to employ Blockchain for ensuring data transmission in a secured way. Ahmed et al. [6] focused on analyzing numerous solutions of artificial intelligence (AI) and machine learning (ML) to path a way for innovative data-centric healthcare discoveries. Sharma [7] presented the architecture of AI along with a few open challenges. They also presented an automatic configuration technique to automatically deploy a Software Defined Network (SDN) on the Internet of Vehicles (IoV). Lalmuanawma, Hussain, and Chhakchhuak [8] extensively reviewed the role of ML and AI as a single technique to screen, predict, forecast, trace contacts, and develop drugs for COVID and related infections. Sedik et al. [9] provided the best COVID-19 detection technique based on deep learning (DL). The output of the simulation reveals that the proposed DL modalities could be one of the best and accepted solutions for rapid screening of coronavirus. Turabieh et al. [10] proposed a dynamic Layer-Recurrent Neural Networks (L-RRN) technique to get back any missing values from applications of IoMT. The proposed method helps in predicting the missing data quickly for saving cost as well as time. Khan et al. [11] presented PART also known as a partial tree (association rule learner) that has a modern set of features for detecting brain tumors of different grades ranging from grade 1 to 4. Kilic [12] provided a summary of AI and ML and its relation with cardiovascular healthcare. Song et al. [13] introduced deep reinforcement learning-based techniques. They also effectively utilized techniques based on neural networks to proficiently realize the deep reinforcement learning strategies for spectrum sensing and access. Girardi et al. [14] discussed a few problems related to health data protection and management exchanged via medical devices. Noura [15] proposed cryptographic methods or solutions to guarantee improved resistance against modern and traditional types of attacks. Masud et al. [16] proposed a lightweight protocol that makes use of physical unclonable functions (PUFs) to activate the devices on a network for verifying the user legitimacy (doctor) and sensor node (SN). It protects the SN installed in hostile areas from attacks, tampering, and cloning. Liaqat et al. [17] proposed a hybrid DL-based software-defined network which is enabled IoMT architecture to influence Convolutional Neural Network (CNN) and Cuda deep neural network. The Cuda deep neural network is a Long Short Term Memory network. The leveraging is done on both neural networks to efficiently detect malware botnets. Cecil et al. [18] discussed an IoMT-driven architecture for surgical training, particularly for the wider Next Generation framework contexts. Askari et al. [19] proposed a fair scheduling algorithm named non-orthogonal multiple access that mutually considers the state of a channel, delays, and power consumption. Badotra et al. [20] aimed to offer the requirement for software-defined networks in IoT applications (particularly healthcare ones). The authors also discussed in detail the benefits, issues, research direction, and architecture of the IoT. Mohd. et al. [21] highlighted the architecture of IoMT, its technologies, and applications, along with its security developments for fighting against coronavirus. Razdan and Sharma [22] highlighted IoMT and various smart E-Healthcare in which various wireless technologies have been explained in detail.

1.4 IOMT AND SMART E-HEALTHCARE

1.4.1 IOMT AND ENABLING WIRELESS TECHNOLOGIES

IoT frameworks comprise sensors and gadgets associated with utilizing an organization of cloud environments over the fast network between every module. The crude information gathered at these gadgets/sensors is sent straightforwardly to the immense stockpiling presented by cloud administrations. This information is additionally cleaned and afterward broken down to acquire further experiences into it. This requires extra programming, apparatuses, and applications which will additionally aid perception, investigation, handling, and the executives of the information.

Figure 1.2 shows few remote advancements like Radio Frequency Identification (RFID), Near Field Communications (NFC), Bluetooth, Long Term Evolution (LTE), and 5G/6G (and then some) between connected with a few gadgets, for example, cell phones, observing gadgets, sensors, savvy wearable, and other clinical gadgets [22]. As of now, the utilization of 5G/6G or past is common in IOMT because of their high data transfer capacity and super low inertness benefits.

1.4.2 SMART E-HEALTHCARE

Smart clinics are emergency clinics that are based on insightful robotized and streamlined modules (AI/ML-driven) on the information and communication technology (ICT) framework to further develop patient consideration techniques and to add new abilities. There are a few uses of smart emergency clinics, for example, telemedicine, telehealth, far-off robot medical procedures. Telemedicine is to give clinical consideration in a far-off area, while telehealth is to give non-clinical consideration a good ways off. In distant robot medical procedures, clinical robots do a medical procedure through guidance from a specialist sitting far away.

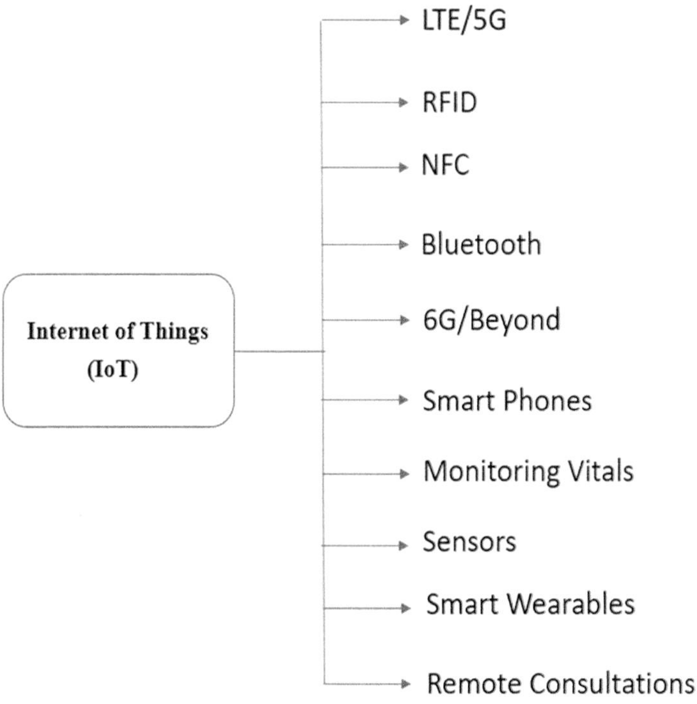

FIGURE 1.2 Advancements in IoT.

Information could be named unstructured assuming that it is gathered disconnected on paper as clinical notes by the employee. Assuming the information is gathered in an organized structure from the gadgets and sensors by utilizing predefined information fields for clients to enter, then, at that point, it turns out to be not difficult to process in additional frameworks like Customer Relational Management (CRM) System.

The CRM brings to utilizes the tools for investigating information and afterward assigning it to its predefined focus in the ecosystem [22]. The fundamental information and data from EHR frameworks are forwarded to the CRM framework and it processes this patient information. This handle information creates further triggers to patients and clinical faculty in the environment. The patients get outbound correspondence from clinics and health specialists as customized patient regimens. The specialists and other clinical staff get advised with regards to the updates and different cautions from a similar CRM software in the environment.

1.5 ARCHITECTURE OF IOMT

The design for IoMT's comprises three layers. There are three layers: (1) things layer, (2) fog layer, and (3) cloud layer [23–25]. Medical care specialists in this architecture type can likewise convey straightforwardly through the router between the thing layer and fog layer and the neighborhood handling servers at the fog layer. The things layer comprises patient checking gadgets, sensors, actuators, clinical records, drug store controls, nourishment routine generator, and so on. This layer is straightforwardly in touch with the clients of the environment. The information from components, for example, wearables, patient-checking information, remote consideration information is gathered at this layer. The gadgets utilized at this ought to be safely positioned to guarantee trustworthiness in the information gathered. The nearby routers in the ecosystem are answerable for interfacing these gadgets to the fog layer [26–28]. The information is additionally handled at the fog and at the cloud layer to produce significant data. Further, to decrease the deferral, the medical care specialists can get the patient information through this router. The other layers are defined in brief as the following sections.

The fog layer works between the cloud and the things layer. This layer comprises nearby servers and entryway gadgets for a sparsely dispersed fog organizing structure [29–31]. The local power processing is harnessed by the lower layer gadgets for continuous reaction to their clients. These servers are likewise used to oversee and control the security and trustworthiness of the framework. The door gadgets at this layer are answerable for diverting this information from these servers to the cloud layer for additional handling. Further, to lessen the delays, the medical services specialists can get the patient information through this switch.

The cloud layer comprises information storage and computational resources for the information to be broken down and infer decision-making frameworks dependent on it. The cloud additionally offers a tremendous reach to consolidate huge hospitals and medical services frameworks to deal with their everyday tasks effortlessly. This layer comprises cloud assets where the information created from the clinical foundation will be put away and scientific work could have proceeded as considered significant later on.

1.6 EMERGING TECHNOLOGIES IN IOMT

During the past few years, the manufacturing and usage of medical devices (IoT-based) has increased rapidly [32–35]. However, this fast-growing technology has also attracted many cybercriminals because these medical devices which are IoT-based stores, gather, and transmit important healthcare-related information via hardware or sensors. Six types of devices are at the highest risk including smartpens (used as entry points to patient data), infusion, and insulin pumps (control of infusions can be lost as a result of hacking), implantable cardiac devices (pacemakers may be switched off by the hackers), alterations of HR and blood glucose levels could be done, and security cameras could be used for compromising data.

There are myriads of interesting innovative advancements in medicine being used in a way that more quality time can now be spent with the patients. The pandemic (COVID-19) in 2020 forced healthcare industries to test many medical technologies on a huge scale. Advanced telemedicine is one such technology that has helped to overcome the situation of the COVID-19 pandemic in many ways. Numerous telehealth regulatory barriers are removed. The healthcare industries in today's era know the way to assess and improve services related to telehealthcare. Much healthcare focuses on integrating existing physical ones with telehealth services. The primary care and urgent appointments are continuously increasing through virtual appointments [36–37].

Another milestone achieved during the pandemic is the development of drugs, for example, the production of multiple vaccines to fight against coronavirus in less than a year is the greatest achievement. Another advancement is Data-Driven Healthcare. Due to the increase in data collection related to healthcare, applications for refining the treatment choice have also considerably improved during the last few years. However, one of the biggest disadvantages is, data related to one organization/industry is not easily processed by other industries. But, the issue of interoperability improved at the end of 2020. Google Cloud introduced its interoperability healthcare program in November 2020.

Further, nanomedicines is medical nanotechnology application having applications in sensing, diagnosis, imaging, and sending via medical gadgets. Scientists are discovering novel methods to make use of nanomedicine for targeting individual cells.

With almost zero delays, another fastest technology of 2021 is fifth-generation sensors or medical gadgets that can collect and send data rapidly which in turn improves the monitoring of patients. Scientists believe that robotics along with fifth-generation would revolutionize the way medical treatments are performed. Tricorders have also been used which are palm-sized devices that can accurately and quickly diagnose while monitoring a wide array of vital signs. Google and Alexa home has brought a tremendous change in people's life or other words we can say that these technologies have changed the way people and technology interact nowadays. The artificial pacemaker is also a critical medical piece being used by millions of patients. It can correct or prevent heart arrhythmias.

1.6.1 BLOCKCHAIN TECHNOLOGY

Nodes in the network keep track of their transactions in a distributed ledger called the blockchain. It's no secret that blockchain technology is a game-changer for healthcare security because of its ability to solve several problems that have arisen in traditional medical records. Information transmitted between any two nodes in a network is stored and may be used for cross-references on the blockchain, which comprises blocks or nodes linked by a network. Using this technology, it is possible to pinpoint the actual location of network trolls and other bad actors. This discarding of blocks that are not identifiable in the network paves the door for blockchain being regarded as a trusted technique in information exchange systems such as the IoT. The blockchain is a distributed ledger that eliminates the need for a central authority to manage the network's transactions. The blocks of data that make up a blockchain record each contain a unique identifier. According to the previous statement, these blocks include information about the blocks immediately before them in the chain, which is protected by cryptographic protocols. Other users can read the data contained in these blocks, but the data itself is impenetrable.

Smart contracts that don't need a central authority to be triggered may be easily processed on the blockchain. These contracts are self-executable and thus do not require any kind of supervision from anyone. Because Ethereum supported blockchain-based "smart contracts," this type of service is readily available. Blockchain's role in healthcare necessitates the division of infrastructure into smaller modules to implement solutions based on it. The IoMT framework may then use these modules in conjunction with the necessary IoMT devices. The distributed nature of the system will allow for the decentralization of network power. While the flow of data in the healthcare environment is ever-increasing, the advantage of using blockchain technology comes with a trust aspect. The blockchain promises to meet the healthcare infrastructure's ever-increasing need for data exchanges. Some clinical studies throughout the globe are now testing the use of blockchain for electronic health records (EHRs).

1.6.2 VIRTUAL REALITY

Virtual reality (VR) is a type of technology that offers a 3-D appealing multisensory atmosphere by modifying the reality experience. The head-mounted display is worn in clinical VR by the users with nearby-proximity screens which give a feeling of being 'transported to 3-D life-like world.' This reality type is a mixture of, extinction learnings, distractions, cognitive-behavioral principle, gate-control theories, along the spotlight attention theory. VR is also used in mental healthcare as well as anxiety disorder, stroke management, and many more. It helps as a supportive instrument to treat cancer patients to influence psychological function. They restrict symptoms (psychological) related to cancers and facilitate the patient's emotional well-being. VR could help offer comforting care. It further reduces the undesirable effects of the current disease through video calls. As well as simulation of the real offering-togetherness feel of people without traveling.

1.6.3 AUGMENTED REALITY

Augmented reality could be proved helpful by helping in visualizing invisible ideas and explanations by navigating in the virtual world. A Telehealth VR system developed by XR-Health for reducing anxiety as well as the stress of patients who are quarantined for being involved in physical as well as cognitive exercises. A VR was introduced for studying the structures, enzymes, molecular dynamics, and protein of coronavirus. Fundamental VR offers simulators in which practices may be performed to improve surgical methods. Oxford VR is to alleviate fear and mental disorder symptoms.

1.6.4 PARALLEL COMPUTING

Parallel computing methods are the source for distributed computing methods which involves examples such as cloud, grid computing, fog, and edge computing. Applications of IoMTs involve real-time communications which are the emergent basis of big data. Thus, segregation and determination of data which requires local maintenance need sharing across servers on the cloud. The traditional method of centralized cloud computing for processing the data architecture in the case of IoMT systems has been shifted to the distributed fog computing system. Hierarchical layers are formed among the cloud servers and the hardware components by Fog computing. The amount of stored data is also reduced around the cloud servers which further helps in reducing the bandwidth of the network along with the response time. Further, it also leads to reduced internet and network latencies in cloud computing. Moreover, as data is locally stored, therefore, Fog computing improves the security of data.

1.6.5 5G NETWORKING

The quick popularity of applications based on IoMTs has resulted in the development of 5G (5th Generation) networks. Current 5G networks include Radio Access Technology (RAT), improvement of antennas, higher frequency usage, and rearchitecting of networks. 5G amalgamates IoMTs, cloud computing, big data, and AI. It also activated the IoMTs application which is powerful sufficient for supporting several medical instruments concurrently. Some of the examples of 5G technology include telemedicine, teleconsultation, remote surgery, and intelligence medicine.

1.6.6 BIG DATA VISUALIZATION AND ANALYTICS

The accumulation of huge data from IoMTs devices requires analyzing correctness for making the decision. One effective solution to this is Big Data Visualization and Analytics.

1.6.7 AI

AI in healthcare, which makes use of clinical, laboratories, and demographic data for screening, diagnosing, and predicting different diseases. Integration of AI and cloud-fog technology is also one of the emerging technologies today. AI also facilitates large-level screenings, monitoring, allocating resources, and predicting interactions with proposed therapies.

1.6.8 BLOCK-CHAINING

In smart healthcare, a huge amount of data sharing remains essential amid medical instruments and healthcare workers, which results in the fragmentation of data resulting in inadequate information decryption which further hampers the therapeutic processes. To mitigate this, the Block-chaining method was developed for establishing connections among the data repository existing in the networks. It is a rising records list in which records are linked with one another using hashing (a cryptographic method). Every record consists of a hash (cryptographic) of the preceding record, for chaining the records and making them impervious to modification.

A smart contracts technique is followed by Block-chaining in which uniqueness is achieved and consequently authorizations are established for data accessibility accumulated in the Block-chaining.

1.7 SECURITY CONCERNS IN IOT

The IoT can mean different "things" to different industries. Broadly speaking, any object or device connected over the internet and powered with the ability to collect, transfer, and analyze data over a network can be labeled a "thing" in the emerging digital ecosystem known as IoT. Narrowing it down specifically to the healthcare industry, any medical device that connects to a healthcare provider's network is subclassified as the IoMT. These include but are not limited to medical imaging systems, remote patient monitoring (RPM) devices, smart thermometers, infusion pumps, medical device gateways, and biosensors packaged into wearables for use in apparel or implanted inside the human body. IoMT has been consistently making its way into healthcare and promises a future where it will be not only prevalent, but also fully extended beyond the four walls of the clinic and hospital.

A more humane approach to healthcare and healing is made possible by the IoMT, which is changing the healthcare sector in more ways than just the healthcare industry itself. Some several digital dangers and weaknesses must be considered. A few of the factors contributing to the high number of cyber-attacks in IoMT include: (1) The primary function of Medical Things is the exchange of sensitive patient data. Due to the linkage of a large number of devices and heterogeneous networks, there is incompatibility and complexity in the system. (2) Healthcare producers are rushing to adopt IoMT arrangements without thinking about security concerns since the industry is so new. Because of this, new concerns about data privacy, integrity, and accessibility (CIA) have surfaced. This raises the risk of wireless sensor network (WSN) security breaches for the IoT. (3) The security and availability of the application as a whole, as well as issues with authorization and authentication, are also key concerns. Computable resources are used up quickly by certain security calculations. These are just a few of the reasons why IoT systems are vulnerable to malicious assaults. An error in a patient's data report or analysis might result in their death if these related devices are left unsecured, and this isn't just about patient data. Because of this, it is critical to address how to differentiate and protect medical equipment against threats. However, many IoTs vulnerabilities are particularly suited to IoMTs due to the sensitive nature of health information. Data breach, Man in the Middle attack, replay assault, network communication unscrambling, Denial of Service (DoS), and security and privacy risks are just a few of the digital attacks that may be used against a system. It is possible to classify vulnerabilities in IoMT by layer. Hackers infiltrated Singapore's health system, stealing the personal information of 1.5 million patients, and compromising the prescription records of 160,000 others, according to help net security. Various internet security risks plague the healthcare business on an almost daily basis. They include anything from malware that jeopardizes system integrity and patient privacy to DDoS assaults that disrupt medical institutions' capacity to provide treatment. Healthcare providers are increasingly concerned about the safety of sensitive data, such as health information, as it travels through the IoMT and their ability to use the framework without interruption. In comparison to other sectors of critical infrastructure, the healthcare sector faces unique challenges because of the nature of the motivations driving these attacks. Human life is directly impacted by financial misfortune and privacy breaches. Small or major attacks on the healthcare system are genuinely a threat, even if they come out of the blue. We must discover a means to securely and properly control IoMT as they become a fundamental element of healthcare. On IoMT, the two most significant assaults are highlighted. The Man in the Center (MitM) Defeats in the IoMT a MitM attack occurs when an attacker eavesdrops on or alters the communication between two systems by intercepting it. By using MitM attacks, cybercriminals may get access to a victim's account passwords and other personal information. They can also monitor the target's conversations and degrade data. To trick the recipient into believing they are still receiving a genuine communication, an attacker might pretend to be the original shipper. In the context of IoHT, one can imagine a scenario in which a malicious party may want to fake vital readings sent from the patient monitor to the Central Monitoring Station (CMS) for the staff remotely monitoring a patient to see incorrect real-time information about vital readings.

In the health industry, this is a serious hazard, since the implications might be fatal. MitM attacks are more likely to occur since the great majority of medical devices transfer data that is either decoded or encrypted with inadequate encryption to the server 6. Attacks against the IoT attacks such as DoS may prevent genuine users from accessing the health framework again. This attack overwhelms the smart devices with service requests, causing them to become unusable. DoS assaults come in a variety of forms. To bring down a network, it is used by hackers. Healthcare providers that need a network connection to provide proper patient treatment may be faced with a significant problem. The healthcare industry is one of the most critical in terms of network reliability. This makes it possible to use EHRs and run life-saving apps. EHRs are electronic health records. Many of today's IoT gadgets are designed to be user-friendly and accommodating, which allows attackers to launch DoS assaults that might bring large enterprises to a standstill. Life-threatening vacations are possible in the healthcare business. As DoS assaults knock networks down, hospitals lose control of IoMT equipment, access to eHealth information, and the ability to conduct research projects, among other things. All network traffic, both inbound and outbound, must be closely scrutinized and monitored so that malicious activity may be quickly identified. The dangers of using IoT health apps for health-related applications, IoT devices are critical. To assist the functioning of healthcare apps, IoT devices collect quantifiable and analyzable healthcare data. For this reason, the security of IoT healthcare apps is critical for healthcare organizations, and for this reason, many researchers have proposed various security systems. The security of IoT devices is endangered by a wide range of threats.

Optimum use of resources sensors has an important role in healthcare. Sensor innovation has made it simple to collect healthcare data that can be measured and analyzed. Wearable medical technologies have become more popular in recent years. Using wearable sensors, healthcare providers may collect a wide range of data without affecting patients. Nevertheless, the use of energy by wearable gadgets is a serious concern. Because wearable devices are compact and may be used to collect data on the structure of an individual's body, they have become more popular. They continually collect medical data from the body. Currently, the battery isn't capable of storing and transmitting health data to healthcare apps. A wearable device's battery also needs regular recharging. IoT devices in healthcare have real problems to deal with. Literature on the subject of privacy offers numerous definitions. Privacy protections as well as ethical health research provides valuable advantages to humanity. Health researchers are important for improving human well-being along with healthcare. Caring patients involved in researches from harm and protecting the rights becomes crucial to ethical research. Main reason to protect personal confidentiality is for protecting the individual interests. On the contrary, the main justification for gathering personally identifiable health information for health researches is to benefit societies. However, it is imperative to stress that privacies also have values at the societal level as it allows complex activities which further includes public health activities and research to be carried out in ways that guard one's dignity. Simultaneously, health research can profit individuals, for instance, when it enables access to newer therapies, better diagnostics, and more efficient methods to avoid illness and deliver care. The patient's privacy protection is a critical component of every healthcare system's information security strategy. Therefore, a large number of international organizations have come up with definitions of privacy. It is defined by the Organization for Economic Cooperation and Development (OECD) as "any information of an identified or identifiable person." IoT devices collect healthcare data. They collect data using remote access technologies that raise privacy and security concerns. The sensor sends the information it collects to a database or cover over the internet. Devices in the IoT are linked to the internet and interact with one another through it. Health information is at risk because of security flaws in internet and IoT devices. Health information is also acquired from a variety of healthcare facilities. Various health units exchange health data. Data should be kept private in each unit. Because of the importance of the information included in healthcare records. They all want to get their hands on your medical records. In light of this, data privacy must be protected. Because it ensures data security and privacy, trust management is essential for IoT devices and applications. It's because all gadgets connect to the network and provide information to programs. As a result, devices connected to the internet may be relied upon to protect user data and personal information. To modify data, attackers might use device IoT apps. The trustworthiness of data collection is not a joke because of the massive amounts of data that are acquired from gadgets. IoT health applications make good use of big data to make informed decisions about patients and improve the standard of treatment they get. This includes data interfacing, analysis, and mining for IoT healthcare services. IoT devices might be used by attackers to corrupt large amounts of data. As a result, scientists have a better understanding of IoT trust management difficulties. The network and application layers of IoT should be used to control the trust. Anti-DoS assaults (DoS) The goal of denial-of-service (DoS) attacks against IoT devices and apps is to prevent them from assisting. The IoT is a network of interconnected devices that exchange data and coordinate actions. IoT apps must connect to a network and get data from connected devices to function. IoT applications are vulnerable to DoS attacks due to the lack of a machine or network asset. IoT devices are restricted in memory, bandwidth, battery life, and circular space because of their small size. As a result, they are vulnerable to DoS assaults. Threats of violence Because IoT devices collect data in an unsecured environment, physical security is not a joke. The IoT devices are also compact and incorporated into many types of equipment such as televisions (TVs), vehicles, air conditioning, and broilers. Because of this, these devices are vulnerable to theft or tampering with their settings. Data supplied by IoT devices may be altered by hackers. Several physical assaults against IoT devices have been discovered, such as suspicious theft, software manipulation, and hardware tampering. IoT healthcare apps rely on data for their data manipulation in data manipulation. All healthcare systems make use of data. As a result, cyberattacks aim to compromise the safety and privacy of personally identifiable information (PII). Data is stolen, data is manipulated, and data is damaged as a result of attacks. Because it includes information that may be used to identify a specific individual, health data is both critical and sensitive for countries throughout the world. In the IoT, data travels in many directions. Devices in the IoT (data generators, recipients, and aggregation points) are connected to the internet, the cloud, and machines in the cloud (application services, big data repositories, and analytics). For nefarious purposes, attackers take health data and injure victims. When IoT devices create and convey data, they steal it. To steer fatalities in the direction of the attackers' desires, attackers alter or modify data. Because of data tampering, doctors may make diagnoses and treatments that are incorrect. Furthermore, IoT health applications have a serious problem with data loss. IoMT devices are susceptible to network/wireless assaults due to their dependency on an open wireless connection. IoMT devices are vulnerable to data interception and eavesdropping due to the lack of safety safeguards implemented by the manufacturers, or because of inadequate security authentication methods. Talented attackers may circumvent security measures to acquire patient data since most IoMT devices are unable to detect or prevent assaults. Consequently, when infecting devices with harmful code or malware, attackers might benefit from an increased advantage. Researchers found that medical gadgets are susceptible to zombie and botnet assaults in their examination. For example, an assault may potentially alter or manipulate the dosage of a medicine, which could result in serious health consequences or even death. As a result, an enemy may identify the patient's record on the blockchain network since most IoMT devices make use of blockchain advancement. Medical records are prone to falsification in healthcare systems, which may lead to inappropriate medicine delivery and patient forecasts, which can lead to an allergic response. Attackers may also issue bogus medical warnings and inflict significant financial loss as a result of security flaws. As dependable and powerful as the IoMT (IoMT) technologies are, they confront serious security issues. As a result, security is seen as a major concern in biomedical research due to the ease with which sensitive physiological data about patients may be leaked or misused. Control of access to prevent unauthorized access to assets by unlawful clients while identifying the appropriate authorization levels for approved customers, access control is used. Clients of IoMT systems may restrict access to IoMT devices via the use of access control, which allows each client to set their access levels. It is vital to point out that IoMT devices may be protected by access control. Figures for confirming an individual's identity to prevent unauthorized workers from accessing critical medical information, identity authentication is used to verify the client's identity. By encoding data in a code or other structure, the information can only be accessed by those who have the correct decoding key – a secret code. This security method is widely used in the biomedical industry since it ensures the integrity of medical data.

1.8 CONCLUSION

Numerous advances are being sent into the medical local area to assist with making a more secure and better future for the world. IoMT gadgets and different advances related are having a major influence to assist with relieving the pandemic. The research in this chapter clarifies the numerous manners by which to utilize IoMT and different innovations to accomplish observation, checking, discovery, anticipation, and relief of the COVID-19 pandemic. Regardless of whether it's conveying IoMT applications for remote patients at home, following patient experimental outcomes around the world, moving secure medical data, or looking for prevention to sicknesses, these innovations are showing this can be refined. Intriguing forward leaps into these advancements will just develop as the medical field and innovations hold combining. These gadgets and applications are most successful when they are consolidated together to bridle their full ability to work far more than their capacities. IoMT gadgets succeed when utilizing large data analytics to test and examine results. Different advancements, for example, AI with huge data consolidated can even assist with tracking down new and creative ways of tracking down medications to battle the pandemic. Computer-based intelligence can have the best calculation on the planet yet without major data to dissect results it's exceptionally trivial. Perhaps the main breakthrough is the headway of telecommunications like 5G, making it feasible for constant data analytics handling and the exchange of data across the organization at lightning-quick speed. Once more, it's not necessary to focus on utilizing one innovation over the other, however, to consolidate numerous applications to get the best reaction. Apart from this, we have highlighted various applications of IoMT along with the various types of attacks and security holes or attacks that are possible. Then, finally, we have given some security measures to overcome different attacks in IoMT.

KEYWORDS

- **blood pressure**
- **electronic health records**
- **food and drug administration**
- **heart rate**
- **internet of medical things**
- **internet of things**

REFERENCES

1. Wu, J., Guo, S., Huang, H., Liu, W., & Xiang, Y., (2018). Information and communications technologies for sustainable development goals: State-of-the-art, needs, and perspectives. *IEEE Commun. Surv. Tutor., 20*, 2389–2406.
2. Naresh, V. S., Pericherla, S. S., Rama, M. P. S., & Reddi, S., (2020). Internet of things in healthcare: Architecture, applications, challenges, and solutions. *Comput. Syst. Sci. Eng., 35*, 411–421.
3. Wang, Q., Zhu, X., Ni, Y., Gu, L., & Zhu, H., (2020). Blockchain for the IoT and industrial IoT: A review. *Internet Things, 10*, 100081.
4. Taylor, P. J., Dargahi, T., Dehghantanha, A., Parizi, R. M., & Raymond, C. K. K., (2020). A systematic literature review of blockchain cyber security. *Digit. Commun. Netw., 6*, 147–156.
5. Dilawar, N., Rizwan, M., Ahmad, F., & Akram, S., (2019). Blockchain: Securing the internet of medical things (IoMT). *Int. J. Adv. Comput. Sci. Appl., 10*, 1.
6. Ahmed, Z., Mohamed, K., Zeeshan, S., & Dong, X., (2020). Artificial intelligence with multi-functional machine learning platform development for better healthcare and precision medicine. *Database*.

7. Sharma, S., (2021). Towards artificial intelligence assisted software-defined networking for internet of vehicles. *Intelligent Technologies for Internet of Vehicles, Internet of Things*.
8. Lalmuanawma, S., Hussain, J., & Chhakchhuak, L., (2020). Applications of machine learning and artificial intelligence for COVID-19 (SARS-CoV-2) pandemic: A review. *Chaos Solitons Fractals, 139*, 110059.
9. Sedik, A., Hammad, M., Abd El-Samie, F. E., Gupta, B. B., & El-Latif, A. A. A., (2021). Efficient deep learning approach for augmented detection of coronavirus disease. *Neural Comput. Appl.,* 1–18.
10. Turabieh, H., Abu, S. A., & Abu-El-Rub, N., (2018). Dynamic L-RNN recovery of missing data in IoMT applications. *Future Gener. Comput. Syst., 89*, 575–583.
11. Khan, S. R., Sikandar, M., Almogren, A., Ud Din, I., Guerrieri, A., & Fortino, G., (2020). IoMT-based computational approach for detecting brain tumor. *Future Gener. Comput. Syst., 109*, 360–367.
12. Kilic, A., (2020). Artificial intelligence and machine learning in cardiovascular health care. *Ann. Thorac. Surg., 109*, 1323–1329.
13. Song, H., Bai, J., Yi, Y., Wu, J., & Liu, L., (2020). Artificial intelligence-enabled internet of things: Network architecture and spectrum access. *IEEE Comput. Intell. Mag., 15*(1), 44–51.
14. Girardi, F., De Gennaro, G., Colizzi, L., & Convertini, N., (2020). Improving the healthcare effectiveness: The possible role of EHR, IoMT and blockchain. *Electronics, 9*, 884.
15. Noura, M., (2019). *Efficient and Secure Cryptographic Solutions for Medical Data*. Theses, Univ. Bourgogne Franche-Comté.
16. Masud, M., Singh, G. G., Alqahtani, S., Muhammad, G., Gupta, B., Kumar, P., & Ghoneim, A., (2020). A lightweight and robust secure key establishment protocol for the internet of medical things in COVID-19 patients care. *IEEE Internet Things J.,* 15694–15703.
17. Liaqat, S., Akhunzada, A., Shaikh, F. S., Giannetsos, A., & Jan M. A., (2020). SDN orchestration to combat evolving cyber threats in the internet of medical things (IoMT). *Comput. Commun., 160*, 697–705.
18. Cecil, J., Gupta, A., Pirela-Cruz, M., & Ramanathan, P., (2018). An IoMT based cyber training framework for orthopedic surgery using next generation internet technologies. *Inform. Med. Unlocked., 12,* 128–137.
19. Askari, Z., Abouei, J., Jaseemuddin, M., & Anpalagan, A., (2021). Energy-efficient and real-time NOMA scheduling in IoMT-based three-tier WBANs. *IEEE Internet Things J.,* 13975–13990.
20. Badotra, S., Nagpal, D., Narayan, P. S., Tanwar, S., & Bajaj, S., (2020). IoT-enabled healthcare network with SDN. In: *2020 8th International Conference on Reliability, Infocom Technologies and Optimization (Trends and Future Directions) (ICRITO)* (pp. 38–42). Noida, India.
21. Mohd, A. A., Hassan, W. H., Sameen, S., Attarbashi, Z. S., Alizadeh, M., & Latiff, L. A., (2021). IoMT amid COVID-19 pandemic: Application, architecture, technology, and security. *J. Netw. Comput. Appl., 174*, 102886.
22. Razdan, S., & Sharma, S., (2021). Internet of medical things (IoMT): Overview, emerging technologies, and case studies. *IETE Technical Review*, 1–14.

23. Singh, P. D., Kaur, R., Singh, K. D., & Dhiman, G., (2021). A novel ensemble-based classifier for detecting the COVID-19 disease for infected patients. *Information Systems Frontiers*, 1–17.
24. Singh, K. D., & Sood, S. K., (2021). QoS-aware optical fog-assisted cyber-physical system in the 5g ready heterogeneous network. *Wireless Personal Communications, 116*(4), 3331–3350.
25. Singh, P. D., Kaur, R., Singh, K. D., Dhiman, G., & Soni, M., (2021). Fog-centric IoT based smart healthcare support service for monitoring and controlling an epidemic of swine flu virus. *Informatics in Medicine Unlocked, 26*, 100636.
26. Kaur, R., Singh, P. D., Kaur, R., & Singh, K. D., (2021). A delay-sensitive cyber-physical system framework for smart health applications. In: *Advances in Clean Energy Technologies* (pp. 475–486). Springer, Singapore.
27. Kaur, S., Singh, K. D., Singh, P., & Kaur, R., (2021). Ensemble model to predict credit card fraud detection using random forest and generative adversarial networks. In: *Emerging Technologies in Data Mining and Information Security* (pp. 87–97). Springer, Singapore.
28. Sood, S. K., & Singh, K. D., (2021). Identification of a malicious optical edge device in the SDN-based optical fog/cloud computing network. *Journal of Optical Communications, 42*(1), 91–102.
29. Angurala, M., Bala, M., Bamber, S. S., Kaur, R., & Singh, P., (2020). An internet of things assisted drone based approach to reduce rapid spread of COVID-19. *Journal of Safety Science and Resilience, 1*(1), 31–35.
30. Sood, S. K., & Singh, K. D., (2019). HMM-based secure framework for optical fog devices in the optical fog/cloud network. *Journal of Optical Communications*.
31. Singh, P. D., Kaur, R., Dhiman, G., & Bojja, G. R., (2021). BOSS: A new QoS aware blockchain assisted framework for secure and smart healthcare as a service. *Expert Systems*, e12838.
32. Seth, J., Nand, P., Singh, P., & Kaur, R., (2020). Particle swarm optimization assisted support vector machine based diagnostic system for lung cancer prediction at the early stage. *PalArch's Journal of Archaeology of Egypt/Egyptology, 17*(9), 6202–6212.
33. Sood, S. K., & Singh, K. D., (2019). Optical fog-assisted smart learning framework to enhance students' employability in engineering education. *Computer Applications in Engineering Education, 27*(5), 1030–1042.
34. Sadek, I., Rehman, S. U., Codjo, J., & Abdulrazak, B., (2019). Privacy and security of IoT based healthcare systems: Concerns, solutions, and recommendations. In: *International Conference on Smart Homes and Health Telematics* (pp. 3–17). Springer, Cham.
35. Singh, K. D., & Sood, S. K., (2020). Optical fog-assisted cyber-physical system for intelligent surveillance in the education system. *Computer Applications in Engineering Education, 28*(3), 692–704.
36. Singh, P. D., Dhiman, G., & Sharma, R., (2022). Internet of things for sustaining a smart and secure healthcare system. *Sustainable Computing: Informatics and Systems, 33*, 100622.
37. Kouicem, D. E., Bouabdallah, A., & Lakhlef, H., (2018). Internet of things security: A top-down survey. *Computer Networks, 141*, 199–221.

CHAPTER 2

Distributed Bragg Reflector Biosensor for Medical Applications

RANJITH B. GOWDA[1] and PREETA SHARAN[2]

[1]Department of Electronics and Communication Engineering, Dayananda Sagar University, Bangalore, Karnataka, India

[2]Department of Electronics and Communication Engineering, The Oxford College of Engineering, Bangalore, Karnataka, India, E-mail: sharanpreeta@gmail.com

ABSTRACT

Distributed Bragg reflector (DBR) is a one-dimensional (1D) photonic-crystal (PC) structure having two different materials arranged alternatively. It has the property of refractive-index variation periodically in a single direction. Advancements in modern technology are introducing many new methods for the diagnosis of major life-threatening diseases. An accurate diagnosis of the disease in its early stages will reduce the factor of risk at higher stages and also saves the life of a person. Various methodologies have been researched and implemented successfully in recent years. Optical sensing technology is growing tremendously for the detection of many diseases in their early stages. Major benefits of optical sensors include small in size and weight, very high accuracy, no electromagnetic interference, small sample requirement, high Q factor, good sensitivity, and can be easily integrated with lab-on-chip (LOC) devices.

Refractive index (RI) is one of the key optical parameters which can be used for bio-sensing applications. Using this method, many diseases,

Computational Health Informatics for Biomedical Applications. Aryan Chaudhary and Sardar M. N. Islam (Naz) (Eds.)

malignant cells, bacteria, viruses, foreign bodies, etc., can be effectively detected in their initial stages. The medical diagnosis uses initial laboratory tests like enzyme-linked immunosorbent assay (ELISA), reverse transcription–polymerase chain reaction (RT-PCR), micro-immunofluorescence (MIF), etc., for the detection and identification of bacteria, virus, fungi, or any other immigrant cells. These tests regularly use blood, urine, or saliva as a bio-analyte for diagnosis purposes. These bio-analytes have distinguished or unique optical properties like RI, light absorption capability, light scattering, and so on. As the normal and infected cells in the biosample exhibit different RI, it can be easily detected using optical detection techniques. There exists a number of optical detection techniques like surface plasmon resonance (SPR), optical reflectometric interference, bioluminescent, evanescent wave fluorescence, ellipsometry, interferometry, and many more. The optical biosensors can be designed and analyzed using advanced technologies such as micro-electro-mechanical systems (MEMS), micro-electronics, biochemistry, microbiology, molecular biology, nanotechnology, and so on. Fabrication of the optical biosensor is also possible by using advanced micro-machining techniques. Biosensors use different signal transductions like piezoelectric or magnetic or resistive, optical, thermometric, electrochemical, etc., for the detection of target bodies. The optical sensing method has the benefit of label-free sensing, i.e., it does not require any chemical reagents for the detection of target species. Whereas, the normal clinical laboratory tests use chemical stains and reagents for the identification of disease-causing infectants. In an optical sensor, a beam of input light or the optical field interacts with the sample under test and gives the signature at the output. The optical signature can be either the change in amplitude, frequency, wavelength, or intensity of the input light. By comparing the input transmitted signal and the output received signals, with and without bio-analyte, the conclusion can be drawn for the presence or for the absence of diseases in the sample under test. It is also possible to detect the various stages or cycles of infected cells using these optical methods. In the human body, the cells or tissue from the specific organ have unique RI values. The body fluids, viz, blood, saliva, and urine is also exhibiting a different index of refraction in the presence and absence of the malignant cells. Therefore, the RI of a cell or bio-analyte can be effectively used for the identification of various diseases.

In this chapter, a 1D photonic crystal-based DBR is used for the detection of foreign bodies like cancer cells or the infected stages of blood cells for quick medical diagnosis. The one-dimensional (1D) Bragg Reflector sensor can also be fabricated by using electron beam lithography or chemical vapor deposition (CVD) techniques.

2.1 INTRODUCTION

Photonic crystals (PC) are materials that exhibit periodic variation in the dielectric constant in certain directions. These materials are transparent to the incident light. It allows the incident light with a certain wavelength by rejecting other wavelength regions. The range of wavelength over which there is no transmission of light is called the stop band region [1, 2]. It can be one-, two- or three-dimensional PC if it exhibits dielectric variation in single, double or in three directions, respectively, as shown in Figure 2.1.

FIGURE 2.1 One-, two-, and three-dimensional photonic crystal structures.

The incident light with a particular wavelength can be made to propagate through these structures by creating a defect in its structure. The defect can be a point defect, line defect or surface defect based on removing either one point, line, or a certain area in the PC. Materials that have a periodic dielectric change in a single direction are referred to as 1D PC. Due to their simplicity in design, analysis, and fabrication process, they are widely used to develop optical sensors. Biosensor design is one of the emerging and quickly growing areas in biological and chemical sensing applications. Many biosensors have been developed in the past decades and can be found elsewhere in the literature [3, 4]. Biosensors can be used to detect target protein molecules [5], bacteria [6], viruses

[7], nucleic acids [8], enzymes [9], and hormones [10]. Normal diagnostic methods like blood and urine testing, ELISA, RT-PCR, etc., has drawbacks like more time consuming, less accuracy, more sample requirement, and the need for skilled laboratory technicians to test sample. Whereas the biosensors have the advantages like more accuracy, less time consuming, a requirement of less sample, fast, and quick response, less in size and weight. Because of these advantages, biosensors are the prominent devices that can be effectively used as point-of-care devices in medical domains.

2.2 MATHEMATICAL MODELING

1D photonic crystal can be analyzed efficiently with the help of the transfer matrix method (TMM) [11]. In this method, each layer of 1D PC can be represented by a transfer matrix to find transmission and reflection of light by each layer. In order to analyze the behavior of electromagnetic wave (EMW) inside each layer of 1D photonic crystals, Maxwell's equations for EMW are considered as follows:

$$\nabla . D = \rho_v$$
$$\nabla . B = 0$$
$$\nabla \times E = -\frac{\partial B}{\partial t}$$
$$\nabla \times H = J + \frac{\partial D}{\partial t} \tag{1}$$

where; E is the electric field vector; H is the magnetic flux vector; D is the electric flux density; and B is the magnetic flux density. The quantities ρ and J are taken as the sources E and H vectors, respectively.

The electric and magnetic flux densities are denoted as:

$$D = \in E = \in_0 E + P$$
$$B = \mu H = \mu_0 H + M$$

where; ε_0 and μ_0 represent the electric and magnetic permittivity; P and M represent the electric and magnetic polarization, respectively.

For dielectric slab, P = 0 as there is a linear electric polarization and also are non-magnetic; M = 0. Electric and magnetic fields can be mathematically represented as:

$$E(a,t) = E(a)e^{-j(\omega t - k)}$$
$$H(a,t) = H(a)e^{-j(\omega t - k)} \quad (2)$$

Using these two equations, Maxwell's equation can be re-written as:

$$\nabla \times H = -j\omega \in E$$
$$\nabla \times E = -j\omega\mu H \quad (3)$$

Here,

$$\frac{\partial}{\partial t} \leftrightarrow -j\omega$$
$$\frac{\partial}{\partial x} \leftrightarrow -jk$$

Hence, the cross product of electric field can be written as:

$$\nabla \times \nabla \times E = \omega^2 \mu \in E \quad (4)$$

Using the identity of vector operation,

$$\nabla \times \nabla \times E = \nabla (\nabla . E) - \nabla^2 E$$

We can write,

$$\nabla^2 E + \frac{\omega^2}{c^2} \in_r \mu_r E = 0$$
$$\nabla^2 H + \frac{\omega^2}{c^2} \in_r \mu_r H = 0 \quad (5)$$

The above equation is called the Helmholtz equation with:

$$c^2 = \frac{1}{\in_0 \mu_0}$$
$$\in_r = \frac{\in}{\in_0}$$
$$\mu_r = \frac{\mu}{\mu_0}$$

For non-magnetic materials, $\mu_r = 1$ and equation (3) can be written as:

$$\nabla \times H = -j\omega \in_r \in_0 E$$
$$\nabla \times E = -j\omega\mu_0 H \quad (6)$$

The above equations can be simplified, and we get:

$$\nabla \times \frac{1}{\epsilon_r} \nabla \times H = \omega^2 \epsilon_0 \mu_0 H$$

or

$$\nabla \times \frac{1}{\epsilon_r} \nabla \times H = \frac{\omega^2}{c^2} H \tag{7}$$

The above master equation is used to analyze the photonic crystal materials and their properties.

2.3 DIELECTRIC SLAB STRUCTURE

Dielectric slab consists of alternate layers of two different materials, which are repeated periodically over a certain length. Let, n_H is the RI of material 1, and n_L is the RI of material 2 such that $n_H > n_L$, i.e., material 1 is having a higher RI than material 2. The geometrical length of material 1 and material 2 layers are taken as l_H and $L_{L,}$ respectively, and N represents the total number of dielectric pairs or periods considered over its length. Figure 2.2 shows a dielectric slab structure with a period of 8.

When a plane wave is made to incident on this type of multilayer structure, it changes its velocity inside the structure due to different dielectric constants of the material layers. Some portion of the light is transmitted, and some portion of the light is reflected back. This process repeats for each layer, and it gives rise to a strong reflection and is called Bragg's reflection [12]. This Bragg's reflection creates a stopband region in the specified range of input light wavelength.

The plane wave propagation in the dielectric slab is mathematically given by [11]:

$$E_1(x) = P_1 e^{jk_1 x} + Q_1 e^{-jk_1 x}$$

$$E_1(y) = R_1 e^{jk_2 y} + S_1 e^{-jk_2 y} \tag{8}$$

where; P_1 and Q_1 are the amplitudes of the incident and reflected light in layer 1; R_1 and S_1 are the amplitudes of the incident and reflected light in layer 2; k_1 and k_2 represents wave number and are given by:

$$k_1 = \frac{\omega}{c} n_1 \cos\theta_1$$

$$k_2 = \frac{\omega}{c} n_2 \cos\theta_2$$

where; θ_1 and θ_2 represents the incident angle of light in layer 1 and 2, respectively.

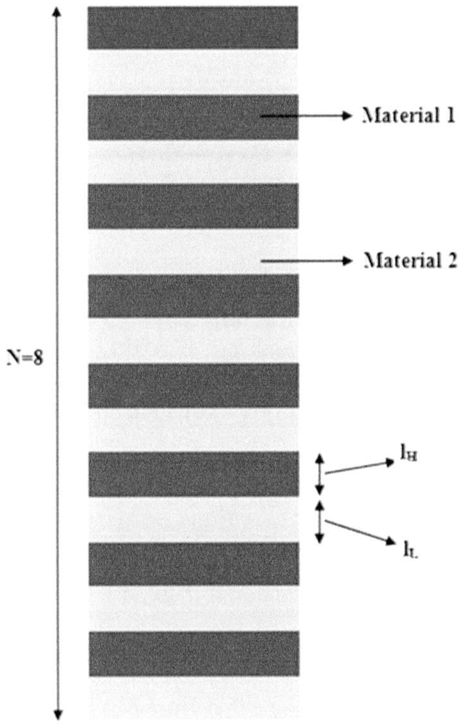

FIGURE 2.2 Dielectric slab consisting of alternate materials with eight periods.

In terms of transfer matrix:

$$\begin{pmatrix} R_1 \\ S_1 \end{pmatrix} = M_{12} \begin{pmatrix} P_1 \\ Q_1 \end{pmatrix} \tag{9}$$

With $M_{12} = \begin{pmatrix} 0.5\left(1+\frac{k_1}{k_2}\right)e^{jk_1 l_H} & 0.5\left(1-\frac{k_1}{k_2}\right)e^{-jk_1 l_H} \\ 0.5\left(1-\frac{k_1}{k_2}\right)e^{jk_1 l_H} & 0.5\left(1+\frac{k_1}{k_2}\right)e^{-jk_1 l_H} \end{pmatrix}$

The EM waves are assumed to be continuous at the interface boundary between two layers and then,

$$\begin{pmatrix} P_2 \\ Q_2 \end{pmatrix} = M_{21} \begin{pmatrix} R_1 \\ S_1 \end{pmatrix} \quad (10)$$

From equations (9) and (10), we can write:

$$\begin{pmatrix} P_2 \\ Q_2 \end{pmatrix} = M_{12} M_{21} \begin{pmatrix} R_1 \\ S_1 \end{pmatrix}$$

$$\begin{pmatrix} P_2 \\ Q_2 \end{pmatrix} = M_{i,j} \begin{pmatrix} R_1 \\ S_1 \end{pmatrix}$$

where; $M_{i,j} = M_{12} M_{21}$

For an N layer structure, we get:

$$M = \begin{pmatrix} M_{11} & M_{12} \\ M_{21} & M_{22} \end{pmatrix} = \left(L_1 L_2\right)^N$$

The reflection (R) and transmission (T) of such structures can be written as:

$$R = \left(\frac{| M_{12} |^2}{| M_{21} |^2 + \left(\frac{\sin\left(Ka\right)}{\sin\left(NKa\right)} \right)^2} \right) \quad (11)$$

where, K is the Bloch wave number, and it has the relation:

$$cos\left(Ka\right) = 0.5(M_{11} + M_{22}$$

$$T = 1 - R \quad (12)$$

2.4 DEFECTED BRAGG REFLECTOR STRUCTURE AS A BIO-SENSOR

Distributed-Bragg-Reflector (DBR) is a 1D PC multilayer structure consisting of dielectric slabs arranged periodically. By proper choice of the materials for a dielectric slab, light property inside the structure can be precisely controlled [13, 14]. In this work, initially, a multilayer structure consisting of Germanium (material 1) and zinc sulfide (material 2) with

8 periods is considered to find the region of the stopband region. The structure of the proposed sensor is shown in Figure 2.3.

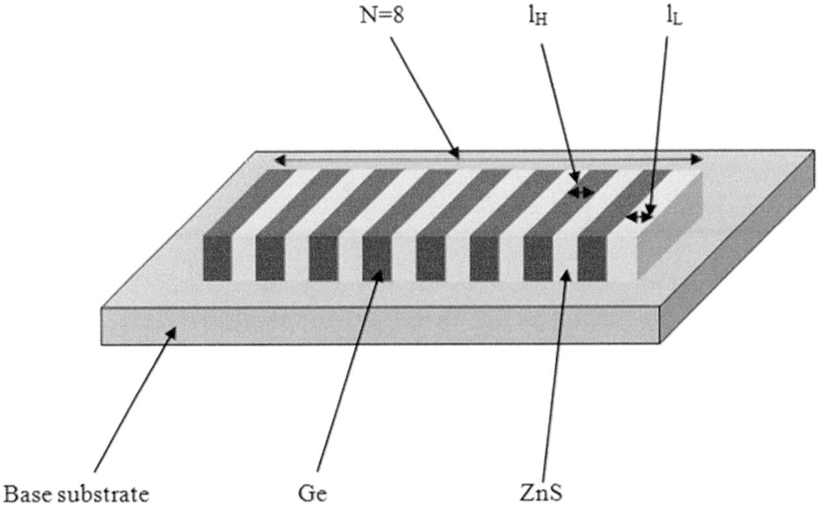

FIGURE 2.3 Proposed DBR sensor multilayer structure with eight periods.

The Germanium material has an RI of 4, and that of zinc sulfide is 2.19. These two materials are optically transparent to the incident light with a mid-infrared frequency, 3–25 μm [15]. 1D DBR can be easily fabricated using CVD or E-beam deposition method [16]. The incident light beam is chosen to have a wavelength of 10.6 μm which is in the mid-infrared region. Devices operating in this range have higher sensitivity due to better molecular interactions [17, 18].

The optical length of both the materials is chosen to satisfy the following condition:

$$l_H n_H = l_L n_L = \frac{\lambda}{4}$$

where; l_H is the geometrical length of material 1; l_L is the geometrical length of material 2; n_H is the RI of material 1; n_L is the RI of material 2; is the wavelength of incident light.

This satisfies the condition of Bragg reflection and prohibits a certain range of frequency to propagate in the structure called stopband. The transmission spectrum for the above structure is shown in Figure 2.4.

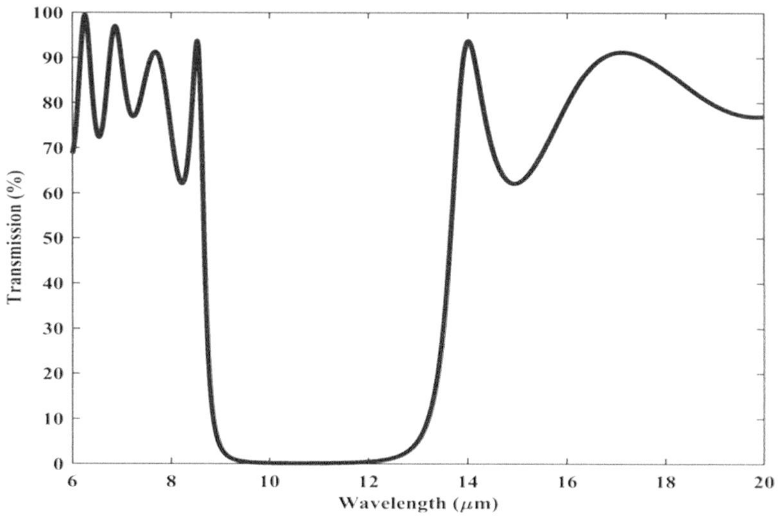

FIGURE 2.4 Transmission spectra for the proposed DBR sensor multilayer structure with eight periods.

The transmission spectrum shows that the range of wavelength from 9.4 μm to 12.3 μm is not propagating through the structure and is referred to as the stopband region. But it is possible to get a transmission in the stopband region by altering the periodicity of the structure. This can be made possible by having a defect layer in the design (Figure 2.5).

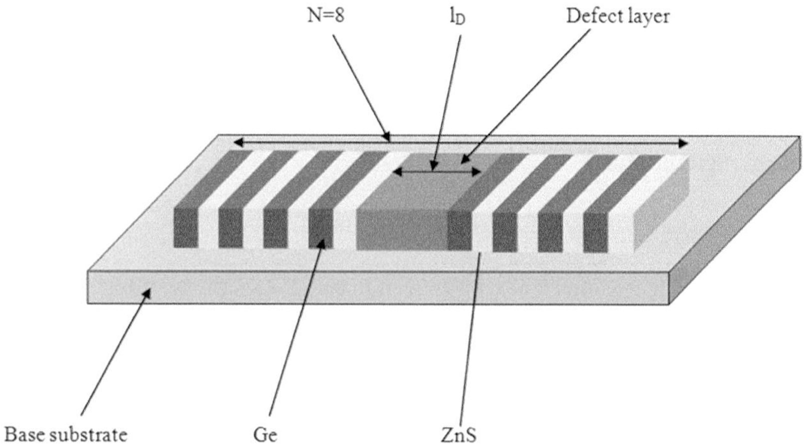

FIGURE 2.5 Proposed DBR with defect layer structure with eight periods.

The obtained transmission spectrum for this structure is shown in Figure 2.6.

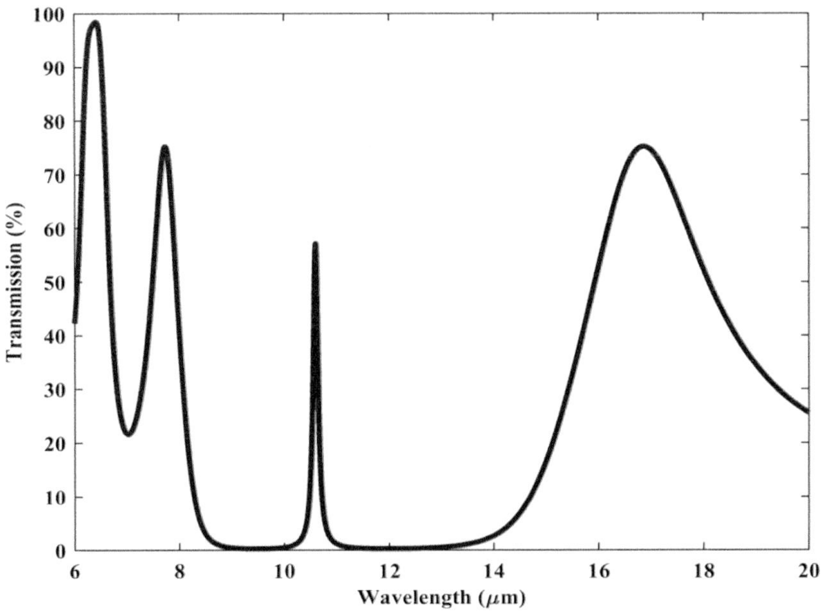

FIGURE 2.6 Transmission spectra for the proposed defected DBR structure.

Figure 2.6 shows that an incident light with the wavelength 10.6 μm is propagating in the region of the stop band. This is due to the creation of a defect layer in the mid of the multilayer structure. The incident light with the wavelength 10.6 μm is propagating through the structure as the defect is an air cavity with the RI of 1.

In this work, we are intended to design a DBR biosensor for the detection of oral cancer cells in the given sample. Cancer is one of the quickly growing and death-causing diseases which occupies a majority of the mortality rate across the world. Hence, early detection of cancer cells is a challenging task in the medical domain. Biosensors play a major role in the diagnosis of diseases. Cancer in the early stage can be treated well and can be cured. Hence initial stage cancer cells detection and its identification may avoid the death of the patient. Cancerous cells detection in the given sample can be identified by using the RI method. The normal cells and cancerous cells have unique RI values and can be effectively used to sense malignant cells. The cancer cells show higher RI distribution as compared to normal cells [19, 20].

The presence of oral-cancerous cells in the given sample can be detected by using the proposed biosensor. The RI of normal and oral cancerous cells is tabulated in Table 2.1 [21].

TABLE 2.1 RI Variation of Normal and Oral Cancerous Cells

Cells	Refractive-Index
Normal	1.343
	1.344
	1.345
	1.348
	1.351
Cancer	1.369
	1.371
	1.372
	1.377
	1.378

Figure 2.7 represents the bar chart of Normal and Cancer cells RI variation. From the Table 2.1, it is clear that the variation of RI profile for normal cells is lesser than that of the cancer cells. The reason for this is because of the higher protein concentration in the cancer cells [19]. Using this as the sensing parameter, it is possible to differentiate normal cells from the cancer cells.

In Figure 2.5, the defect region of the proposed optical sensor is filled with the sample to be analyzed. A microfluidic channel can be created for the sample flow, which fills the defect region and is represented in Figure 2.8. Once the region of defect is filled with the sample to be analyzed, a light beam of wavelength 10.6 μm is made to an incident on the DBR structure. As the light beam propagates through the structure, it interacts with the sample at the defect region and a beam of light is transmitted through the structure. A shift in the wavelength of the transmission spectrum is observed. Based on this wavelength shift, it is possible to identify whether the sample contains cancer cells or only normal cells. The following transmission spectrum plot shows the wavelength region for normal and cancer cells (Figure 2.9).

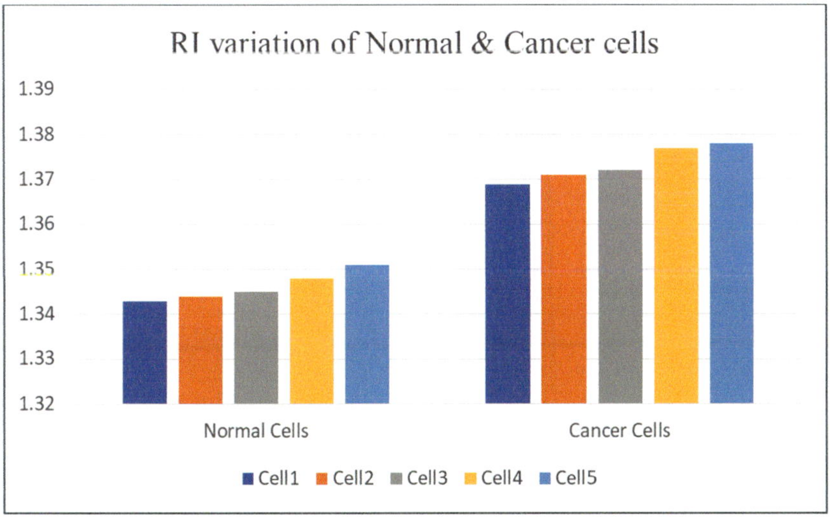

FIGURE 2.7 Refractive-index profile of normal and cancer cells.

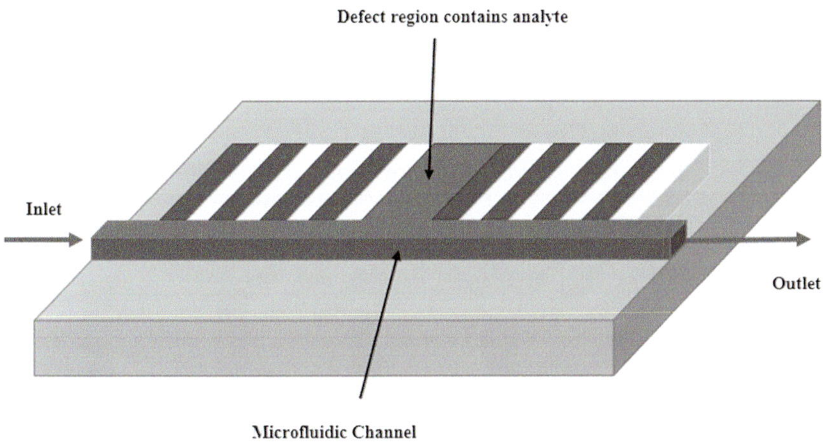

FIGURE 2.8 Proposed sensor with a microfluidic channel for the sample flow.

The Figure 2.7 graphs conclude that there is a considerable shift in the resonant mode for the normal and cancer cells. This shift can be measured significantly to detect the presence of cancer cells in the given test sample. The bar chart representation of the resonant mode shift clearly indicates the recognizable wavelength shift between normal and cancer cells (Figure 2.10).

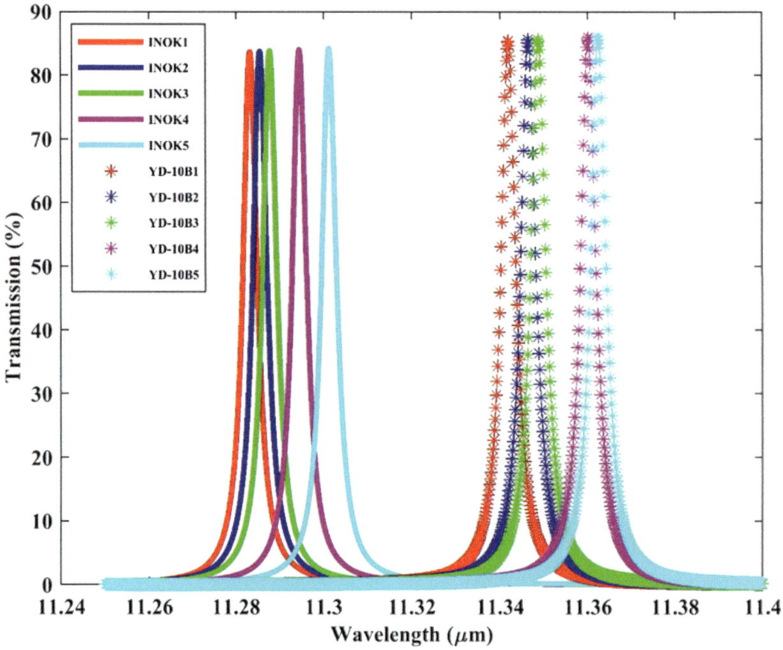

FIGURE 2.9 Wavelength region of resonant modes for normal and cancer cells.

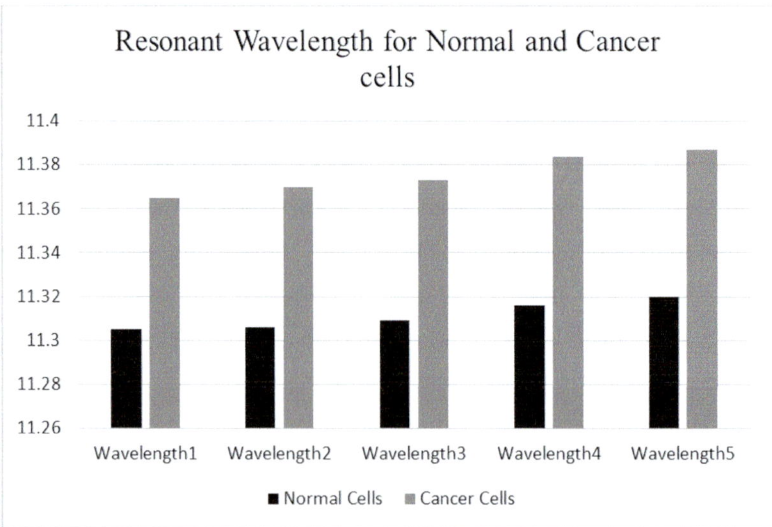

FIGURE 2.10 Bar chart representation of resonant mode shift between normal and cancer cells.

2.5 SENSOR DESIGN PARAMETERS

The two main sensor design parameters of an optical sensor are sensitivity (S) and quality factor (Q). These two parameters describe how good the proposed sensor is for the detection of malignant cells in the given test sample.

2.5.1 SENSITIVITY (S)

This parameter represents how good the proposed sensor is to detect the presence of cancer cells in the sample.

Mathematically it is given by:

$$S = \frac{\Delta\lambda}{\Delta n} nm / RIU$$

where; $\Delta\lambda$ represents the change in resonant wavelength; Δn represents the change in RI.

Change in the resonant wavelength for the corresponding change in RI of the cells accurately describes the sensitivity of the sensor. The sensitivity variation of the proposed sensor for the sample under test is represented graphically in Figure 2.11. From the chart, it is concluded that the highest sensitivity of 2,482 nm/RIU is achieved with the proposed design.

2.5.2 QUALITY FACTOR (Q)

This parameter represents the sensor's ability to accurately distinguish normal and cancer cells in the test sample.

Mathematically it is given by:

$$Q = \frac{\lambda}{FWHM}$$

where; λ represents the resonant wavelength; FWHM represents full-width Half-maximum of resonant wavelength.

Figure 2.12 represents the Q factor variation of the proposed sensor for the cells under consideration. The highest Q factor of 141.5 is achieved for the proposed sensor design.

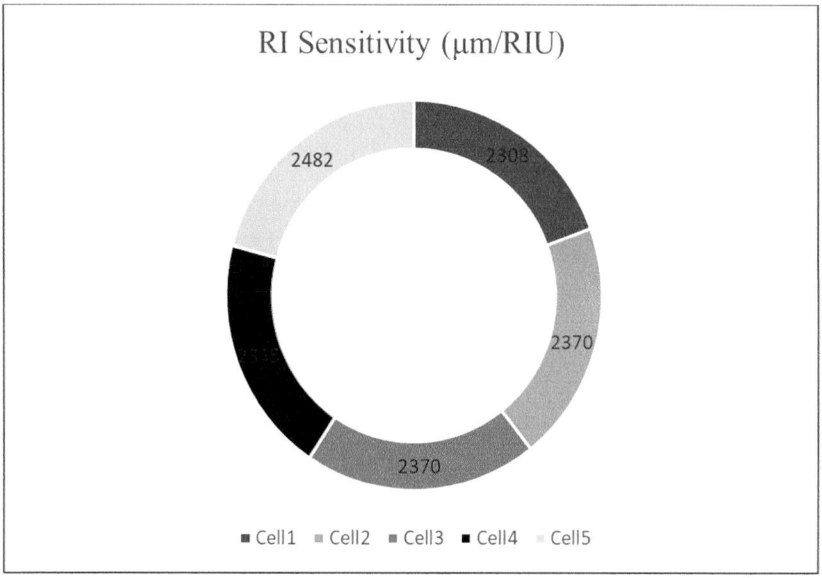

FIGURE 2.11 Sensitivity of the proposed sensor for cells under consideration.

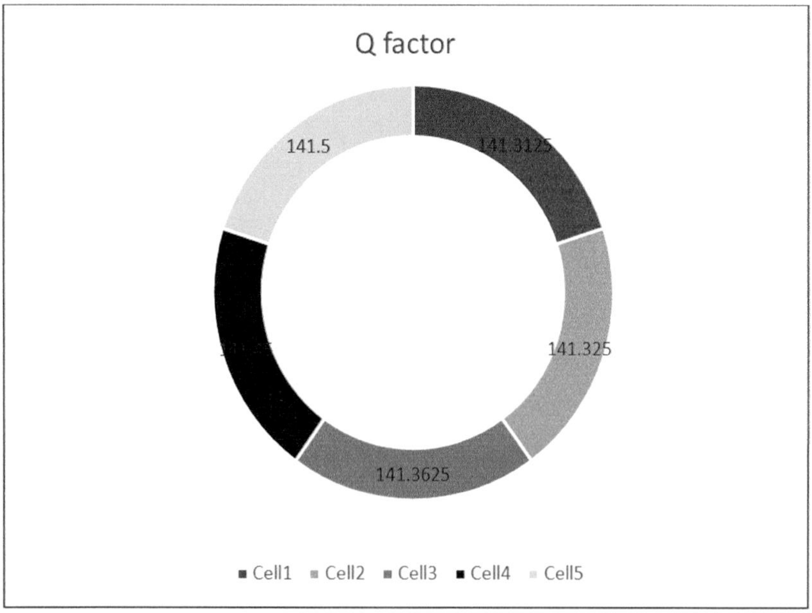

FIGURE 2.12 Q-factor of the proposed sensor for cells under consideration.

2.6 CONCLUSION

In this work, an optical sensor using 1D DBR is designed and proposed theoretically to sense oral cancerous cells in the given test sample using the TMM. 1D DBR is designed using mid-infrared optically transparent materials like Germanium and zinc sulfide. The proposed sensor operated at the mid-infrared range to achieve high sensing parameters. The sensing parameters of an optical sensor like Q factor and sensitivity are calculated and obtained a very high sensitivity and Q factor of 2,482 nm/RIU and 141.5, respectively. From the results obtained, it is concluded that the sensor proposed in this work can be used to quickly and accurately detect oral cancerous cells in the given sample under test and can be suitably used in point-of-care applications.

KEYWORDS

- **distributed Bragg reflector**
- **enzyme-linked immunosorbent assay**
- **micro-immunofluorescence**
- **photonic-crystal**
- **reverse transcription–polymerase chain reaction**
- **zinc sulfide**

REFERENCES

1. Noda, S., & Toshihiko, B., (2003). *Roadmap on Photonic Crystals* (Vol. 1). Springer Science & Business Media.
2. Yablonovitch, E., (1993). Photonic band-gap structures. *J. Opt. Soc. Am. B, 10*, 283–295.
3. Inan, H., et al., (2017). Photonic crystals: Emerging biosensors and their promise for point-of-care applications. *Chemical Society Reviews, 46*(2), 366–388. doi: 10.1039/c6cs00206d.
4. Monošík, R., Streanský, M., & Šturdík, E., (2012). *Acta Chim. Slovaca., 5*, 109–120.
5. Washburn, A. L., Luchansky, M. S., Bowman, A. L., & Bailey, R. C., (2010). *Anal Chem., 82*, 69–72.
6. Yang, L., & Bashir, R., (2008). *Biotechnol Adv., 26*, 135–150.
7. Xu, J., Suarez, D., & Gottfried, D. S., (2007). *Anal. Bioanal. Chem., 389*, 1193–1199.

8. Ozsoz, M., Erdem, A., Kara, P., Kerman, K., & Ozkan, D., (2003). *Electroanalysis, 15*, 613–619.
9. Myung, S., Yin, P. T., Kim, C., Park, J., Solanki, A., Reyes, P. I., Lu, Y., et al., (2012). *Adv. Mater., 24*, 6081–6087.
10. Kirsch, J., Siltanen, C., Zhou, Q., Revzin, A., & Simonian, A., (2013). *Chem. Soc. Rev., 42*, 8733–8768.
11. Born, M., & Wolf, E., (1986). *Principles of Optics – M. Born, E. Wolf*.pdf.
12. Pochi, Y., & Amnon Y., (1976). Bragg reflection waveguides. *Optics Communications, 19*, 427–430.
13. Joannopoulos, J. D., Johnson, S. G., Winn, J. N., & Meade, R. D., (2008). *Photonic Crystals: Molding the Flow of Light* (2nd edn.). Princeton University Press, Singapore.
14. Sibilia, C., Benson, T. M., Marciniak, M., & Szoplik, T., (2008). *Photonic Crystals: Physics and Technology* (1st edn.). Springer Milan, Milano.
15. Rogalin, V. E., Kaplunov, I. A., & Kropotov, G. I., (2018). Optical materials for the THz range. *Opt. Spectrosc. 125*(6),1053–1064.
16. Persano, L., Camposeo, A., Del, C. P., Mele, E., Cingolani, R., & Pisignano, D., (2006). *Very High-quality Distributed Bragg Reflectors for Organic Lasing Applications by Reactive Electron-Beam Deposition, 14*(5), 167–171.
17. Hemati, T., & Weng, B., (2019). The mid-infrared photonic crystals for gas sensing applications. *Photonic Crystals – A Glimpse of the Current Research Trends*.
18. Seddon, A. B., (2011). A prospective for new mid-infrared medical endoscopy using chalcogenide glasses. *Int. J. Appl. Glass Sci.,* 177–191.
19. Lue, N., Choi, W., Popescu, G., Yaqoob, Z., Badizadegan, K., Dasari, R. R., & Feld, M. S., (2009). Live cell refractometry using Hilbert phase microscopy and confocal reflectance microscopy. *J. Phys. Chem. A, 113*, 13327–13330.
20. Aly, A. H., & Zaky, Z. A., (2019). Ultra-sensitive photonic crystal cancer cells sensor with a high-quality factor. *Cryogenics, 104*, 102991.
21. Choi, W. J., Jeon, D. I., Ahn, S., Yoon, J., Kim, S., & Lee, B. H., (2010). Full-field optical coherence microscopy for identifying live cancer cells by quantitative measurement of refractive index distribution. *Opt. Express, 18*(22), 23285–23295.

CHAPTER 3

Photonic MEMS Sensor for Biomedical Applications

ANUP M. UPADHYAYA[1] and PREETA SHARAN[2]

[1]*Department of Mechanical Engineering the Oxford College of Engineering, Bangalore, Karnataka, India*

[2]*Department of Electronics and Communication Engineering, The Oxford College of Engineering, Bangalore, Karnataka, India*

ABSTRACT

Biosensor chips are considered an easy and simple detection and diagnosis system for different diseases. Enzyme-linked immunosorbent assay (ELISA) is the one decease detection technique used widely. In this, fluorescent labeling is given to antigen and antibodies. In another optical sensor, such as surface plasmon resonance (SPR), optical fiber sensors use the light propagation technique to observe the change in resonance wavelength due to a change in the refractive index (RI) of the medium. Silicon photonics-based optical MEMS sensors use different biosamples for disease detection. A Microcantilever beam with nanometer/micrometer dimension is a well-adopted sensor structure, which undergoes deflection and generates surface stress in the microcantilever. By monitoring the adsorption in the microcantilever and generated surface stress, microcantilever can be used for the detection of different chemicals and biomolecules causing dangerous diseases. Detection schemes in microcantilevers include electrical, optical techniques. Integrating the photonic chip in the microcantilever can be

Computational Health Informatics for Biomedical Applications. Aryan Chaudhary and Sardar M. N. Islam (Naz) (Eds.)

considered as a photonic MEMS sensor. Photonic crystal sensing patches are integrated with many optical sensing systems such as Mach Zehnder interferometer, and microcantilever. Optical imaging and waveguide are used for interrogating optical sensors for large distance applications.

Photonic MEMS sensor mainly based on the principle of photoelasticity. The photonic sensing layer undergoes mechanical deformation due to applied pressure. This mechanical deformation influences the change in the RI of the sensing layer due to molecular rearrangements. This change in the RI of the sensing layer will change the overall wavelength shift obtained at the monitor. Photonic crystal-based MEMS technology considered fewer system integrations compared to conventional optomechanical sensors. Conventional optomechanical systems such as atomic force microscopy, photoacoustic microscopy, accelerometer, mass sensing can be replaced with photonic integrated circuit-based photonic sensing layers in the near future. Photo-sensitive-based microcantilevers can have a high-resolution range while integrating the photonic crystal sensing method with a different mechanical system. Integrated optical readout techniques are becoming more influential in sensing technology due to the requirement of reduction in cost and size of the sensor. Recently nanocavities sensing patches have shown high potential and resolution for monitoring displacement, pressure, and strain. The construction process of photonic MEMS sensor involves optical waveguides or sensing patches, transduction system as mechanical, output monitors, etc. The main advantage of the photonic MEMS system is mechanical movement, and optical read outputs techniques are uncoupled and made independent of electric circuits considering the final fabrication of the photonic MEMS sensing system. Using the RI modulation system, the photonic sensor offered more reliability and flexibility in large distance sensing applications.

FBG-based MEMS sensing system utilizes the principle of light reflection from fiber Bragg grating sensing structure. A significant improvement over sensor specification and measuring system can be achieved with the help of a highly sensitive FBG sensor. Photonic MEMS are used for monitoring many other parameters such as shape and distribution, strength, and stiffness. Other parameters of MOEMS researchers' interest are electrical, thermal, biological optical, and chemical elements. Most precision sensors of force, pressure, strain, and acceleration are considered in the present situation having limitations over sensitivity, electromagnetic interference, and compactness. Scientists are showing more interest in the development of nano and micro-scale sensors with ultra-high sensitivity and Q factor using laser light sources.

3.1 INTRODUCTION TO PHOTONIC MEMS

The development of photonic MEMS sensors, fiber Bragg grating sensors, photonic crystal pressure sensors, Mach Zehnder interferometer, and microcantilever sensors has received a lot of attention in the past decade. Integration of FBG's, photonic crystal pressure sensor with photonic sensing layer has a crucial role in the packaging of photonic MEMS structure. Each sensing configuration has a unique integration approach.

3.1.1 PHOTONIC INTEGRATED MACH ZEHNDER INTERFEROMETER

Photonic crystal-based MZI structure is designed by line defect and integrated with the silicon slab with oxide layer as substrate. As we can see from Figure 3.1, the sensor has light input and output ports. Approximate length, width, and height are shown in Figure 3.1. This sensing structure consists of two parts, one end of the sensing arm will deflect for applied pressure, such that it deforms in a downward direction. Other ends of the sensing arm do not deflect, such that it is fixed in an integrated position. This end of the sensing arm has some bulk material for support.

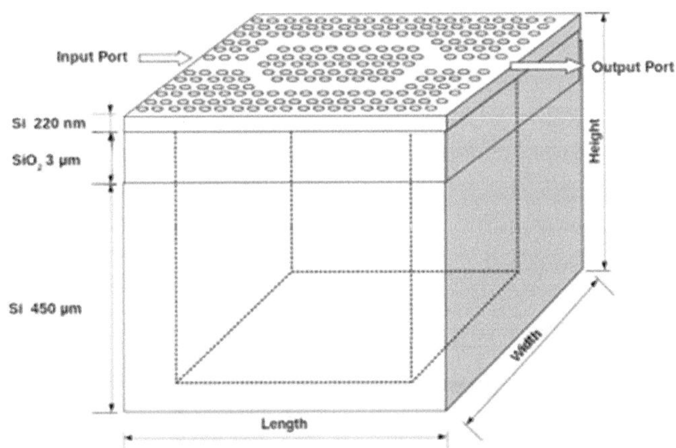

FIGURE 3.1 MZI pressure sensor.

Source: Reprinted from Ref. [1]. https://creativecommons.org/licenses/by-nc-nd/4.0/

3.1.2 PHOTONIC INTEGRATED MICROCANTILEVERS

Integrating the photonic crystal with a microcantilever is a novel readout technique in the optical sensor category. Microcantilever has a fixed end and a free end. Fixed end of microcantilever integrated with optical sensing system and readout techniques. The free end of the microcantilever is integrated with a microfluidic channel for sensing biological pathogens and tissues. Microcantilever integrated with microfluidic channels will be introduced with biosamples of various conditions. Microfluidic channels surface will adhere with a coating material which will adsorb the biosample introducing the sensing structure. Biomolecules attaching with the sensing layer will deflect the microcantilever for increased pressure at the tip of the microcantilever. The photonic sensing structure is attached to the based end of the microcantilever will undergo deformation. Due to the deformation in the sensing structure, there will be geometrical variation in the position of sensing holes or micropillars. Research has been carried out in various ways for variation in the shape of the microcantilever and for its sensitivity. T shape, V shape is the popular structure used in microcantilever integrated photonic circuits. Micromachining and die saw technologies are used, fabricating the microcantilever chip and integrating it with the photonic sensing layer. Figure 3.2 shows the photonic integrated microcantilever sensing configuration for biomolecules detection. Figure 3.3 shows the light propagation in the integrated photonic sensing layer.

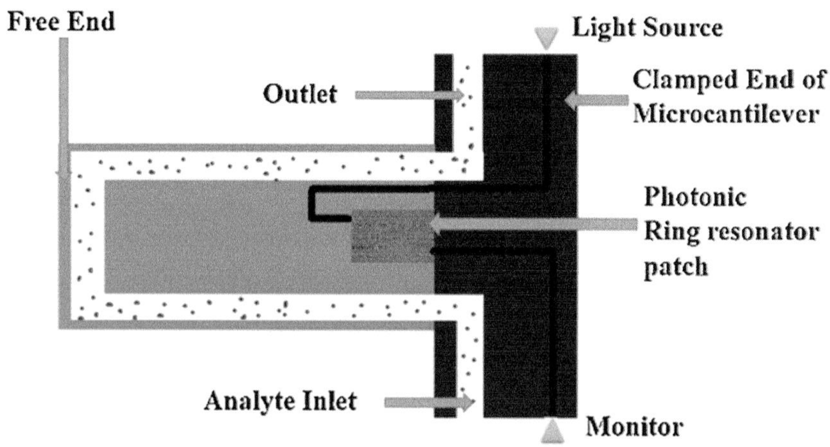

FIGURE 3.2 Photonic integrated microcantilever [2].

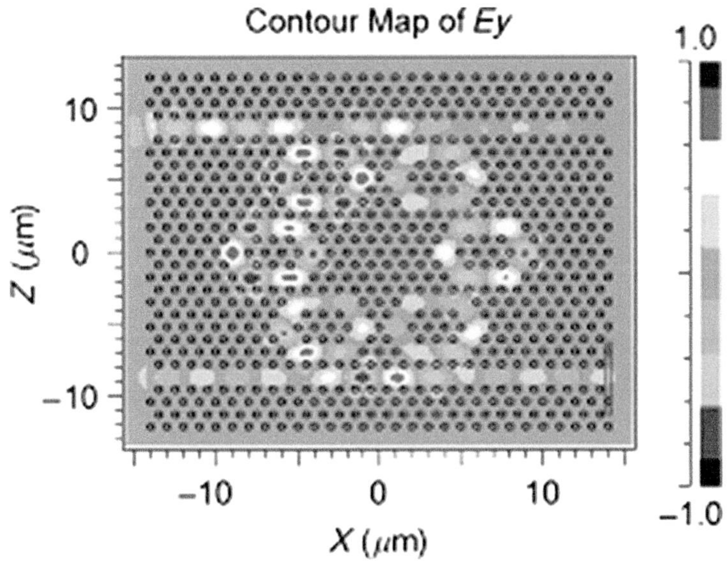

FIGURE 3.3 Light propagation in photonic sensing layer for microcantilever integration.

3.1.3 OPTICAL PRESSURE SENSORS IN SILICON ON INSULATORS

This silicon on insulator structures was fabricated with the process of UV lithography techniques. Silicon on insulator layer is deposited with silicon dioxide layer of thickness 2 μm. Simulation and analysis of silicon insulator waveguides can be carried out in various types of simulating tools. The most widely used simulating tools for simulation of silicon on insulators in the market are COMSOL multiphysics ANSYS lumerical. Silicon on insulator structure can be used in both pressure and biosensing application. Determining the confident field factor is a critical factor in the SOI structure. Figure 3.4 shows silicon on insulator structure for photonic MEMS application.

3.1.4 FIBER BRAGG GRATING SENSOR

Fiber Bragg grating sensors are formed by periodic variations of the RI of the fiber. FBG sensor has high sensitivity due to strong reflected signal. This periodic arrangement of grating lines can be generated by various techniques. Fiber Bragg grating mainly works on the principle of Bragg law. Electromagnetic radiation undergoes Bragg reflection has the same order of magnitude with atomic spacing. The observation of change in the RI of fiber was initially noticed in the germane silica fiber. UV lasers are used to inscribe the gratings on fibers. Important parameters which can be monitored with optical fiber Bragg grating are strain, vibration, electric current, leakage impedance, pressure, temperature, gas concentration and distance between two objects.

$$\lambda_B = 2n_{eff}\Lambda\lambda_B = 2n_{eff}\Lambda$$

where; λ_B is the Bragg wavelength of FBG function; Λ is the period – the distance between grating lines; $_{Neff}$ is the effective refractive index.

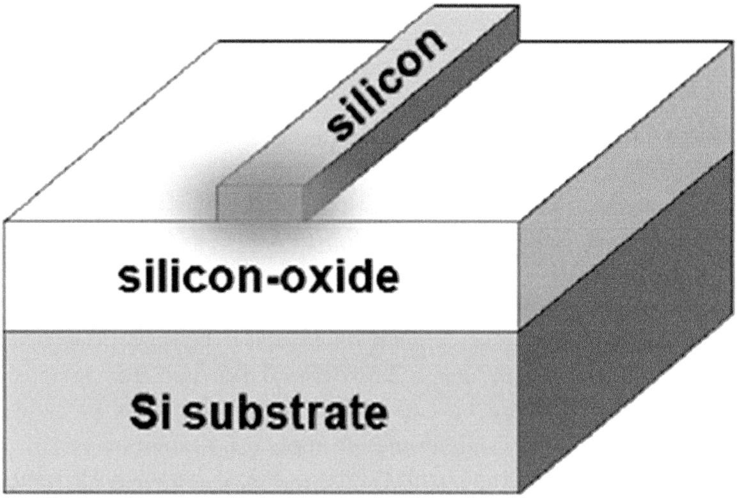

FIGURE 3.4 Silicon on insulator structure.

Source: Reprinted from Ref. [3]. Open access.

A small amount of light reflected back for each change in refraction. All the reflected light signals combine at one coherently to make one large reflection. It is referred to as the Bragg condition. Light signals at wavelengths other than Bragg wavelength have transparent signals and generate fewer signals. The light propagated through the grating and with less attenuation. The shift in the central wavelength of reflected light satisfies the Bragg condition.

Fiber Bragg grating is widely used in real-time applications, railway monitoring, biomedical applications, and general industrial applications as temperature sensors, pressure sensors, and accelerometer sensors. Figures 3.5 and 3.6 show the principle of the fiber Bragg grating sensor.

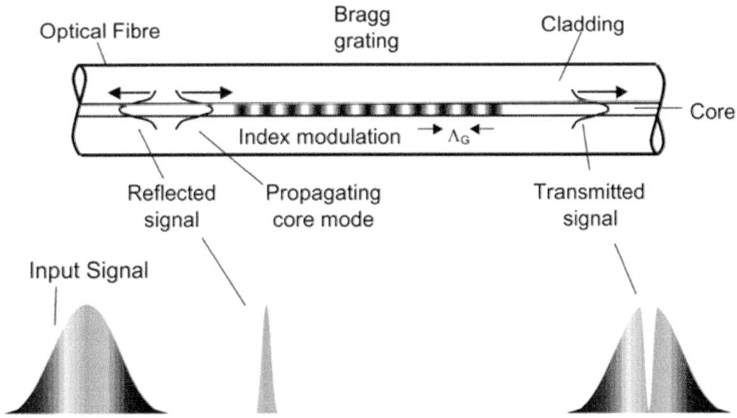

FIGURE 3.5 Fiber Bragg grating principle.

Source: Reprinted from Ref. [4]. © 2012 by the authors; licensee MDPI, Basel, Switzerland. https://creativecommons.org/licenses/by/3.0/

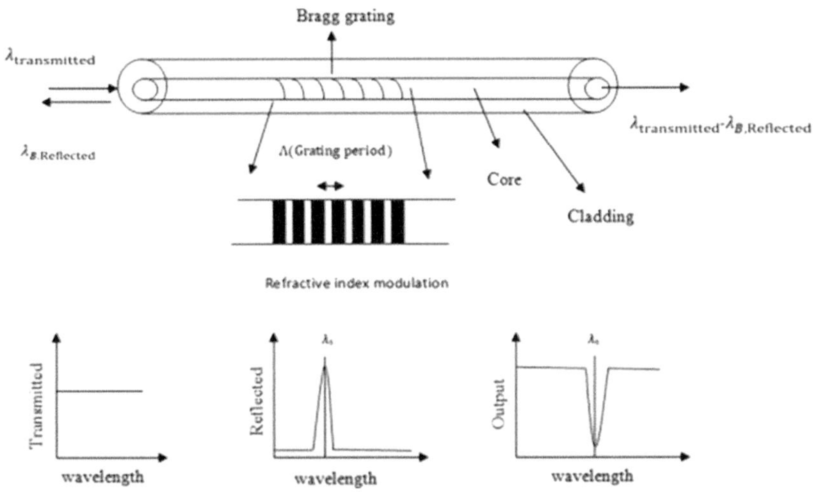

FIGURE 3.6 Fiber Bragg grating wavelength shift

Source: Reprinted from Ref. [5]. © 2020 Society of Photo-Optical Instrumentation Engineers (SPIE). Open access..

3.2 REFRACTIVE INDEX SENSOR AND PHOTONIC STRUCTURES

Sensing has been identified as a promising application for photonic technology. Every year a large number of photonic sensing structures have been undergoing testing for their ability to sense. Apart from the photonic crystal sensing structure, there are a number of RI sensing systems for various bio and mechanical sensing applications.

3.3 BIOSENSING APPLICATIONS OF PHOTONIC SENSOR

3.3.1 PHOTONIC INTEGRATED MICROCANTILEVER SENSOR (PIMS) – DIFFERENT SHAPES

Microcantilever is integrated with various photonic crystal sensing layers. Using the regular rectangular sensor photonic crystal sensing layer is integrated on the base of the microcantilever (Figure 3.7). Microcantilever is designed in CAD software tools. Numerical analysis of different shapes such as trapezoidal, T shape, and V shape microcantilever has been carried out. Optimization of regular microcantilever is an important step in photonic integrated microcantilever. Regular microcantilever is optimized for tip width and thickness. Tip width varied in such a way that, microcantilever has undergone different such as trapezoidal and triangular shapes (Figure 3.8). MIT electromagnetic equation propagation tool has been used to design the sensing structure. The sensor is investigated for its performance in air and water medium. The maximum sensitivity of 1.93 nm/KPa. The applied pressure range in the sensing structure was 100 KPa to 250 KPa [6].

3.3.2 PHOTONIC INTEGRATED MICROCANTILEVER SENSORS FOR CANCER DETECTION

Microcantilever with two different shapes, such as rectangular and v shape, has been designed and analyzed for cancer detection application. Photonic Integrated Microcantilever undergoes defection due to pressure applied. Microfluidic channels are integrated in the microcantilevers (Figure 3.9).

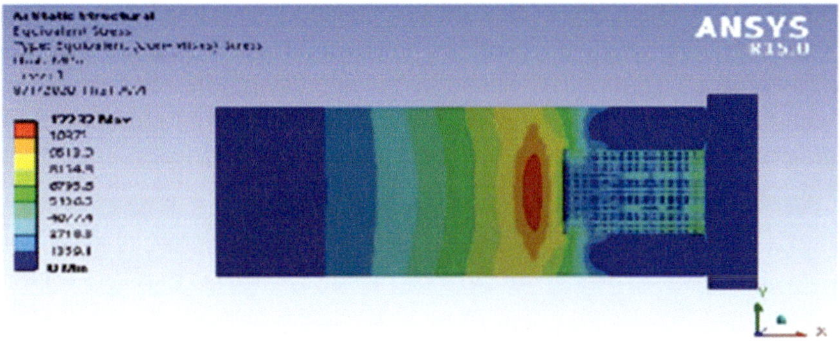

FIGURE 3.7 FEA analysis of photonic integrated rectangular microcantilever.

Source: Reprinted with permission from Ref. [6]. © 2020 John Wiley.

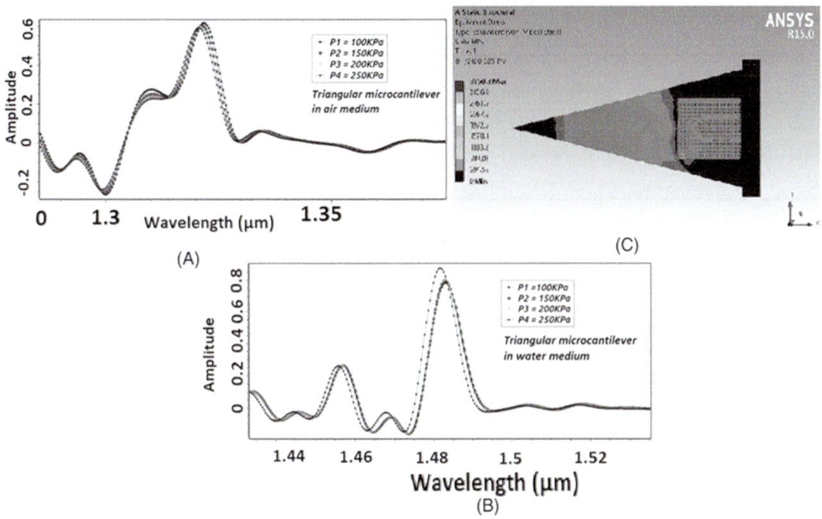

FIGURE 3.8 Wavelength shift for triangular microcantilever.

Source: Reprinted with permission from Ref. [6]. © 2020 John Wiley.

Carcinoembryonic antigen (CEA) molecules are responsible for colon cancer in humans. Biofluids containing CEA molecules are introduced in the microfluidic channel. The tip of the microcantilever undergoes deflection due to the CEA molecules binding to the surface of the microcantilever due to adsorption. Pressure on the microcantilever is calculated to form the weight of the CEA attaching to the surface of the microcantilever. The pressure sensitivity of this microcantilever obtained was 60 nm/MPa. The transmission efficiency of the sensor obtained was around 90%. As the number of biomolecules increases, microcantilever deflection increases [7]. Figures 3.10 and 3.11 show the rectangular profile microcantilever with integrated photonic structure and transmission spectrum of the triangular microcantilever.

FIGURE 3.9 Photonic integrated rectangular microcantilever.

Source: Reprinted with permission from Ref. [2]. © 2020 Elsevier GmbH.

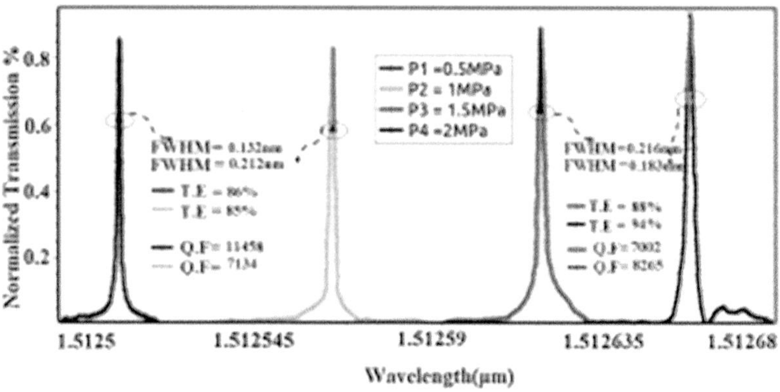

FIGURE 3.10 Transmission spectrum for triangular microcantilever [7].

3.3.3 PHOTONIC MEMS-BASED RING RESONATORS

Ring resonators optical waveguides in which one is a closed-loop connected with input and output waveguides. Due to the nature of constructive interference and the total internal reflection process, ring resonators are considered to achieve high-quality signals. Because only some of the wavelengths get resonated within the optical waveguides. Ring resonators can be designed with many numbers, such as single ring resonators, double-ring resonators, and triple ring resonators, etc. (Figure 3.12).

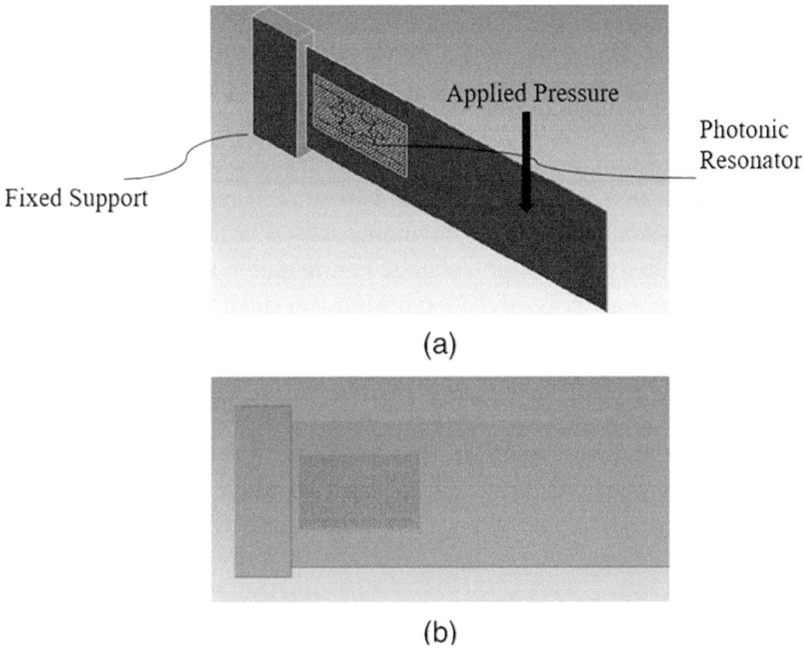

FIGURE 3.11 CAD models of integrated microcantilever [8].

Source: Reprinted with permission from Ref. [8].

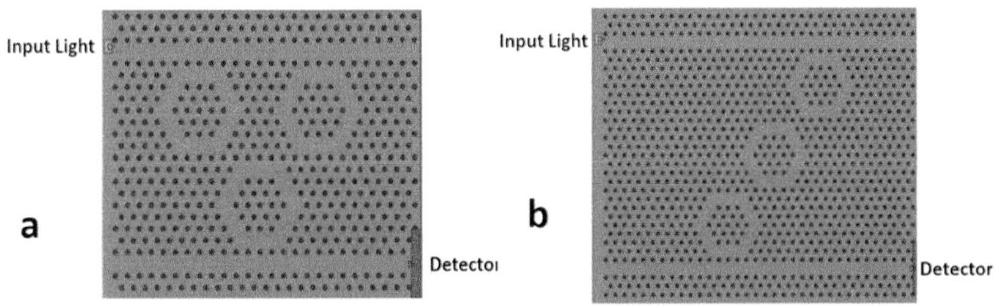

FIGURE 3.12 Triple ring resonators in parallel (a); and series process (b).

Source: Reprinted with permission from Ref. [9].

Series and parallel arrangement of triple ring resonators are proposed in the proposed work. Sensing structures are incorporated in the hexagonal lattice configuration, sensing layer made with silicon material. Specific distance re maintained between double-ring parallel and triple ring parallel structures. Similarly, vertical distance is maintained between the double-ring series and triple ring resonator structure. During the propagation of light in the sensing structure due to the microcantilever deflection, the RI of the sensing structure changes. This, in turn, changes the overall RI of the structure. This property of light that changes in RI for mechanical deformation of sensing structure is called photoelasticity. Peak resonant wavelength charges for change in RI. As the applied pressure increases, the strain generated at the base of the microcantilever increases. The bandgap range of each sensing structure changes for a change in orientation rings in the structure. High-pressure sensitivity of 49.7 nm/MPa and quality factor 3260. Triple ring resonator in parallel configuration was found to behave more sensitivity compared to the other three configurations [7]. Sensing structures can be fabricated and used for biosensing applications in the near future.

3.3.4 MICROMECHANICAL OPTICAL SENSOR FOR BIOSENSING APPLICATION

Line defect is used with photonic crystal microcavities to create optical waveguides. A micromechanical ultra-sensitive membrane is constructed to get a bandgap structure. The structure has shown a sharp resonant peak wavelength with high sensitivity and Q factor. Suspended photonic crystal structure. The sensing structure is optimized with pressure and temperature. The temperature of the sensing structure has been changed from 21°C to 31°C. Suspended bridge-type photonic sensing structure undergoes deflection due to applied pressure in air and water medium. Q factor 2356 was obtained for the above sensor in micro holes configuration in air and water medium. The sensing structure is designed in the FEA tool and optical simulator to get the required mechanical and optical properties of the sensing structure. Input light is transmitted through the line defect created in the sensing structure. Overall strain is calculated from vertical and horizontal deformation in the sensing structure. This deformation taking place in the sensing structure changes the position, size, and shape of micro holes. Pressure is applied, and the temperature of the fluid is changed gradually. The high sensitivity of 0.34 nm/RIU is obtained during the sensing structure analysis. The bandgap of the sensing structure changes for a change in temperature and pressure. Bandpass frequency has to be identified for varying pressure and temperature. The sensitivity of the sensing structure can also be further increased by rearranging the defect created or microcavities. The proposed sensor has tremendous application biosensing, such as water impurities detection as a chemical sensor, biomolecule detection as a biosensor, and cancer tissue marker detection. Photonic MEMS is evolving multidisciplinary field. Developing sensors of this kind helps society in a different way in terms of cost and ease of availability. With substantial progress made with the optical MEMS fabrication process, the proposed sensor can be fabricated in different methods such as reactive ion beam etching, bulk, and surface micromachining, photolithography, electronic bel lithography, die saw Technology, Lithography Galvanoformung Abformung (LIGA) [11] (Figures 3.13 and 3.14).

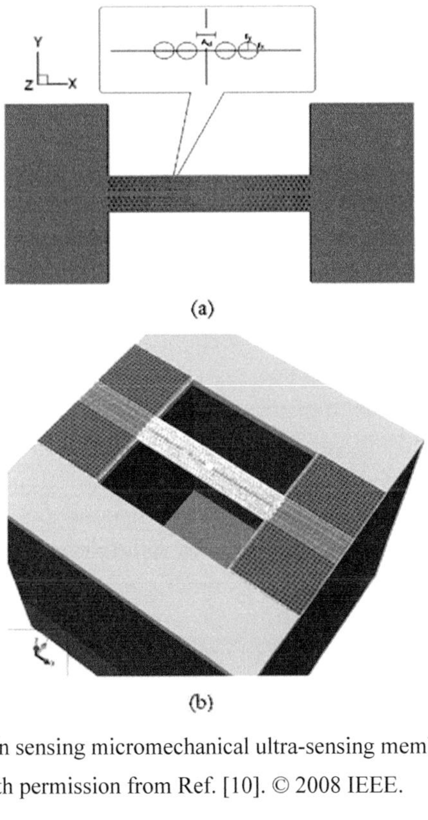

FIGURE 3.13 Strain sensing micromechanical ultra-sensing membrane [10].

Source: Reprinted with permission from Ref. [10]. © 2008 IEEE.

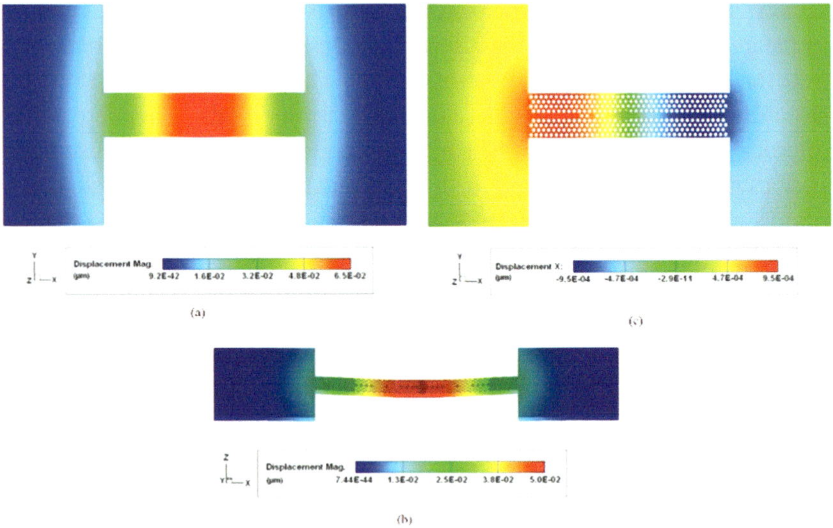

FIGURE 3.14 FEA simulation photonic integrated bridge structure [10].

Source: Reprinted with permission from Ref. [10]. © 2008 IEEE.

3.3.5 MOEMS DISPLACEMENT SENSOR FOR MUSCLE ACTIVITY DETECTION

Muscles play an important role in managing good health. Muscle helps in keeping the posture, joint, and all internal organs of the human body in shape and healthy way. The sensor is designed with the help of a photonic crystal. Micropillar sensing configuration with hexagonal lattice is considered in the design process. The radius of micropillar, lattice size, number of micropillars in X and Y direction, lattice configuration are the input parameters for sensing layer construction. Rectangular slab structures are designed within the crystal structure. Slabs are moved to a particular distance. Later sensing layer is modeled in the FEA tool, and strain data are collected. Sensitivity and Q factor are optimized for change input parameters. As the pressure changes in the sensing structures, the peak resonant wavelength is changed. As the Q factor increases, there is more probability of obtaining high sensitivity sensor configuration [11] (Figure 3.15).

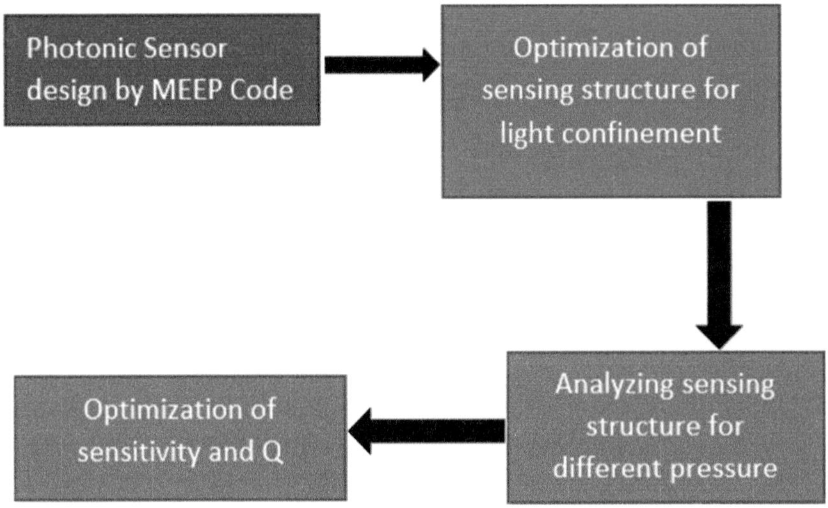

FIGURE 3.15 Block diagram of photonic displacement sensor operation.
Source: Reprinted with permission from Ref. [12]. © 2021 Springer.

3.3.6 COMPARISON OF OPTICAL RING RESONATOR STRUCTURE WITH RING COUNT

Three different types of optical ring resonator structures are designed and analyzed with optical and FEA tools for sensing performance investigation. Each sensing structure has one ring, two rings, and three-ring resonators. The Sensitivity and Q factor of each ring resonator structure has been carried out. Q factor, sensitivity, and minimum detectability of sensing structure have been investigated. FEA modeling each circular ring resonator structure has been constructed using the Ansys tool. Pressure is applied in the range of 1 MPa to 6 MPa, increment of 1 MPa. Maximum stress and strain developed in the overall structure have been considered for the sensing layer. Minimum observable deformation of sensor such as 1.007 μm. The high sensitivity of 450 nm/RIU is achieved with this specific sensing configuration. A Maximum Q factor of 8913 is obtained for a single ring resonator circular diaphragm. The sensitivity obtained for two and the three-ring resonator-based circular diaphragm is 316 nm/RIU and 415 nm/RIU, respectively. For no-load conditions, one ring resonator structure has given a feasible peak amplitude of 1.9196 V at a peak resonant wavelength of 1.53 nm.

Figure 3.16 shows the FEA simulation of a three-ring resonator-based circular diaphragm. During the application of pressure at the center of the circular diaphragm, deflection is observed more at the center of the circular diagram compared to the side. More stress is observed at the center of the circular diaphragm. In the FEA simulation in Figure 3.16, red indicates the high-stress concentration region, and blue indicates the low-stress concentration region. Figure 3.17 shows the three different types of circular diaphragm modeled in the Ansys workbench tool.

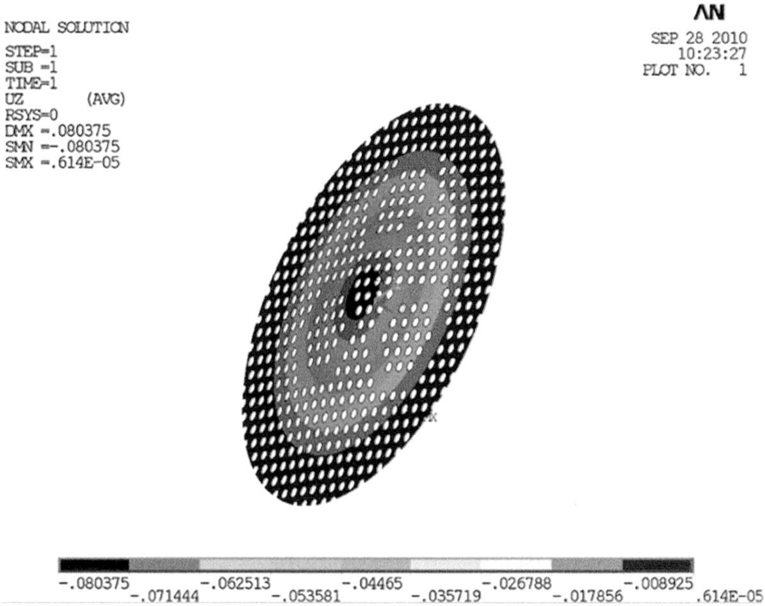

FIGURE 3.16 FEA simulation of ring resonator structure.

Source: Reprinted with permission from Ref. [13]. © 2011 Elsevier.

Figure 3.18 depicts the peak resonant wavelength shift in the double-ring resonator. The wavelength is shifted from 1 μm to 1.45 μm.

3.3.7 FIBER BRAGG GRATING-BASED OPTOMECHANICAL SENSOR

Fiber Bragg grating sensor is giving a new look to biomedical technology in terms of ease of handling, sensitivity, and easily accessible data. Plenty of FBG sensors applications is at an advanced level for usage purposes. FBG sensor is constructed with polydimethylsiloxane polymer on the

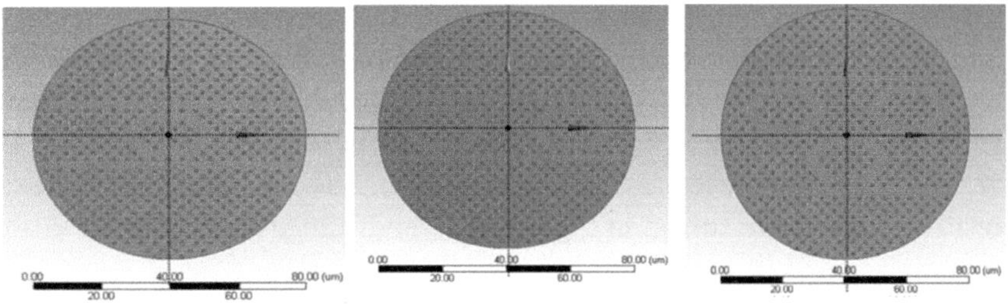

FIGURE 3.17 Comparison of three different ring resonator structures [13].

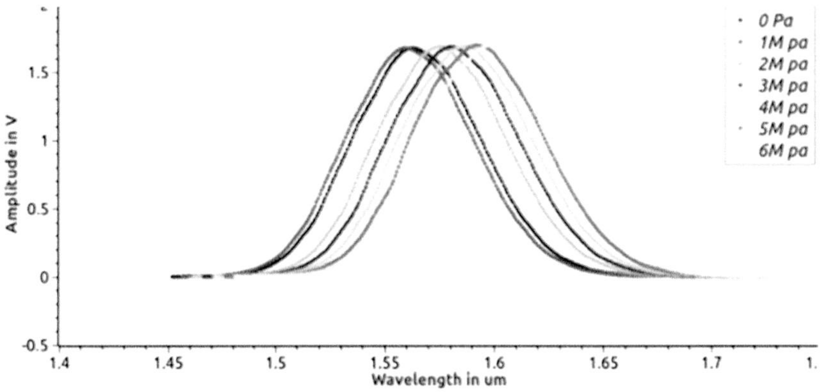

FIGURE 3.18 Wavelength shift for double-ring resonator [14].

Source: Reprinted with permission from Ref. [14]. © 2019 IEEE.

sensing FBG for monitoring cardiac vibrations. Elastic and thermal property of FBG sensor. Real-time experimental with FBG sensor with cardiac vibration is validated with sensor performance [15]. The need for an FBG sensor was also extended for pulse rate monitoring with a fingertip. Vital sign of the human body is measured with the help of packaged FBG sensor. Plastic FBG sensor with young's modulus of silica glass considered in the real-time experiment process [16]. FBG is also used for joint movement monitoring during human activities. FBG sensor is embedded within a silicon tube for joint movement monitoring. As the finger movement is taking place, the joint in the finger is observed, and FBG sensor output is collected. Movement in the knee, arm, and finger joint is considered [16]. Cancer detection has been made easy these days with sophisticated technology. Early detection of human breast cancer is a major requirement in the present era. Much advanced imaging technology has been in the market. But these methods are costly and uncomfortable and take a lot of time to give output. The detailed thermal view is obtained with the help of COMSOL Multiphysics. Temperature variation of 0.3°C is detected at the presence of tumors location [18]. Different applications of FBG sensors are shown in Figures 3.19 and 3.20 [17].

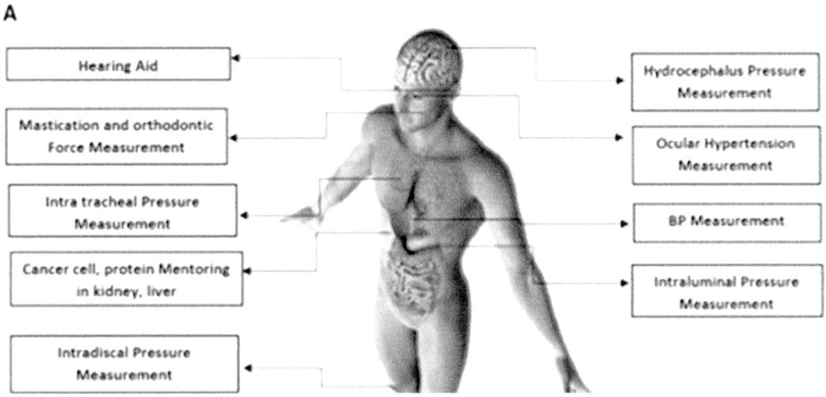

FIGURE 3.19 FBG sensor applications.

Source: Reprinted from Ref. [18]. https://creativecommons.org/licenses/by/4.0/

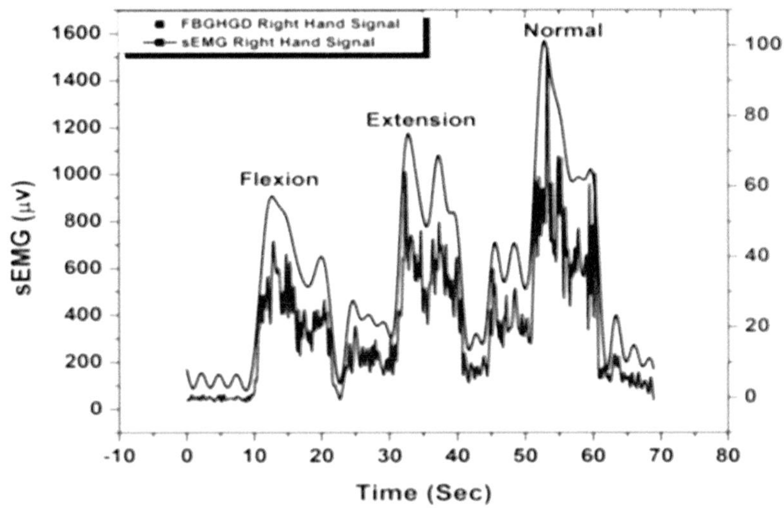

FIGURE 3.20 Handgrip pressure monitoring using FBG sensor [19].

FBG sensor is also used in monitoring handgrip pressure in flexion, extension, and normal condition. Sensor signal obtained from the FBG sensor is compared with EMG signals. Obtained results are converted to force/kilograms using a calibration factor in the sensor output. Wrist angular strength has major significance in monitoring human hand strength. Along with the measuring entire grip by hand, FBG is also monitored by foot pressure for neutral foot, cavus foot, supinated foot. Foot pressure from different numbers of persons is considered [7]. The smart insole is developed to monitor foot pressure in diabetic patients. Performance and functional elements of Fiber Bragg Grating are evaluated for different experimental conditions of Fiber Bragg Grating Sensors. FBG sensors are placed in the different pressure points of the insole. Six sensors are placed in the insole structure to capture the pressure range in each region of the foot section. As shown in Figure 3.21, during the working foot is divided into different parts such as medial forefoot, central forefoot, lateral forefoot, medial midfoot, lateral midfoot, and heel parts. Figure 3.20 represents the transmission spectrum for the handgrip pressure monitoring.

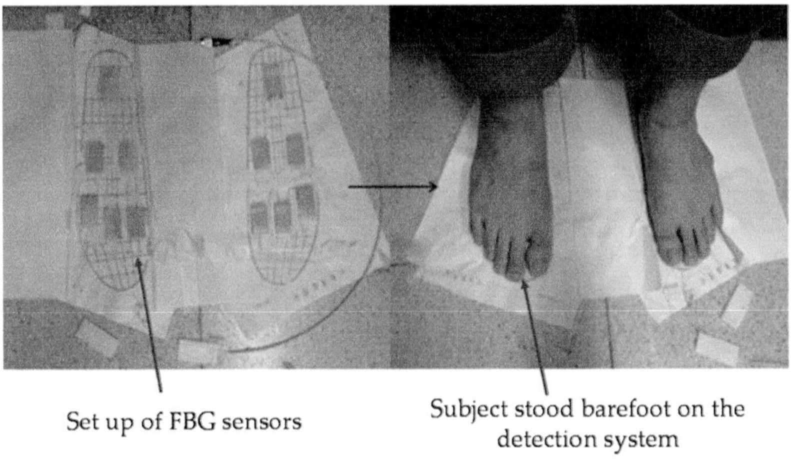

FIGURE 3.21 Set-up of FBG sensors and test in action.

Source: Reprinted from Ref. [7]. © 2016 by the authors; licensee MDPI, Basel, Switzerland. https://creativecommons.org/licenses/by/4.0/

3.4 CONCLUSION

In this chapter, different types of photonic MEMS sensors are designed and discussed for different biomedical applications. The sensing mechanism, fabrication process, sensing principle of each type of sensor are discussed. Major sensors types such as Mach Zehnder interferometer, silicon insulator structures, microcantilevers with integrated photonic sensor and integrated ring resonator structures, and Fiber Bragg Grating Sensor structures. The main challenges when it comes to the above sensor's methods are sophisticated fabrication techniques and the process of characterization of the sensor. Demodulating the wavelength of fiber Bragg grating sensor is the second challenge faced by fiber optic (FP) sensors. Conversion of optical wavelength to the required form of output is an additional task for researchers. Fiber Bragg Grating Technology has been revolutionized the sensing system in general industrial applications and especially in biomedical applications. Most of the FBG sensor-based biomechanical system developed is ready for real-time implementation. Photonic MEMS sensor has shown promising possible realization of lab on chop techniques with the integration of micromechanical and microelectronic components.

KEYWORDS

- **carcinoembryonic antigen**
- **enzyme-linked immunosorbent assay**
- **fiber Bragg grating**
- **micro-electro-mechanical system**
- **microelectronic components**
- **photonic integrated microcantilever sensor**

REFERENCES

1. Venkateswara, R. K., Basavaprasad, Indira, B., & Srinivas, T., (2021). A highly sensitive photonic crystal Mach-Zehnder-interferometer based pressure-sensor. *Results in Optics, 5*, 100118. ISSN 2666-9501, https://doi.org/10.1016/j.rio.2021.100118.
2. Upadhyaya, A. M., Srivasta, M., & Sharan, P., (2021). Integrated MOEMS based cantilever sensor for early detection of cancer. *Optik, 227*, 165321, ISSN 0030-4026, https://doi.org/10.1016/j.ijleo.2020.165321.
3. Jing, Z., Grigorij, M., Joan, J., Sulakshna, K., et al., (2019). III-V-on-Si photonic integrated circuits realized using micro-transfer-printing. *APL Photonics, 4*, 110803. https://doi.org/10.1063/1.5120004.

4. Mihailov, S. J., (2012). Fiber Bragg grating sensors for harsh environments. *Sensors, 12*(2), 1898–1918. https://doi.org/10.3390/s120201898.
5. Jasjot, K. S., Neena, G., & Divya, D., (2020). Fiber Bragg grating sensors for monitoring of physical parameters: A comprehensive review. *Optical Engineering, 59*(6), 060901. https://doi.org/10.1117/1.OE.59.6.060901.
6. Upadhyaya, A. M., Srivastava, M. C., & Sharan, P., (2021). Performance analysis of optomechanical-based microcantilever sensor with various geometrical shapes. *Microwave and Optical Technology Letter, 63*, 1319–1327. https://doi.org/10.1002/mop.32652.
7. Liang, T. C., Lin, J. J., & Guo, L. Y., (2016). Plantar pressure detection with fiber Bragg gratings sensing system. *Sensors (Basel, Switzerland), 16*(10), 1766. https://doi.org/10.3390/s16101766.
8. Upadhyaya, A. M., Srivasta, M., Sharan, P., & Roy, S. K., (2021). Silicon nanostructure-based photonic MEMS sensor for biosensing application. *Journal of Nanophotonics, 15*(2), 026001. https://doi.org/10.1117/1.JNP.15.026001.
9. Patil, P. P., Kamath, S. P., Upadhyaya, A. M., & Sharan, P., (2021). Design and analysis of photonic MEMS-based microring resonators for pressure sensing application. *Journal of Micromechanics and Microengineering, 31*(11), 115004. IOP Science. 10.1088/1361-6439/ac2bb1.
10. Lee, C., Radhakrishan, R., Chen, C. C., Li, J., Thillaigovindan, J., & Balasubramanian, N., (2008). Design and modeling of a nanomechanical sensor using silicon photonic crystals. *J. Lightwave Technol., 26*, 839–846.
11. Upadhyaya, A. M., Srivastava, M. C., Sharan, P., Yashaswini, P. R., & Srikant, P. C., (2019). Micromechanical deformation sensor based on ultra-sensitive photonic crystal membrane. *2019 Workshop on Recent Advances in Photonics (WRAP)*, (pp. 1-3), doi: 10.1109/WRAP47485.2019.9013699.
12. Sharan, P., Sandhya, K. V., Barya, R., et al., (2021). Design and analysis of moems based displacement sensor for detection of muscle activity in human body. *Int. J. Inf. Tecnol., 13*, 397–402. https://doi.org/10.1007/s41870-020-00533-6.
13. Bo, L., & Chengkuo, L., (2011). NEMS diaphragm sensors integrated with triple-nano-ring resonator. *Sensors and Actuators A: Physical, 172*(1), 61–68. ISSN 0924-4247, https://doi.org/10.1016/j.sna.2011.02.028.
14. Upadhyaya, A. M., Srivastava, M., Sharan, P., & Srinivas, T., (2019). Performance analysis of optical MEMS based pressure sensor using ring resonators structure on circular diaphragm. In: TENCON *2019 IEEE Region 10 Conference (TENCON)* (pp. 668–672). doi: 10.1109/TENCON.2019.8929302.
15. Mohapatra, A. G., Tripathy, P. K., Mohanty, M., & Khanna, A., (2022). Fiber Bragg grating (FBG) sensor for the monitoring of cardiac parameters in healthcare facilities. In: Gupta, D., Khanna, A., Kansal, V., Fortino, G., & Hassanien A. E., (eds.), *Proceedings of Second Doctoral Symposium on Computational Intelligence: Advances in Intelligent Systems and Computing* (Vol. 1374). Springer, Singapore. https://doi.org/10.1007/978-981-16-3346-1_57.
16. Li, L., et al., (2021). Embedded FBG-based sensor for joint movement monitoring. In: *IEEE Sensors Journal* (Vol. 21, No. 23, pp. 26793–26798). doi: 10.1109/JSEN.2021.3120995.

17. Prasad, A., Pant, S., Srivatzen, S., & Asokan, S., (2021). A non-invasive breast cancer detection system using FBG thermal sensor array: A feasibility study. In: *IEEE Sensors Journal* (Vol. 21, No. 21, pp. 24106–24113). doi: 10.1109/JSEN.2021.3111684.
18. Upadhyaya, A. M., Hasan, M. K., Abdel-Khalek, S., et al., (2021). A comprehensive review on the optical micro-electromechanical sensors for the biomedical application. *Front Public Health, 9,* 759032. doi:10.3389/fpubh.2021.759032.
19. Chethana, K., Honnungar, R. V., Ibrar, J. M. A., & Asokan, (2020). Comparison studies on hand grip measurement techniques using FBG sensor and sEMG. In: *2020 IEEE International Conference on Electronics, Computing and Communication Technologies (CONECCT)* (pp. 1–6). doi: 10.1109/CONECCT50063.2020.9198583.

CHAPTER 4

Chaotic and Nonlinear Features as EEG Biomarkers for the Diagnosis of Neuropathologies

JISU ELSA JACOB

Department of Electronics and Communication Engineering, Sree Chitra Thirunal College of Engineering, Thiruvananthapuram, Kerala, India, E-mail: jisuelsa@sctce.ac.in

ABSTRACT

Electroencephalogram signals or EEG signals are highly complex, nonlinear signals which clearly reveal the information regarding the brain working and crucial clues of various neuropathologies. Linear features are not sufficient for classifying disease case EEGs from that of normal healthy subjects. This chapter summarizes the emphasis of various chaotic and nonlinear features for EEG for the diagnosis of various neurological diseases.

EEG signal is decomposed using discrete wavelet transform (DWT) to generate EEG sub-bands. Chaotic analysis of signal calculates correlation dimension (CD) and Lyapunov exponent, which gives a measure of the complexity and chaoticity of the signal after reconstructing the phase space. The fractal dimension (FD) also gives a measure of the signal complexity and shows the significant difference between the disease group and the normal group. Entropy gives a measure of randomness or irregularity in the signal.

Computational Health Informatics for Biomedical Applications. Aryan Chaudhary and Sardar M. N. Islam (Naz) (Eds.)

All these nonlinear features have proven to be good biomarkers for the diagnosis of epilepsy, Alzheimer's disease, encephalopathy, depression, etc. Thus, chaotic and nonlinear features have proven to be good biomarkers of EEG in the diagnosis of various neurological diseases.

4.1 INTRODUCTION

Biomedical signals are very crucial in delivering information about the pathological conditions and the condition of various organs of the body. Being an inexpensive and invasive method of data acquisition, monitoring biomedical signals for health monitoring is very important.

The electrical activity of the brain is recorded in EEG signals (Figure 4.1). Electrodes are placed across the scalp covering various regions of the brain. Each electrode records the electrical potential with respect to a reference electrode which is usually placed at the earlobe. A commonly used method for EEG recording is the International 10–20 electrode system (as shown in Figure 4.2). The recorded signals have very low amplitudes in the range of a few microvolts, which makes signal amplification and noise removal very crucial. This is the major focus and scope of research in biomedical signal processing.

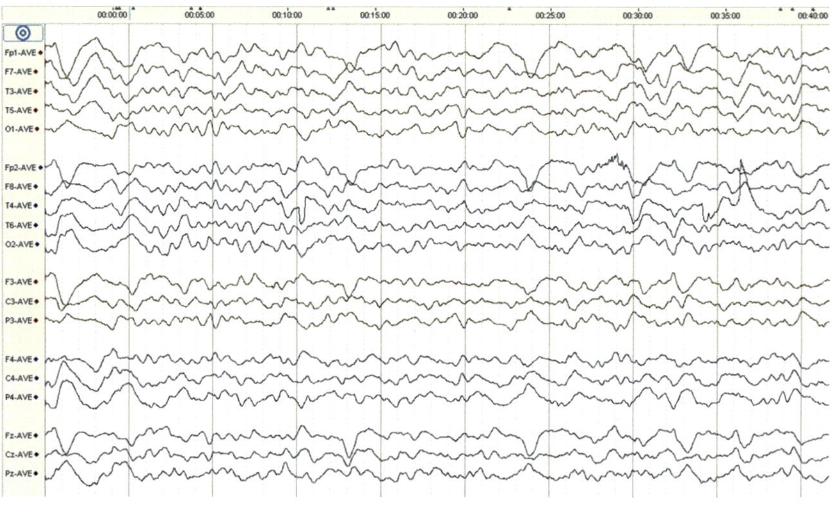

FIGURE 4.1 A sample EEG recording.

The standard EEG frequency bands are the delta (0.5–4 Hz), theta (4 to 7 Hz), alpha (8 to 13 Hz), and beta (14 to 30 Hz) bands [1], denoted by δ, θ, α and β waves. EEG signals with frequencies greater than 30 Hz are known as gamma waves. Time and frequency domain analysis techniques are well employed for EEG signal analysis. Even though they give a lot of information about the signal like the power of various sub-bands, level of slowing, presence of various particular patterns as evidence of some diseases, etc., in general, an EEG signal has complex behavior with nonlinear dynamic properties, and it can be represented after digitization as a sequence of time samples.

International 10–20 electrode system is a common method of electrode placement adopted for EEG recording. Various electrodes cover various areas of the brain, namely frontal, parietal, temporal, central, and occipital, due to which electrodes are also named on its basis along with a number to represent the location in the brain hemisphere. Electrodes placed on the right hemisphere are given even numbers 2, 4, 6, 8, while those placed on the left hemisphere are given odd numbers 1, 3, 5, 7. Figure 4.2 shows the electrode placements on the scalp of a person according to the International 10–20 electrode system for measuring EEG signals.

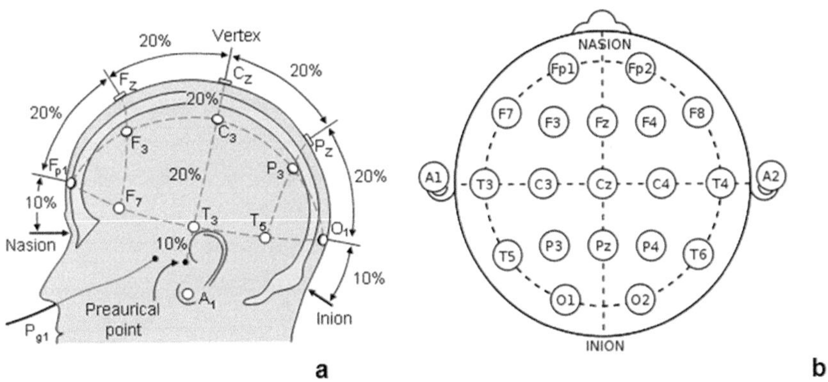

FIGURE 4.2 Measurement of (a) electrode placements on the scalp; (b) international 10–20 electrode placement system.

EEG signal is comprised of some waves corresponding to different frequency ranges called delta, theta, alpha, beta, and gamma waves. Alpha waves are observed in EEGs corresponding to the relaxed state of a person, particularly in the eyes-closed state. Its frequency ranges from 8–13 Hz.

They are called alpha waves, as they were identified first in EEGs. Beta waves in a frequency range of 13–30 Hz correspond to logical thinking, problem analyzing and solving. It is prevalent in the frontal and parietal regions.

Infants and sleeping adults have delta waves in EEG corresponding to a frequency of 0.5 to 4 Hz. Theta waves in the frequency range of 4–7 Hz are seen in EEGs of sleeping and resting adults and normal infants in an awake state. High-frequency components of EEG in the frequency range higher than 30 Hz are termed as Gamma Band activity. It is related to various cerebral functions, such as perception, attention, memory, consciousness, and motor control. Generally, low-frequency EEG signals, i.e., δ, θ and α sub-bands, are assessed for disease diagnosis. Various EEG sub-bands are shown in Figure 4.3.

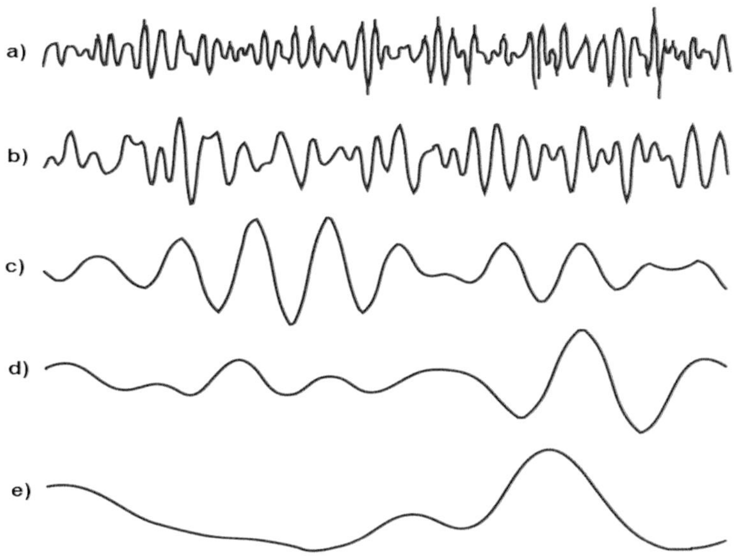

FIGURE 4.3 Sub-bands of EEG signal (a) gamma wave (>30 Hz); (b) beta wave (13–30 Hz); (c) alpha wave (8–13 Hz); (d) theta wave (4–7 Hz); (e) delta wave (0.5–4 Hz).

4.2 DIFFERENT DOMAINS OF ANALYSIS

EEG signals can be analyzed in various domains. As a signal is in the time domain, it can be assessed in the time domain. Other possibilities are frequency domain or time-frequency domain analysis or nonlinear analysis techniques.

4.2.1 TIME-DOMAIN ANALYSIS

In time-domain analysis, EEG signal is analyzed in time-domain with features like line length and energy. Line length is calculated as the vertical distance between consecutive samples of the signal. It represents the amplitude-frequency characteristic of the EEG signal, and it calculates the average absolute difference between consecutive samples in the signal [2, 3].

Energy is another time-domain feature that can be calculated directly from the signal in a straightforward manner [4] by taking the sum of squares of the signal samples. Other temporal features included maximum value, the minimum value of signal, etc. Time-domain features are easy to calculate as raw EEG signals are in the time domain. But complex, the non-stationary signal cannot be studied in detail with just time-domain features. Root mean square (RMS) is also used in studies as a measure of the magnitude of the varying quantity.

4.2.2 FREQUENCY-DOMAIN ANALYSIS

EEG signals can be converted to frequency domain to get more meaningful features in frequency domain. Transforms like Fourier transform is used to convert to the frequency domain. Power spectral density (PSD) can be calculated for EEG signals to compare between the disease group and normal healthy subjects. PSD is calculated as the Fourier transform of the autocorrelation of the EEG times series signal [4]. The advantage of parameters in the frequency domain is that they are less susceptible to signal quality variations that occur due to the placement of electrodes or characteristics of patients and other human subjects [5].

Peak frequency is one of the features reported in many studies as having good discriminative capabilities for neurological diseases. It describes the frequency of the highest peak in the PSD of the signal. The median frequency is calculated as the middle point or middle of the frequency power spectrum where the sum of the points on each side is equal [4].

4.2.3 TIME-FREQUENCY DOMAIN ANALYSIS

Both time and frequency domain analysis extracts very important information regarding the signals and the underlying system. Other than specific patterns or frequency domain features, time-frequency domain features are found to be much frequency components or relative ENEEG is analyzed by the frequency components of the signal or percentage of various EEG sub-bands. Frequency domain analysis or spectrum analysis of EEG signal is done using Fourier transform. Generalized EEG slowing is studied by neurologists for assessing encephalopathy. But, time-domain information is not preserved in such cases. Time-frequency representation includes both natural variables, i.e., time and frequency, together. Thus, time-frequency analysis like wavelet decomposition is more effective in extracting and assessing the sub-bands of EEG.

DWT is done for separating the EEG signal into sub-bands: delta (0.5–4 Hz), theta (4–7 Hz), alpha (7–13 Hz) and beta (13–30 Hz). Based on the highest frequency fm present in the signal, levels of decomposition are chosen. Mother wavelet is chosen such that it has a maximum correlation with the signal to be decomposed. Figure 4.4 illustrates the use of DWT for extracting the EEG sub-bands. If the sampling rate is 500 Hz, the highest frequency content is taken as 250 Hz. Accordingly, waves are obtained with six levels of decomposition (DWT), and A6, D6, D5, and D4 correspond to delta, theta, alpha, and beta waves, respectively. Generally, Daubechies (db4) is chosen as the mother wavelet for EEG decomposition. Many studies have applied this technique for extracting sub-bands and further calculating the sub-band energies [6] as well as calculating nonlinear features of sub-bands [7]. Hasan Ocak generated EEG sub-bands by applying DWT on EEG signals, calculated the entropies of those sub-bands, and obtained a high accuracy of 96% [8].

Various other decomposition techniques are also used in biosignal analysis. Empirical mode decomposition is another nonlinear decomposition technique where the local properties of the signal are extracted in an efficient manner [9, 10]. It is an adaptive and data-dependent method that is suitable for decomposing EEG signals. EMD generates a set of symmetric, amplitude, and frequency-modulated (AM-FM) components called intrinsic mode functions (IMFs), which represent the oscillatory modes in the data [11].

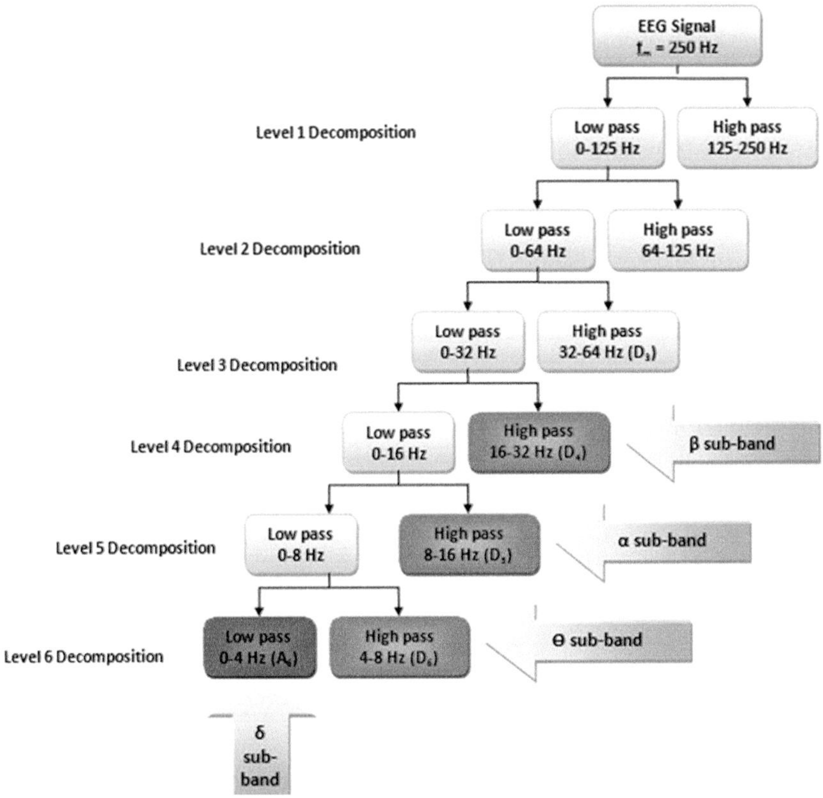

FIGURE 4.4 Discrete wavelet transforms for generating EEG sub-bands.

The EMD technique decomposes the signal into various IMFs, where IMF should obey two conditions: Firstly, the number of zero crossings and extremes must be the same or differ at most by 1.

The second condition is that, at all points, the mean value of an envelope formed by local maxima and that of local minima should be zero.

A signal *s(t)* can be decomposed into various IMFs and residue signal as:

$$s(t) = \sum_{i-1}^{M} IMF_i \tag{1}$$

where; IMF_i are the M number of IMFs; and *r(t)* is the residue signal.

The major steps in EMD are as follows:

i. Find the number of extremas (both maxima and minima) in the signal *s(t)*.
ii. Generate the upper envelope $e_{up}(t)$ and lower envelope $e_{low}(t)$ by connecting maxima and minima separately using cubic spline interpolation.
iii. The mean is calculated as:

$$\mu(t) = (e_{up}(t) + e_{low}(t))/2 \tag{2}$$

iv. The detail of the signal is extracted as:

$$s'(t) = s(t) - \mu(t) \tag{3}$$

v. Check if $s'(t)$ is an IMF by checking the two conditions described earlier:
   a. If $s'(t)$ is an IMF, $IMF_1 = s'(t)$ and residue $r(t)$ is taken as:

$$r(t) = s(t) - IMF_1(t) \tag{4}$$

   Replace $s(t)$ with $r(t)$.
   b. If $s'(t)$ is not an IMF, replace $s(t)$ with $s'(t)$.
iv. Repeat the above steps until a constant residue is obtained, i.e., a function with only one maxima and one minima from which further IMFs cannot be generated.

4.3 CHAOTIC ANALYSIS

Another method of analysis is chaotic analysis which tried to explore the chaotic nature of EEG signal and try to measure various features of its chaotic behavior. The two major features include CD and Lyapunov exponent.

A dynamical system is a mathematical model of a system that evolves in time. In this case, even if the time series is apparently irregular in the time domain, the linearity is manifest as sharp peaks in the frequency domain. On the other hand, if the time series is from a nonlinear dynamical system, with or without small added noise, or a random process (large noise), then the time series is generally irregular in both the time domain and the frequency domain. However, there are other methods of time series analysis, based on phase space reconstructions, which can reveal structure in time series from nonlinear dynamical systems when compared with time series from random processes.

Brain signals, being chaotic in nature, can be studied by chaotic analysis. The first step is phase space reconstruction done based on Taken's method. According to Theiler, the brain can be considered as a deterministic system, as it can be completely studied by analyzing the electrical output of a single neuron. He highlighted the possibility of applying delay-time embedding to study the self-organizing performance of a complex dynamical system in a system where the complete state of the system at any given time is unknown. This study pointed out that chaotic analysis is sufficient to study the behavior of the brain.

4.3.1 INTRODUCTION TO CHAOTIC ANALYSIS

In the field of nonlinear analysis and chaotic analysis of dynamic systems, they are often portrayed in their state space, using the one-dimensional (1D) time series. A set of trajectories evolving in this state-space conveys how the states of the dynamic system are evolving over time. In the case of dissipative deterministic dynamical systems where energy loses with time, various trajectories will converge to a subspace of the total state space, which is called attractor, when the system is observed for a long time [12]. The subspace is called an attractor because it is supposed to converge (or attract) trajectories from all possible initial conditions. The dynamics corresponding to a strange attractor are called deterministic chaos.

Considering an infinite time, the dynamical attractor of the system can be traced, which can be considered as the subset of the total phase space to which all trajectories will converge. Most of the information regarding the dynamical system is included within the attractor [13].

The EEG signals available as a 1D time series can be used to reconstruct the attractor of the underlying dynamic process of the brain. By reconstructing the attractor, it is possible to explore all the characteristics of the dynamic system it is representing. This is made possible with Taken's embedding theorem [14].

One dimensional EEG signal x(n) is mapped to m-dimensional space as:

$$X_m(n) = \left[x(n), x(n+\tau), x(n+2\tau), \ldots, x(n+(m-1)\tau)\right] \tag{5}$$

where; τ is, the time delay and m are the embedding dimension. CD and largest Lyapunov exponent (LLE) are the features that can be computed in this method as complexity measures. CD represents the complexity, whereas LLE represents the chaoticity of the signal and the system that generates it.

4.3.2 PHASE SPACE RECONSTRUCTION

Taken's embedding theorem [14] proved that it is possible to reconstruct the attractor from the time-series signals, and the reconstructed attractor reveals all the properties of the original attractor of the dynamical system. State-space reconstruction from a time series is a very efficient approach for analyzing the chaotic nature of a system. This is a very important step in exploring the features and characteristics of the system from the time series. After successfully reconstructing the phase space, features like CD and Lyapunov exponent can be calculated, which can give a quantitative measure of the system complexity and chaoticity [15].

4.3.2.1 FINDING OPTIMUM VALUE FOR TIME DELAY

The mutual information (MI) method is used to calculate the optimum value of time delay [16]. The MI is calculated for every point with a time lag. Thus, for various values of time lag, average MI is calculated, and the graph is plotted between time lag and average MI.

$$s = -\sum_{k=0}^{n} p_{ij}(\tau) ln \frac{p_{ij}}{p_i p_j} \tag{6}$$

Figure 4.5 shows the plot of mutual information versus time lag. The time lag at which mutual information reached the first minimum value is taken as the optimum time delay. It is marked with the arrow in Figure 4.5.

4.3.2.2 FINDING OPTIMUM VALUE FOR EMBEDDING DIMENSION

For calculating the optimum value-form, False nearest neighbor (FNN) is commonly used, which was proposed by Kennel [17]. In this method, the minimum sufficient value of m, i.e., m_0 is selected such that the reconstructed attractor in state space of dimension m_0 will be a one-to-one image of the original attractor. Only for a dimension $m > m_0$, the topological properties of the attractor will be preserved. The basic idea is that many points may seem near due to projection at lower values of m. As m increases, they will no longer be neighbors, which are referred to as false neighbors.

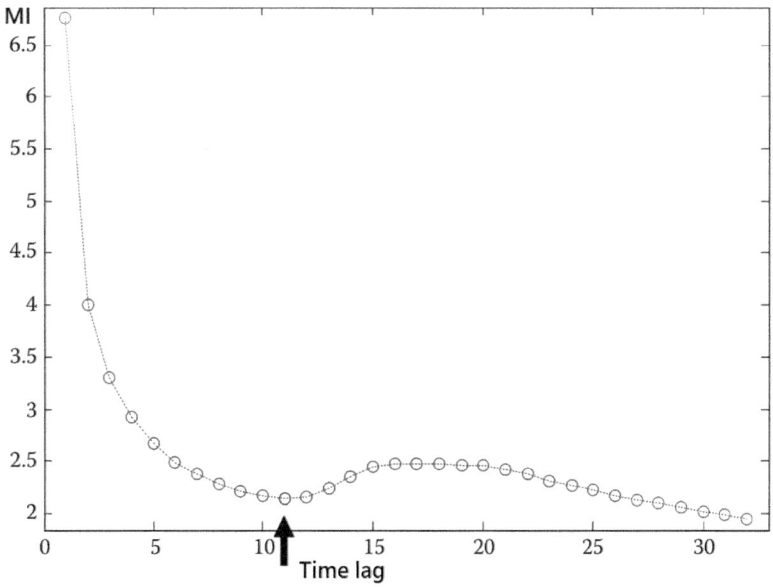

FIGURE 4.5 Optimum value of time lag for state-space reconstruction – mutual information (MI) method.

Consider a point p_j in time series and let p_k be the nearest neighbor in m dimensional space:

$$R = \frac{\left\| p_{j+1} - p_{k+1} \right\|}{\left\| p_j - p_k \right\|} \tag{7}$$

This measure of the ratio of distance R is computed and compared with a threshold value R_t. The point is marked as having a FNN if $R > R_t$. Number of nearest neighbors is calculated for every point on the trajectory is examined, and a number of FNNs is also calculated for increasing values of m. At an optimum value of *m*, when the embedding space is properly constructed with a sufficient value of *m*, the percentage of FNNs will drop to a very small value or almost zero. This is shown in Figure 4.6, where *m* can be taken as 4.

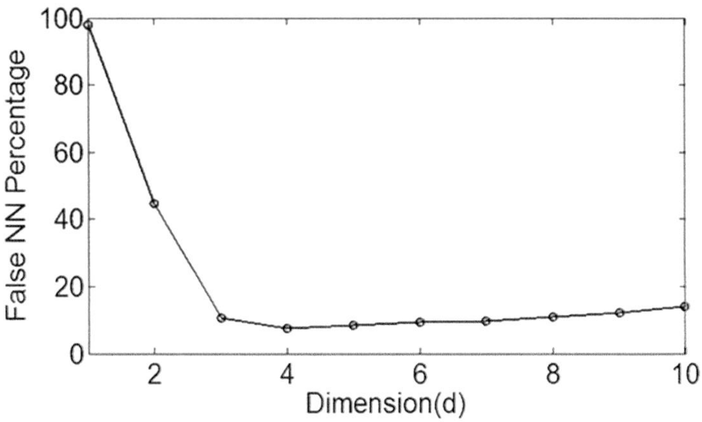

FIGURE 4.6 Optimum value of embedding dimension (m) for state space reconstruction – false nearest neighbor (FNN) method.

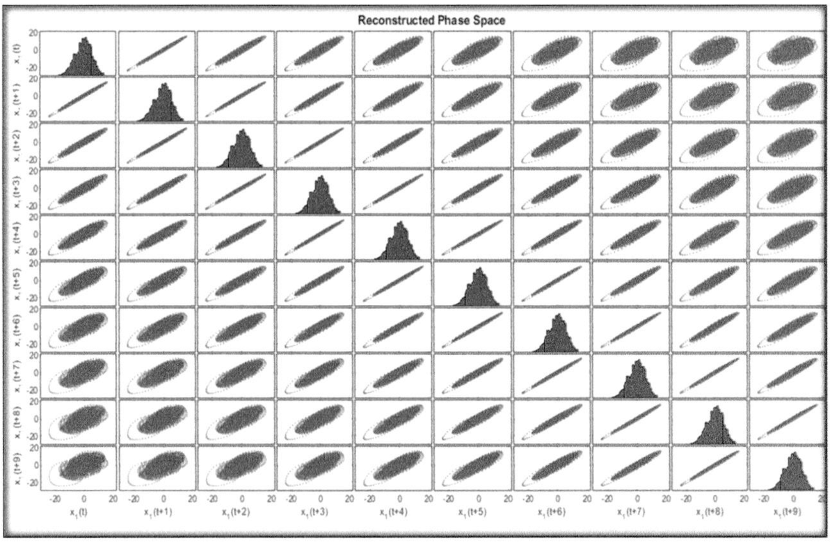

FIGURE 4.7 State space reconstruction (done in MATLAB) (Here, m = 10).

4.3.3 CORRELATION DIMENSION

CD is calculated by Grassberger-Procaccia algorithm [18]. First, the correlation sum is calculated as the probability that two points in phase space are distant r or less. Heaviside function in Eq. (4) gives the number of data points distant r or less.

Correlation sum C(r):

$$C(r) = \frac{1}{N^2} \sum_{i=1}^{N} \sum_{j=1, i \neq j}^{N} \theta\left(r - \left|v_i - v_j\right|\right) \tag{8}$$

Correlation dimension CD is calculated as:

$$CD = \lim_{r \to 0} \frac{logC(r)}{log(r)} \tag{9}$$

where; N is the number of data points in phase space; r is the radial distance around each reference point Xi; v_i, v_j are the points of the trajectory in the phase space; θ is the Heaviside function.

This method is employed for calculating the value of CD for increasing values of m. CD value goes on increasing with increasing value of m. The particular value of m is noted, after which the CD value stops increasing, which is termed the saturation of CD. Figure 4.8 shows the computation procedure of the CD. The value of the CD can be considered as a measure of the complexity of the signal.

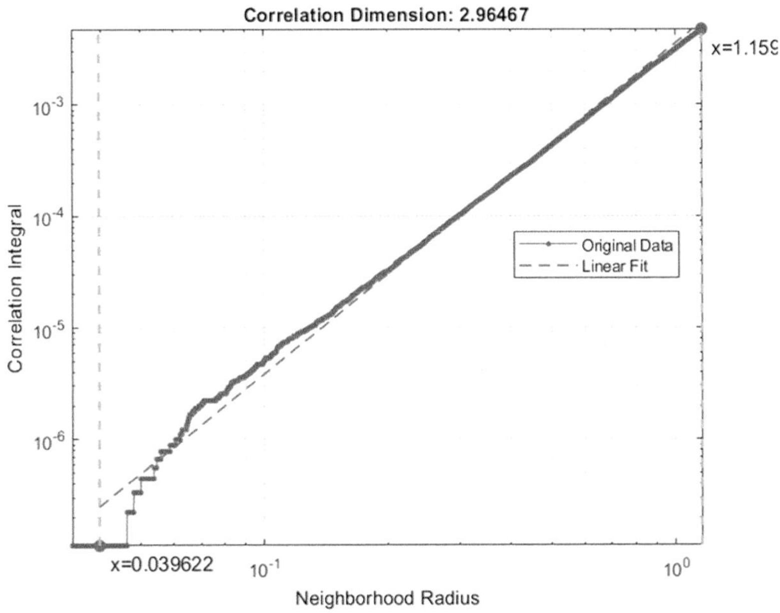

FIGURE 4.8 Computation of correlation dimension.

4.3.4 LYAPUNOV EXPONENT

Lyapunov exponent gives the rate at which trajectories converge or diverge. Wolf's algorithm and Rosenstein's algorithm are commonly employed for the calculation. LLE is calculated as the slope of the average logarithmic divergence of neighboring trajectories. The algorithm was developed by Rosenstein in 1993 [19].

$$y(i) = \frac{1}{\Delta t} < ln\ [d_j(i)] > \tag{10}$$

Positive Lyapunov exponent is a good indicator that the system under consideration is chaotic in nature. Figure 4.9 shows the LLE values of a healthy normal EEG calculated in MATLAB.

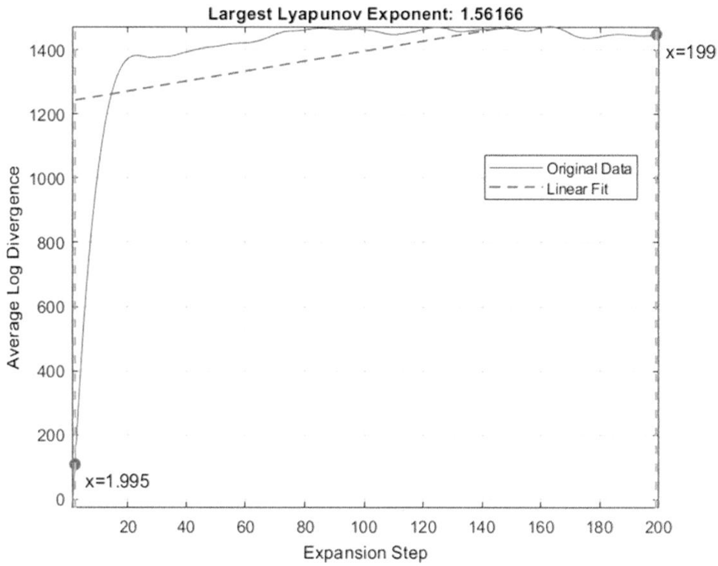

FIGURE 4.9 Computation of Lyapunov exponent.

4.4 NONLINEAR ANALYSIS

4.4.1 ENTROPY

One commonly used and defined concept in the nonlinear technique for getting a qualitative measure for the dynamic characteristics of signals is entropy. It gives a measure of complexity and reflects the degree of randomness of the system from its states defined as a time series. The entropy shows the extent of the unpredictability of the signal and hence, the system. Entropy is defined as a nonlinear feature giving the measure of randomness in the signal. The higher the value of entropy, the higher is the value of randomness and irregularity.

Approximate entropy (ApEn) was defined by Pincus as the logarithmic likelihood that two close sequences remain close in the next instant [20].

$$ApEn(m,r,N) = \frac{1}{N-m+1}\sum_{i=1}^{N-m+1} logC_i^m(r) - \frac{1}{N-m}\sum_{i=1}^{N-m} logC_i^{m+1}(r) \quad (11)$$

where; correlation integral is given by:

$$C_i^m(r) = \frac{1}{N-m+1}\sum_{j=1}^{N-m+1}(r - \| X_i - X_j \|) \quad (12)$$

ApEn is found to be more effective for small datasets and is efficient in discriminating the signal from random signals. Sample entropy (SampEn) was redefined by Richman and Moorman with a modification in ApEn algorithm by excluding self-comparisons making the entropy value more unbiased [21]. Algorithm is evaluated as less robust to noise and entropy value as more independent of data length.

$$SampEn(m,r,N) = -ln[C^{m+1}(r)/C^{m(r)}] \quad (13)$$

where; $C^m(r)$ gives the probability that two sequences match when m points are taken and $C^{m+1}(r)$ gives the probability corresponding to m + 1 points. The SampEn algorithm avoids the comparison to its own data; it is calculated as the negative average of the natural logarithm of conditional probability, and it is unbiased.

4.4.2 FRACTAL DIMENSION

Fractal dimensions (FDs) are a measure of the complexity of the signal. The two commonly used algorithms are Higuchi's algorithm and Katz's algorithm. From the time series, Higuchi [22] developed an algorithm for finding fractal dimensions directly from the time series [22]. From the original time series $X(1), X(2).... X(N)$, new time series are defined as:

$$X_m^k = X(m),\ X(m+k),\ X(m+2k),....\ X\left(m + int\left(\frac{N-m}{k}\right)*k\right) \quad (14)$$

where $m = 1, 2, 3, \ldots, k$

Here, m is the initial time; and k is the interval time. The length of the curve is denoted as *Lm(k)*

$$\mathrm{Lm(k)} = \frac{\sum_{i=1}^{int\left(\frac{N-m}{k}\right)} | X\left(m+ik\right) - X(m+\left(i-1\right)k| * \left(n-1\right)}{k * int\left[\frac{N-m}{k}\right]} \tag{15}$$

Here; N is the total number of samples. The length of the curve is taken as the average value of k values of Lm(k). Higuchi developed the idea that a curve understudy is fractal-like if L(k) is proportional to k-D where D is the FD [22]. Higuchi's FD is calculated as the slope of least-squares linear best fit of the plot *ln (L(k))* versus *ln (1/k)*.

Another algorithm was developed by Katz [23] in which the maximum distance from the initial point (d) is calculated for the time series *x(1), x(2).x(n)*.

$$d = max\ (|x_1 - x_j|) \text{ where j} = 2, 3, \ldots, \text{N} \tag{16}$$

The total length of the time series is:

$$L = \sum_{i=2}^{N} x_i - x_{i-1} \tag{17}$$

The average distance between two successive points is given by:

$$a = \frac{L}{N-1} \tag{18}$$

$$\text{Calculate Katz's FD} = \ FD = \frac{ln\frac{L}{a}}{ln\frac{d}{a}} \tag{19}$$

4.5 APPLICATION TO NEUROLOGICAL DISEASES

4.5.1 CHAOTIC ANALYSIS

Adeli et al. calculated the CD and LLE values for each of the sub-bands of EEG and reported that CD is significant in beta and gamma sub-bands while LLE is significant in alpha sub-band in the three groups [24].

When CD values are calculated for normal, ictal, and interictal phases, CD is found to be highest for normal healthy individuals. CD decreases for an epileptic patient during the normal phase (inter-ictal phase), and it further reduces during the ictal period [25]. Figure 4.1 illustrates this change in CD value which shows that the complexity of brain dynamics decreases during the ictal phase, indicating a more chaotic character of brain signal during normal conditions.

LLE value decreases during the ictal phase when compared to the interictal phase when the patient is not experiencing the seizure. LLE is highest for a normal healthy person indicating a more chaotic character of brain signals during normal conditions. For an epileptic patient, LLE is less showing reduced chaoticity for an epileptic patient. It is further reduced during the ictal phase when the patient undergoes a seizure [25]. These values strongly propose the decrease in complexity of EEG signal during the interictal phase when compared to normal EEG. This result can be exploited for diagnosing Epilepsy during the non-seizure period. CD and LLE were considerably reduced during encephalopathy compared to normal healthy subjects [26].

4.5.2 NONLINEAR ANALYSIS

Entropies have been reported as good biomarkers for the diagnosis of various neurological diseases. It is reported that entropy was lower during seizure of an epileptic patient when compared to that of normal healthy individuals [27–30]. FDs of EEG were also proven to give good accuracy for classifying epilepsy cases from that of normal healthy subjects [31–34]. Lin et al. showed that FD is useful in giving a quantitative measure of the complexity of dynamical signals in biology and medicine like EEG [35]. A fractal is a shape that can retain the structural details for different scales of magnification. Thus, a single number called FD can measure the complexity of the structure of such a set, irrespective of its scale.

EEG signal can be considered as a fractal, and FD can be used to find out the transient deterministic data in the EEG [32, 36]. Some studies have reported that Katz's method may be more sensitive than the other algorithms of FD in locating seizures in EEG of epileptic patients [37]. Some studies have reported a decrease in complexity of brain dynamics during Alzheimer's disease by a decrease in value of FD during disease when compared to that of normal healthy subjects [38, 39].

4.6 CONCLUSION

EEG is a relatively inexpensive and easily available diagnostic tool aiding the clinician in identifying patients with various neuropathologies. Signal processing algorithms need to be applied in EEGs for extracting hidden information. Time-frequency analysis, chaotic analysis and other nonlinear techniques can give promising results for EEG classification. Further addiction of good ML algorithms like SVM, multilayer perceptron and random forest classifiers can improve the accuracy of classification of EEG signals using these features. This area of research allows the development of automated diagnosis or computer-aided diagnosis (CAD) of various neurological diseases based on EEG signals or to assist neurologists and other doctors for easy diagnosis.

But it demands more research in this field and many more specific results in all types of neurological diseases and done on a different population. Much more data and results are needed to validate the result and for implementing it in the clinical scenario. Thus, EEG signal processing is a promising field for signal processing researchers.

KEYWORDS

- **amplitude and frequency-modulated**
- **discrete wavelet transform**
- **false nearest neighbor**
- **largest Lyapunov exponent**
- **power spectral density**
- **signal processing researchers**

REFERENCES

1. Schomer, D. L., & Da Silva, F. L., (2012). *Niedermeyer's Electroencephalography: Basic Principles, Clinical Applications, and Related Fields*. Lippincott Williams & Wilkins.

2. Tessy, E., Shanir, P. M., & Manafuddin, S., (2016). Time domain analysis of epileptic EEG for seizure detection. In: *2016 International Conference on Next Generation Intelligent Systems (ICNGIS)* (pp. 1–4). IEEE.
3. Guo, L., Rivero, D., Dorado, J., Rabunal, J. R., & Pazos, A., (2010). *Automatic Epileptic Seizure Detection in EEGs Based on Line Length Feature and Artificial Neural Networks, 191*(1), 101–109.
4. Fergus, P., Hignett, D., Hussain, A., Al-Jumeily, A., & Abdel-Aziz, K., (2015). *Automatic Epileptic Seizure Detection Using Scalp EEG and Advanced Artificial Intelligence Techniques, 2015.*
5. Maner, W. L., Garfield, R. E., Maul, H., Olson, G., & Saade, G., (2003). *Predicting Term and Preterm Delivery with Transabdominal Uterine Electromyography, 101*(6), 1254–1260.
6. Jacob, J. E., Nair, G. K., Iype, T., & Cherian, A. J. N. R. I., (2018). *Diagnosis of Encephalopathy Based on Energies of EEG Subbands Using Discrete Wavelet Transform and Support Vector Machine, 2018.*
7. Jacob, J. E., & Nair, G., (2019). *EEG Entropies as Estimators for the Diagnosis of Encephalopathy, 101*(3), 463–474.
8. Ocak, H. J. E. S. W. A., (2009). *Automatic Detection of Epileptic Seizures in EEG Using Discrete Wavelet Transform and Approximate Entropy, 36*(2), 2027–2036.
9. Martis, R. J., et al., (2012). *Application of Empirical Mode Decomposition (EMD) for Automated Detection of Epilepsy Using EEG Signals, 22*(06), 1250027.
10. Jacob, J., Gopakumar, K., Iype, T., & Cherian, A. J. N., (2018). *Automated Diagnosis of Encephalopathy Based on Empirical Mode EEG Decomposition, 50*(4), 278–285.
11. Sharma, R., & Pachori, R. B., (2018). Automated classification of focal and non-focal EEG signals based on bivariate empirical mode decomposition. In: *Biomedical Signal and Image Processing in Patient Care* (pp. 13–33). IGI Global.
12. Pritchard, W. S., & Duke, D., (1992). *Measuring Chaos in the Brain: A Tutorial Review of Nonlinear Dynamical EEG Analysis, 67*(1–4), 31–80.
13. Torkamani, S., Butcher, E., Todd, M., & Park, G., (2011). *Detection of System Changes Due to Damage Using a Tuned Hyperchaotic Probe, 20*(2), 025006.
14. Takens, F., (1981). Detecting strange attractors in turbulence. In: *Dynamical Systems and Turbulence, Warwick 1980* (pp. 366–381). Springer.
15. Stergiou, N., (2018). *Nonlinear Analysis for Human Movement Variability*. CRC press.
16. Fraser, A. M., & Swinney, H., (1986). *Independent Coordinates for Strange Attractors from Mutual Information, 33*(2), 1134.
17. Kennel, M. B., Brown, R., & Abarbanel, H., (1992). *Determining Embedding Dimension for Phase-Space Reconstruction Using a Geometrical Construction, 45*(6), 3403.
18. Grassberger, P., & Itamar, P., (1983). *Characterization of Strange Attractors, 50*(5), 346.
19. Rosenstein, M. T., Collins, J. J., & De Luca, C., (1993). *A Practical Method for Calculating Largest Lyapunov Exponents from Small Data Sets, 65*(1, 2), 117–134.
20. Pincus, S., (1995). *Approximate Entropy (ApEn) As a Complexity Measure, 5*(1), 110–117.
21. Richman, J. S., Lake, D. E., & Moorman, J., (2004). *Sample Entropy, 384*, 172–184.
22. Higuchi, T., (1988). *Approach to an Irregular Time Series on the Basis of the Fractal Theory, 31*(2), 277–283.
23. Katz, M. J. (1988). *Fractals and the Analysis of Waveforms, 18*(3), 145–156.

24. Adeli, H., Ghosh-Dastidar, S., & Dadmehr, N., (2007). *A Wavelet-Chaos Methodology for Analysis of EEGs and EEG Subbands to Detect Seizure and Epilepsy, 54*(2), 205–211.
25. Jacob, J. E., Sreelatha, V. V., Iype, T., Nair, G. K., & Yohannan, D., (2016). *Diagnosis of Epilepsy from Interictal EEGs Based on Chaotic and Wavelet Transformation, 89*(1), 131–138.
26. Jacob, J. E., Cherian, A., Gopakumar, K., Iype, T., Yohannan, D. G., & Divya, K., (2018). *Can Chaotic Analysis of Electroencephalogram Aid the Diagnosis of Encephalopathy?*, 2018.
27. Srinivasan, V., Eswaran, C., & Sriraam, N., (2007). *Approximate Entropy-Based Epileptic EEG Detection Using Artificial Neural Networks, 11*(3), 288–295.
28. Amarantidis, L. C., & Abásolo, D. J. E., (2019). *Interpretation of Entropy Algorithms in the Context of Biomedical Signal Analysis and Their Application to EEG Analysis in Epilepsy, 21*(9), 840.
29. Li, P., Karmakar, C., Yearwood, J., Venkatesh, S., Palaniswami, M., & Liu, C. J. P. O., (2018). *Detection of Epileptic Seizure Based on Entropy Analysis of Short-Term EEG, 13*(3), e0193691.
30. Vijith, V., Jacob, J. E., Iype, T., Gopakumar, K., & Yohannan, D. G., (2016). Epileptic seizure detection using nonlinear analysis of EEG. In: *2016 International Conference on Inventive Computation Technologies (ICICT)* (Vol. 3, pp. 1–6). IEEE.
31. Torabi, A., & Daliri, M. R., (2021). *Applying Nonlinear Measures to the Brain Rhythms: An Effective Method for Epilepsy Diagnosis, 21*(1), 1–9.
32. Ahmadlou, M., Adeli, H., & Adeli, A., (2010). *Fractality and a Wavelet-Chaos-Neural Network Methodology for EEG-Based Diagnosis of Autistic Spectrum Disorder, 27*(5), 328–333.
33. Jacob, J. E., Nair, G. K., Cherian, A., & Iype, T., (2019). *Application of Fractal Dimension for EEG Based Diagnosis of Encephalopathy, 100*(2), 429–436.
34. Radhakrishnan, M., Won, D., Manoharan, T. A., Venkatachalam, V., Chavan, R. M., & Nalla, H. D., (2021). *Investigating Electroencephalography Signals of Autism Spectrum Disorder (ASD) Using Higuchi Fractal Dimension, 66*(1), 59–70.
35. Liu, J., Yang, Q., Yao, B., Brown, R., & Yue, G. J. B. C., (2005). *Linear Correlation Between Fractal Dimension of EEG Signal and Handgrip Force, 93*(2), 131–140.
36. Arle, J. E., Simon, R. H. J. E., & Neurophysiology, C., (1990). *An Application of Fractal Dimension to the Detection of Transients in the Electroencephalogram, 75*(4), 296–305.
37. Esteller, R., Vachtsevanos, G., Echauz, J., & Litt, B., (2001). *A Comparison of Waveform Fractal Dimension Algorithms, 48*(2), 177–183.
38. Smits, F. M., Porcaro, C., Cottone, C., Cancelli, A., Rossini, P. M., & Tecchio, F. J. P. O., (2016). *Electroencephalographic Fractal Dimension in Healthy Ageing and Alzheimer's Disease, 11*(2), e0149587.
39. Woyshville, M. J., & Calabrese, J. R. J. B. P., (1994). *Quantification of Occipital EEG Changes in Alzheimer's Disease Utilizing a New Metric: The Fractal Dimension, 35*(6), 381–387.

CHAPTER 5

Application of Artificial Intelligence and Deep Learning in Healthcare

RAMAN CHADHA[1] and ROHIT KUMAR VERMA[2]

[1]UIE, Department of Computer Science and Engineering, Chandigarh University, Gharuan, Punjab, India, E-mail: dr.ramanchadha@gmail.com

[2]Assistant Professor, Department of Computer Science, Himachal Pradesh University Regional Center, Dharamshala, Kangra, Himachal Pradesh, India

ABSTRACT

Utilization of Man-made reasoning (simulated intelligence) has expanded in medical services in numerous areas. Associations from medical services of various sizes, types, and strengths are presently a day's keener on how man-made consciousness has advanced and is helping patient necessities and their consideration, likewise diminishing expenses, and expanding effectiveness. This review investigates the ramifications of computer-based intelligence on medical services the executives and difficulties associated with utilizing artificial intelligence (AI) in medical care alongside the audit of a few examination papers that pre-owned computer-based intelligence models in various areas of medical services like Dermatology, Radiology, Medication plan and so forth Computerized reasoning (man-made intelligence) intend to imitate human intellectual capacities. We study the current status of artificial intelligence (AI) applications in medical services and examine its future. Simulated intelligence can be applied to different

Computational Health Informatics for Biomedical Applications. Aryan Chaudhary and Sardar M. N. Islam (Naz) (Eds.)

sorts of medical services information (organized and unstructured). Significant sickness regions that utilization computer-based intelligence devices incorporate malignancy, nervous system science and cardiology. We then, at that point, audit in more subtleties the computer-based intelligence applications in stroke, in the three significant spaces of early discovery and finding, treatment, just as result expectation and forecast assessment.

The medical services industry faces difficulties in fundamental regions like electronic record the executives, information combination, and PC supported conclusions and infection forecasts. It is important to lessen medical care costs and the development of customized medical services. The quickly growing fields of profound learning and prescient examination have begun to assume an urgent part in the development of a huge volume of medical care information practices and exploration. Profound learning offers a wide scope of instruments, strategies, and systems to address these difficulties. Wellbeing information prescient examination is arising as an extraordinary instrument that can empower more proactive and protection treatment choices. More or less, this chapter center around the system for profound learning information examination to clinical dynamic portrays the review on different profound learning procedures and devices practically speaking just as the utilizations of profound learning in medical care. Over the previous decade, profound learning has made striking progress in different man-made reasoning exploration regions. Developed from the past research on fake neural organizations, this innovation has shown better execution than other AI calculations in regions, for example, picture, and voice acknowledgment, regular language handling, among others.

Lately, the main rush of uses of profound learning in drug research has arisen, and its utility has gone past bioactivity expectations. It has shown guarantee in resolving assorted issues in drug revelation. Models will be examined covering bioactivity forecast, once more atomic plan, amalgamation expectation and organic picture investigation. Profound learning is one of the most intriguing and quickly developing strategies in Man-made brainpower. Consequently, the interesting limit of profound learning models to take in designs from the information separates it from other AI methods. Profound learning is answerable for a critical number of late forward leaps in AI. In any case, profound learning models are exceptionally reliant upon the fundamental information. Thus, consistency, exactness, and culmination of information are fundamental for a profound learning model.

Our contextual analysis is expected to give significant experiences to the profound learning local area just as for information researchers to direct conversation and future examination in applied profound learning with certifiable information. Notwithstanding the enormous contrasts among creating and created nations, access is a significant issue in country wellbeing all throughout the planet. Indeed, even in the nations where most of the populace lives in provincial regions, the assets are packed in the urban communities. All nations experience issues with transport and correspondence, and they all face the test of deficiencies of specialists and other wellbeing experts in provincial and far-off regions.

5.1 INTRODUCTION

The coming of computerized advances in the medical services field is described by consistent difficulties in both application and reasonableness. Unification of dissimilar wellbeing frameworks has been slow, and the reception of a completely incorporated medical care framework in many areas of the planet has not been cultivated. Human science's innate nature and intricacy, just as the variation between individual patients, has reliably shown the significance of the human component in diagnosing and treating infections. Nonetheless, progresses in computerized advances are most likely becoming irreplaceable devices for medical services experts in giving the best consideration to patients.

The improvement of information advances, including the capacity size, computational power, and information move speeds, has empowered the inescapable reception of AI in many fields—medical services included. Because of the multivariate idea of giving quality medical services to an individual, the new patterns in medication have underscored the requirement for a customized medication or "accuracy medication" way to deal with medical care. The objective of customized medication is to utilize a lot of medical services information to find, foresee, and examine indicative choices, which doctors can thusly execute for every individual patient. Such information incorporates yet isn't restricted to hereditary or familial data, clinical imaging information, drug mixes, population-wide persistent wellbeing results, and regular language handling of existing clinical documentation.

We will zero in principally on three of the biggest utilization of AI (ML) in the clinical and biomedical fields. As a quickly developing field, there is a wide scope of expected utilization of AI in the medical services field, which might incorporate helper parts of the field like faculty the executives, protection strategies, administrative issues, and considerably more. Thusly, the points shrouded in this section have been reduced to three normal utilizations of AI.

The first is the utilization of AI in clinical pictures, for example, attractive reverberation imaging (MRIs), modernized pivotal tomography (CAT) checks, ultrasound (US) imaging, and positron discharge tomography (PET) examines. The aftereffect of these imaging modalities is a set or series of pictures that ordinarily requires a radiologist to decipher and make a finding. ML methods have quickly progressed to anticipate and observe pictures that might demonstrate an illness state or difficult issue. The second is the regular language handling of clinical archives. With the push towards electronic clinical records (EMR) in numerous nations, the agreement from numerous medical care experts has been that the cycle is slow, drawn-out, and, as a rule, totally mangled. This can once in a while lead to more unfortunate by and large medical services for the patient role. The major significant difficulty is the measure of actual clinical accounts and certification that occurs in numerous clinics and facilities.

Distinctive designing, manual notes, and plenty of deficient or non-incorporated data have made the change to taking on electronic clinical records not exactly effective. The third AI application envelops the utilization of human hereditary qualities to anticipate sickness and track down reasons for infection. With the approach of cutting-edge sequencing (NGS) methods and the blast of hereditary information, including huge data sets of populace-wide hereditary data, the endeavor to perceive significant data of what hereditary qualities might mean for human wellbeing is presently at the front line of many examinations tries. By seeing how complex sicknesses might show and how hereditary qualities might increment or reduction a distinctive individual's danger can support precaution medical services. This could furnish doctors with more data on the best way to tailor a particular patient's consideration to decrease the danger of getting more intricate sicknesses.

The normal issue present in every one of the three of these points is the means by which to decipher wellbeing information gained from the Internet of Things (IoT) into justifiable, valuable, dependable data for patients and clinicians. How would we decipher a huge number of information sources and boundaries from the information? How would we do this productively? What is the advancement of resolving this issue right now?

5.1.1 ARTIFICIAL INTELLIGENCE AND MACHINE LEARNING (AI&ML)

Man-made intelligence has been complicatedly associated with the rising of cutting-edge enlisting machines. Man-made intelligence has its establishments and beginnings steadily planted ever. Alan Turing's work in breaking the German Enigma machine during World War II transformed into the justification for a great deal of current programming. The Turing Test, which plans to check whether AI has become muddled from human information, is furthermore named after him [1, 2].

At the stature of the Second World War, the Allies had a colossal key snag in the Atlantic. The United States and the United Kingdom expected to set up secure conveyance lines to move the two weapons and troops to England, fully expecting a focal region European assault. In any case, the German U-boats were inconceivably strong at disturbing and sinking countless boats exploring these conveyance ways [3].

Turing and the rest of Bletchley Park were endowed with breaking the coded messages conveyed by The Enigma Machine and at last made The Bombe. This mechanical enlisting contraption successfully decoded the code of The Enigma machine (Figure 5.1). Using the Bombe, they read the German orders delivered off submarines and investigated their boats around these dangers. This was Turing's first sharp machine. Alan Turing would later continue to depict the chance of a thinking machine which would eventually be called AI [4].

Man-made intelligence is a subset of AI, and the term was founded in the last piece of the 1950s by Arthur Samuel, who appropriated a paper on planning PCs to play checkers when he worked with IBM [5]. Reenacted knowledge is best depicted as giving human-like understanding to machines in a manner that clearly impersonates the bearing and treatment of the human still, little voice. ML is the subset of AI that bright lights on empowering machines to learn in a free manner with no human intervention.

By the last piece of the 1960s, experts were by then endeavoring to prepare PCs to play fundamental games, for instance, fit tac-toe [6]. Eventually, the chance of neural associations, which relied upon a speculative model of human neuron affiliation and correspondence, was wandered into fake neural associations (ANNs) [7, 8]. These essential works laid dormant for quite a while as a result of the trouble and horrendous appearance of the systems made. Figuring advancement had not yet advanced to the point of diminishing the computational chance to an even-minded level.

The state-of-the-art PC time provoked extraordinary extensions in both computational power and data accumulating limit. With the introduction of IBM's Deep Blue and Google's AlphaGo in late numerous years, a couple of bounces in AI have shown the restriction of AI to address veritable world, complex issues [9, 10]. Likewise, the assurance of AI has snatched hold in basically every region under the sun.

The inevitable gathering of AI can be generally attributed to the openness of incredibly tremendous datasets and the improvement of computational systems, which diminish overfitting and work on the theory of arranged models. These two factors have been the central purpose of the quick headway and gathering of AI in essentially every field today. This, joined with the extending inescapability of interconnected contraptions or the IoT, has made a rich structure whereupon to collect insightful and modernized systems.

AI is essential for understanding the enormous flood of well-being information today. A framework of frameworks to supplement the expanding IoT foundation will without a doubt depend intensely on these procedures. Many use cases have effectively shown huge guarantees. How do these procedures' function, and how would they give us knowledge into apparently detached data?

5.1.2 MACHINE LEARNING (ML) ALGORITHMS

AI is extensively parted into regulated and solo learning. Calculations falling under the two classifications execute numerical models. Every calculation means to enable PCs to figure out how to play out specific undertakings.

5.1.2.1 SUPERVISED LEARNING

Supervised learning adapting regularly utilizes preparing information known as named information. Preparing information has at least one data source and has a "named" yield. Models utilize these named results to evaluate themselves during preparing, fully intent on working on the expectation of novel information [11]. Ordinarily, administered education models center around order and relapse calculations [12]. Separation problems are very common in medicine. In most clinical cases, a patient's diagnosis involves a specialist who collects the disease by giving him or her a specific plan for symptoms. Recurrence problems will usually look at predicting statistical outcomes as an analyzed length of stay (LOS) in the center of problems given a specific course of data such as significant symptoms, clinical history, and weight.

FIGURE 5.1 Machines of German enigma used in military communications.

Common memorized statistics for this controlled study category are irregular forests (RF), choice tree (DT), Naïve Bayes models, direct backlash and techniques, and vector support devices (SVM). Still, neural organizations also can be prepared with a managed reading. [13]. Rare wood is a type of selected wood, but it is a set of clothes prepared independently. Subsequent predictions of trees are often mid-range to improve outcome and prediction [14]. Each tree is processed using a limited information model by replacement, and in all parts of the applicant that separates the subset of the same key points are selected. This keeps every reader or tree from focusing too much on the best available images of a preparation set that may not have new information. So, create a model guess. Unusual backwoods can have large or large numbers of trees and work best with sound information [15]. A model developed by collecting results from different trees prepared with information will provide expectations that can be tested using experimental information (Figure 5.2).

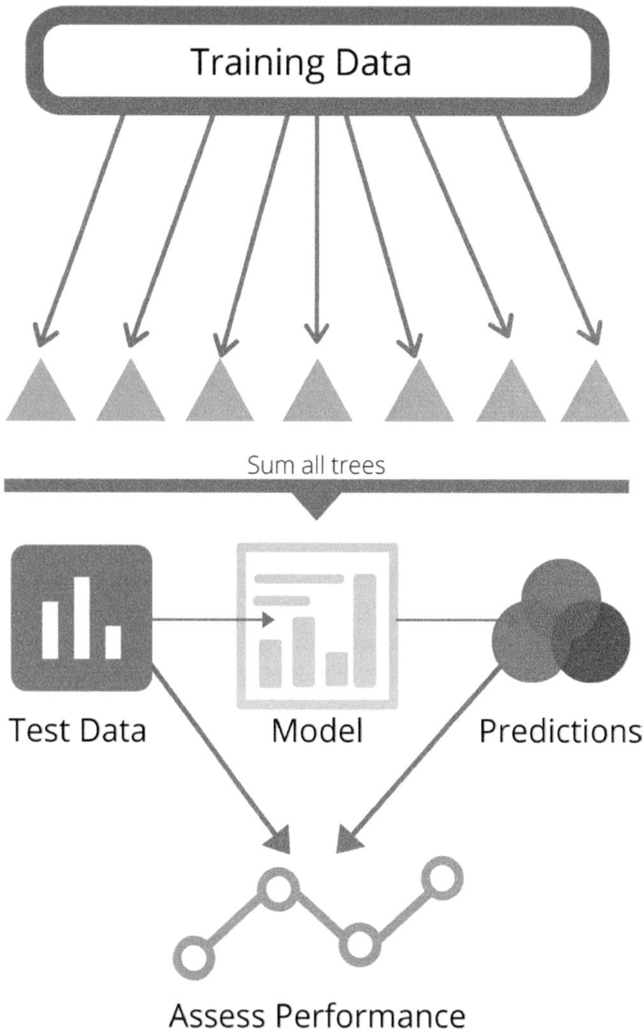

FIGURE 5.2 The application of machine learning in the field medical health science.

Delineation of a work interaction for planning and reviewing a discretionary forest area model. Each green triangle tends to be an independently pre-arranged tree from the planning data. The assumption for each tree is added and is tended to as the model. Test data is then dealt with to the model, i.e., all of the trees and the resulting assumption is made. The assumption is then appeared differently in relation to the primary test data to assess how the model performs.

Typically, better results can be cultivated with point supporting, but tuning is altogether more annoying, and the risk of overfitting is higher. Point helping capacities splendidly with unbalanced data and planning time is out and out faster as a result of the incline plunge nature of the estimation [17, 18].

A system used to additionally foster many coordinated computations is known as slant making a difference. Taking decision trees, for example, the slant supporting machine as it is normally realized plays out a near bunch planning strategy as the discretionary boondocks yet with "feeble understudies." Instead of building the decision trees in equivalent as in the sporadic forest estimation, the trees are created progressively with the screw up of the past tree being used to additionally foster the accompanying tree [16]. These trees are not nearly as significant as the trees of the self-assertive wood, which is the explanation they are assigned "weak" (Figure 5.3).

5.1.2.2 UNSUPERVISED LEARNING

Solo AI utilizes unlabeled information to find designs inside the actual information [19]. These calculations ordinarily dominate at bunching information into applicable gatherings, considering the location of idle attributes, which may not be promptly self-evident. Notwithstanding, they are likewise more computationally concentrated and require a bigger measure of information to perform. The most widely recognized and notable calculations are K-implies bunching and profound learning. However, profound learning can be utilized in an administered way [12, 20]. Such calculations additionally perform affiliation assignments that are like grouping. These calculations are considered solo on the grounds that there is no human contribution with regards to what set of characteristics the groups will be fixated on.

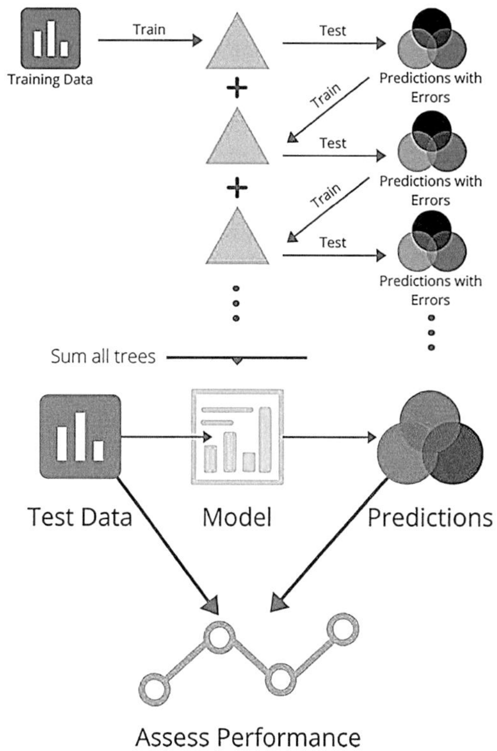

FIGURE 5.3 The use of machine learning in the medical industry.

Source: Reprinted from Ref. [6]. © 2020 The Author(s). Licensee IntechOpen. https://creativecommons.org/licenses/by-nc/4.0/

An outline of the clean method of designing and examining a helping gadget version. Each inexperienced triangle tends to have a predetermined tree from the resulting information on the ensuing tree the usage of the residues or slip-us on the previous tree to break it at its imagination. The fee of each tree is brought and is considered a version. The experimental statistics are then processed to the model, i.e., all timber and the ensuing predictions were accomplished. The gauge is wherein it appears differently in relation to the main test information to test how the version performs.

The ordinary k-implies calculation has some sorts, such as k-medians and ok-medoids; however, the well-known is something comparable for every calculation. The calculation makes use of Euclidian distance to look like the "closest" recognition or mean for a bunch accepting there are ok organizations. It then, at that point, appoints the current facts to highlight that bunch and, in a while, recalculates the center for thc group, clcaning it for the subsequent records point [21]. The best downside to this calculation is that it must be delivered with a regular range of "indicates" or "focuses." Improper willpower of the k worth can result in a helpless grouping.

Profound mastering utilizes neural nets to perform forecasts even on unlabeled statistics, just as order methods. Based on models of human neurons, perceptrons, as they're frequently referred to as, are coordinated into many prepared layers making the company "profound" in nature [20]. Each perceptron has numerous data resources and a solitary result. They have coordinated into layers wherein the results of the past layer fill in as the contributions for the following layer. The records layer calls for one perceptron consistent with the input variable. The resulting is not absolutely set in stone previous to preparation with the aid of a human (Figure 5.4). This is one of the troubles and problems in constructing a viable neural net. The computationally critical nature of figuring each perceptron for a massive neural internet can mean that practice alone can require days to weeks for large informational collections [22].

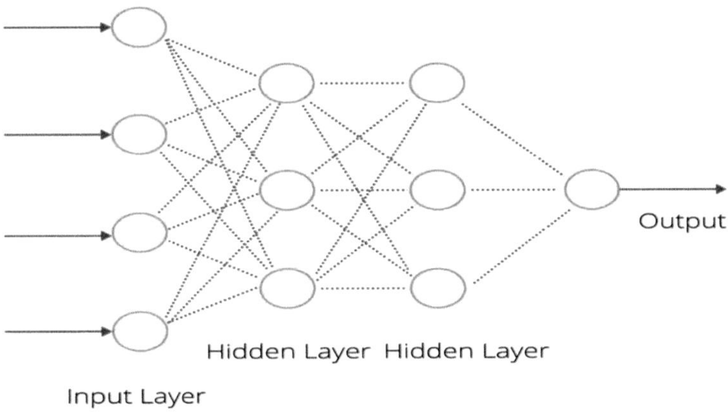

FIGURE 5.4 An overview of neural networks and their applications in medical healthcare sciences.

5.1.2.3 HYPERPARAMETERS

In AI, a model typically has a group of limitations, simply as a bunch of hyperparameters. Boundaries are factors about the version that can be changed throughout the preparation. For instance, limitations may be without a doubt the upsides of the instruction facts, with each piece of record being different alongside one or a number of the obstacles. While hyperparameters are primarily set before getting ready happens and cannot alternate as soon as studying starts evolving. Hyperparameters are commonly set to music esteems like the model's mastering velocity and compel the actual calculation. Various calculations may have various preparations of hyperparameters. For instance, a normal hyperparameter for fake neural groups is the amount of stowed away layers. Also, a distinct yet related hyperparameter is the wide variety of perceptrons in every secret layer. However, a comparable equal in choice timber will be the maximum extreme range of leaves in a tree or the greatest profundity for a tree. Other regular hyperparameters include getting to know the charge, group length, dropout widespread, and halting size.

Appropriately selecting hyperparameters can fundamentally boost up the search for a legitimate summed-up model without forfeiting execution. Nonetheless, watching the suitable set in several cases is an extra amount of expertise than technological know-how. Numerous scientists have endeavored to make hyperparameter searching through a greater effective and reproducible undertaking [23–25]. Once greater, this cycle additionally pretty is based upon the calculation, dataset, and problem you are attempting to address. An AI version can be tuned in a nearly countless degree of numerous approaches of conducting higher execution. Hyperparameters address a technique for duplicating consequences and furthermore fill in as a tool to as it should be approved models.

Illustration of a fundamental neural net with two secret layers of 3 perceptrons every. The quantity of facts assets, variety of stowed away layers, and range of perceptrons in each layer may be changed. Moreover, the institutions amongst layers and perceptrons can likewise be changed.

5.1.2.4 ALGORITHM STANDARDS

Thinking about the velocity of exploration within the discipline, there are steady advances and improvements to a full-size lot of these AI processes. The significant issue to remember is that not all calculations paintings for all usage instances now. Every calculation enjoys benefits and inconveniences. Certain facts might also likewise have an impact on the presentation of character calculations. The time spent executing such models will often be an aftereffect of checking out diverse varieties, obstacles, and hyperparameters inner these calculations to perform the nice-summed-up exhibition.

5.1.2.5 ASSESSMENT OF VERSION PERFORMANCE

The goal of any AI calculation is to apply proper records to make a model that performs out extraordinary on proper conditions and that can be evaluated in a quantitative, reproducible manner. Appraisal of real fashions is a whole subfield in itself, yet we will momentarily speak approximately the rudiments that are appropriate for nearly any AI calculation you'll run over.

5.1.2.6 SENSITIVITY VS. SPECIFICITY MNEMONIC

SnNouts and SpPins is a reminiscence aide to assist you with recalling the distinction between affectability and explicitness.

- **SnNout:** A test with high affectability esteem (Sn) that, whilst bad (N), assists with precluding an infection (out).
- **SpPin:** A test with high particularity esteem (Sp) that, whilst fine (P), assists with administering in contamination (in).

How could I ascertain affectability and particularity esteems?

An ideal test seldom neglects what you are searching for (i.e., it is delicate) and seldom confuses it with something different (for example, it is explicit). Hence, while assessing indicative tests, it is critical to ascertain the affectability and particularity of that test to decide its viability.

The affectability of an asymptomatic check is communicated because of the chance (as a fee) that an example checks wonderfully given that the patient has the illness.

The accompanying circumstance is applied to exercise session a test's affectability:

$$\text{Affectability} = \frac{\text{Number of genuine upsides}}{\left(\text{Number of genuine upsides} + \text{Number of bogus negatives}\right)}$$

$$= \frac{\text{Number of genuine upsides}}{\text{Complete number of people with the ailment}}$$

The particularity of a test is communicated as the likelihood (as a rate) that a test returns an adverse outcome given that that patient doesn't have the illness. The accompanying condition is utilized to compute a test's particularity:

$$\text{Particularity} = \frac{\text{Number of genuine negatives}}{(\text{Number of genuine negatives} + \text{number of bogus upsides})}$$

$$= \frac{\text{Number of genuine negatives}}{\text{The complete number of people without the ailment}}$$

You have another analytic test that you need to assess. You have a board of approval tests where you know for specific whether they are most certainly from infected or sound people for the condition you are trying for. Your example board comprises of 150 up-sides and 400 negatives.

There are four things we will expect to explain in this model:

i. What is the test's affectability? That is, what number of ailing people does it effectively recognize as unhealthy?
ii. What is the test's particularity? That is, what number of sound people does it accurately distinguish as solid?
iii. What is the test's positive prescient worth (PPV)? That is, what is the likelihood that an individual returning a positive outcome is really sick? and
iv. What is the test's negative prescient worth (NPV)? That is, what is the likelihood that an individual returning an adverse outcome is really solid?

In the wake of running the examples through the test, you contrast your outcomes with their known infection status and find:

- Genuine up-sides (test result positive and is really certain) = 144.
- Bogus positive (test result positive, however, is really negative) = 12.
- Genuine negatives (test result negative and is really negative) = 388.
- Bogus negative (test result negative, however, is really positive) = 6.

The trendy dimension for comparing the exhibition of AI models is known as the beneficiary working trademark (ROC). The ROC can be summed up through various from zero to at least one, that is, the planned vicinity underneath-the-ROC bend (AUC). The ROC bend is a 2D plot that moves the advantageous substitute charge instead of the genuine, effective rate. There are four numbers that are utilized to determine the compelling ness of a test: proper fine charge, bogus high-quality rate, real negative price, and bogus negative price.

Genuine fantastic and genuine bad are the proper responses to a take a look at, while bogus fantastic and bogus terrible are misguided responses to the take a look at or model. These numbers may be consolidated further into two numbers referred to as affectability and particularity.

We have as of now tested affectability and particularity, but currently, we can speak approximately how they're utilized to make the ROC.

Preferably a test would have each high affectability and excessive explicitness. Notwithstanding, there's a tradeoff. Specializing in one frequently activates the drawback of the alternative. When setting the threshold low, one will get an excessive obvious high-quality charge (high affectability) and an excessive bogus nice price (low explicitness). On the opposite hand, putting the restriction high will result in a low apparent fine fee (low affectability) and a low bogus advantageous price (excessive particularity).

The ROC and AUC metric is applied to describe the more a part of the characterization errands many AI fashions are endeavoring to do; does this individual have the infection or do they not? Assuming that a test has an excessive affectability and an excessive particularity, it is considered a close to consummate take a look at, and the AUC is near 1 (Figure 5.5). On the off hazard that the test is abnormal, then the AUC is 0. Five. The x-hub is commonly the bogus nice fee (or 1-explicitness). In an excellent international, the bogus wonderful rate is just about as little as it ought to without a doubt be predicted. They-pivot is by and large the real advantageous rate (affectability). The affectability is the component that is commonly amplified. On a commonplace bend, the midpoint of the bend is the most adjusted compromise between sensitivity and particularity, but this isn't always dependably the state of affairs. The AUC esteem is a much less complex, greater summed up manner, to survey the exhibition instead of the transferring tradeoffs among affectability and particularity.

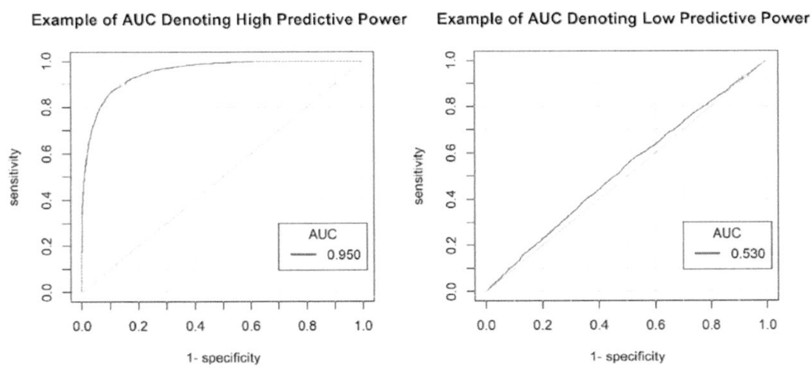

FIGURE 5.5 Analysis of numerous regressions to evaluate the accuracy of prediction models.

One extra method for considering AUC is as a rate the model can appropriately apprehend and isolate a fine outcome from an unfavorable final result. Given a difficult to understand the case, a version with an AUC of zero. Around 75% are able to accurately distinguish whether the case is an advantageous case or a terrible case. This quantity will swiftly assist you in realizing the aftereffects of any version.

5.1.2.7 OVERFITTING

Overfitting is one of the predominant issues when getting ready any model [26]. Basically, while preparing a model on a group of records, over-getting ready the model will paintings at the presentation of the model on that precise dataset yet at the price of losing speculation to different datasets. An overfitted model may not work while applied to new information. It has in no way visible. From a feasible point of view, this type of version isn't particularly helpful in proper software.

When getting ready any AI model, the right final result is a summed-up version. A summed-up version features admirably on an extensive range of cases and a wide variety of datasets, especially records it has never seen. All things considered, several experts are reluctant to provide an extra of trustworthiness to a version or technique that uses a solitary dataset.

An assortment of techniques had been utilized to maintain models from overfitting, and a big range of these are currently typified in the hyperparameters tested before.

Occasions of an AUC show a version which has incredible perceptive power (left) and an AUC implying a model with negative or close unpredictable farsighted energy (right). Specificity is a remarkable quantity which makes a fake fantastic price (for), and its affectability can be examined as an undeniable advantageous rate (tpr).

The wondering is to keep the fashions lower back from converting exorbitantly brief to the dataset it's far being geared upon. This subset of systems is referred to as regularization [27]. One such technique used in neural nets is known as a dropout. This procedure is substantially used to maintain counterfeit neural nets lower back from Overfitting for the duration of the path of motion responsibilities. The method is absolutely crucial. During the making plans cooperation, unpredictable perceptrons and their bearing on affiliations are "dropped" from the affiliation. These "decreased" networks have higher execution stood out from other regularization methods on controlled getting to know endeavors [28]. A technique referred to as cross-endorsement is often used to evaluate the show and help the summarized insightful limit of a model. The most widely perceived method for incorporating AI models is distributing the educational assortment into by way of and huge 80% for planning and 20% for checking out. This package deal is frequently less accommodating for direct models, yet keeping apart is more beneficial for complicated models [29]. During move-endorsement, this cut-up is achieved in self-sustaining spaces of the information to make sure appropriate attention. For instance, if 10-go-over go-endorsement is labored out, the first separated in a long time set with 100 discernments can be the underlying 80 for purchasing prepared and the ultimate 20 for the check, the following split may be the underlying 10 and closing 10 for the test, and the middle 80 for making plans, etc. (Figure 5.6). This made 10 models the usage of a comparative estimation as of overdue geared up and took a stab at extraordinary portions of comparable records. The regular display of these 10 models offers an honest evaluation of the summarized presentation of the estimation on that form of record.

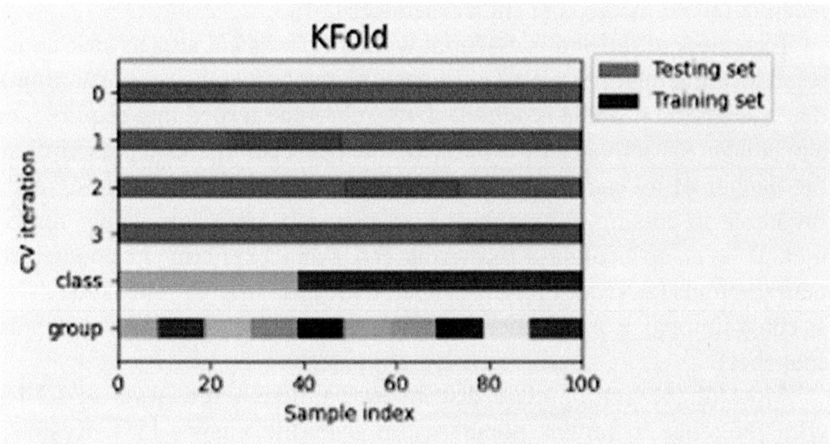

FIGURE 5.6 Illustration of a bunch of cross approval parts.

5.2 MACHINE LEARNING OF MEDICAL IMAGES

Present-day medical pics are superior in nature. To viably use them in medical care, there are a few problems that need to be survived. Clinical imaging depicts an assortment of procedures to make visible portrayals of internal segments of the human body with the stop intention of conclusion, exam, and clinical intercession. This is beneficial in staying away from or diminishing the requirement for the greater established scientific norm of exploratory medical manner. Since establishing any part of the human frame via careful manner expands the hazard of contaminations, strokes, and exclusive inconveniences, medical imaging is currently the popular tool for starting findings within clinical putting.

The modern-day medical norm of comparing clinical snapshots is the utilization of prepared doctors, pathologists, or radiologists who look at the photographs and determine the principal motive force of medical afflictions. This clinical standard is willing to human blunder. It is also outrageous and pricey, frequently requiring years or a few years of involvement to accomplish a degree of know-how that could reliably evaluate these pictures. Taking into consideration that the showcase of appropriate AI abilities within the advanced age changed into shown through Andrew Ng utilizing snapshots pulled from YouTube recordings, it is clear why scientific photographs had been one of the predominant areas tended to in the course of the underlying reception of AI methods in clinical services [54].

Exactness of the end is essential within the medical area as sick-counseled locating ought to set off extreme outcomes and outcomes. Assuming that a scientific process is achieved wherein none turned into required, or a misdiagnosis activates ill-cautioned doses of recommended prescription, the danger of a deadly result increments. In the domain of photographs favorable to cessing, maximum procedures rely on a completely fundamental level on profound mastering (DL) and explicitly in counterfeit neural groups (ANNs). Present-day strategies use improvements to ANNs as convolutional neural agencies (CNNs) to help execution whilst grouping snapshots.

Most of the cutting-edge distributions utilize a few sorts of CNNs with reference to protest popularity in scientific images [55]. Realistic coping with unit (GPU) speed increase has made the structure of profound CNN's greater powerful. However, good-sized problems in making an equipped version sincerely exist. The finest trouble is the requirement for a whole lot of explained medical photograph statistics. The fee to general and make such statistics bases is regularly restrictive since it requires prepared medical doctors' a super opportunity to clarify the photographs. Furthermore, issues along with affected person safety regularly prevent the capacity to make such facts based open-source. Many investigations simply use around 100 1,000 examples in getting ready CNNs. This limited instance size expands the threat of overfitting and diminishes the exactness of the expectations [56].

Concerns regarding the execution of AI into medical conclusions had been raised with regard to appropriate approval of fashions [57]. The fundamental emotions of dread contain appropriately perusing the planned objectives of an AI model, lessening dimensionality of the statistics, and reproducibility of making ready such models on the real world and new scientific facts. Approving outcomes on different datasets can be difficult due to the absence of larger datasets for unique infections, in which the entire of these facts can take more paintings than the real coaching of the model. Clinical imaging records are innately extra tough you bought and are more difficult to save and method. The basis to address the information has essentially not stayed privy to the increment inside the degree of facts.

5.2.1 LESION DETECTION AND COMPUTER-AUTOMATED DETECTION

The famous maximum usage of modern-day AI advancements in remedy is for PC mechanized identification (CAD) explicitly within the discovery of accidents, such as those normally located in mammograms, mind examines, and other frame filters [58]. These strategies use CNNs to expose up the likelihood that a competitor's harm is certainly a sore, regularly using some 2D cuts of 3D rotational outputs of 1 or the other CAT or magnetic resource imaging (MRI) snapshots.

US photos are additionally utilized in preparing and a collection of strategies like the randomized revolution of the photos or focusing up-and-comer injuries inside the focal factor of the picture. Particularly in mammography (MG), CAD techniques have arrived at a degree in which they are applied as a "2nd assessment" for maximum radiologists, appreciably operating at the exactness of screenings without multiplying the price related to utilizing a human because of the "2nd evaluation" (Figure 5.7).

Computer-aided layout is moreover presently parted into discovery and analysis. This qualification is understated but full-size. A harm may be categorized as both innocent or dangerous, primarily based on a health practitioner's records and appraisal. Nonetheless, actual identity is an essential preliminary section in treating a patient.

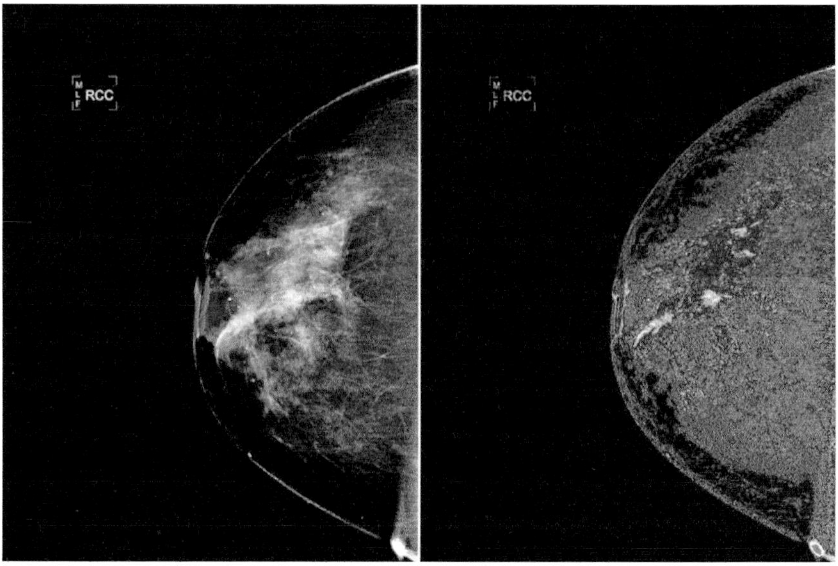

FIGURE 5.7 Illustration of a mammogram NASA programming initially used to improve geology symbolism.

Source: Taken from NASA's official statement, credited to Bartron medical imaging.

PC-supported location is the genuine acknowledgment of expected injuries from a clinical picture. For instance, the discovery and division of glioblastoma is a tough errand because of the intrusive and large nature of these growths. Not in any respect like other cerebrum growths, they may be no longer handily constrained, and evaluating how drugs, for instance, chemotherapy, are appearing is in itself a troublesome task. Profound mastering has supported this with the aid of computerizing the evaluation of glioblastoma MRIs [59].

The PC-supported willpower depicts the probability harm is threatening in nature. These techniques are basically used to paintings at the exactness of willpower and work on early analysis in the clinical setting. Once more, those errands have reliably been accomplished with the aid of AI, specifically in cerebrum-related applications, due to the tough concept of comparing mind wellness. Furthermore, the willpower of Alzheimer's thru medical imaging is a capability application for profound knowledge that displays some assurance [60, 61].

5.3 MACHINE LEARNING IN GENETICS FOR THE PREDICTION AND UNDERSTANDING OF COMPLEX DISEASES

Hereditary records and advances have detonated starting around 2008, making tough challenges in the way to deal with the dramatically increasing records. Progresses in hereditary sequencing tempo, specifically, next-generation sequencing (NGS) advances have dramatically speeded up at which a whole human genome is sequenced while likewise appreciably lessening fees. The human genome is a really complicated design that encodes all the statistics of the human flip of activities and attributes. The genome is profoundly interconnected, and translating the full-size majority of these suggestions is as but a secret to us. The type of genomes amongst people likewise builds the intricacy of comprehending nice collaborations.

Numerous nicely-being drives have zeroed in on obtaining huge instance sizes of human genomes to help with distinguishing measurably critical patterns among various populaces of people. However, the 23 chromosomes of the human genome include around 20,000 features which have been outstanding due to the critical coding successions for the proteins essential in constructing the natural elements of our cells. This quantity is as, but a super wager, and some evaluations display that there is probably upwards of 25,000 traits or as not many as 19,000. A huge wrap of hereditary information that doesn't code for any proteins is excluded from the opinions of the one.

A developing assemblage of writing demonstrates that particular segments of what has been conversationally referred to as hereditary dull take into account, or lacking heritability, exist. These terms allude to the segments of deoxyribonucleic acid (DNA) which don't have any apparent protein-coding capacity but are probably relevant to the degree of exceptional articulation in a character's hereditary code. Degrees of great articulation could provide probable purpose protein over-burden or inadequacy, which could activate a group of clinical problems. Also, number one contrasts inside the bodily production of the way the DNA is certain into chromosomes and in a while in this way unfolded all through both the duplication interaction and interpretation and file method, can likewise have an effect at the degree of exceptional articulation.

As an instance, methylation or acetylation of the DNA backbone can make it more difficult (methylation) or extra sincere (acetylation) to unwind the DNA strand at some point of regular cell methods like replication or protein amassing. Proof of severe duplicates of similar outstanding has furthermore been organized in what is depicted as duplicate range varieties (CNV), which display duplication, tripling, and cancellation activities of unique regions of the genome in a person. Knowledge of this in the particular interconnected and nonlinear connection between all the numerous regions of the human genome is difficult.

With AI, researchers have started to have a look at examples and styles which can be displayed in a further unsurprising way. The usage of the commonly developing measure of hereditary statistics, AI has the functionality of precisely foreseeing who is in the chance of acquiring precise infections like malignant growths and Alzheimer's contamination. Dysfunctional behaviors, for instance, schizophrenia, and bipolar troubles, have likewise been identified to run in families, showing an ability hereditary connection.

A massive assortment of gadgets and ability techniques ought to be brought collectively and normalized to enjoy the elevated facts collection. Records regarding how human hereditary range can add to someone's defenselessness we could help patients and professionals to make the early way of life changes in a precautionary manner. Similarly, it could educate medical doctors concerning which forms of prognostics and diagnostics can be the most pertinent for a selected affected man or woman, placing apart each time and cash while going for walks on quiet results within the prolonged haul. In addition, as AI began with Turing deciphering the puzzler system, we're currently going to utilize AI and AI to disentangle the privileged insights of the human body and genome.

KEYWORDS

- **artificial intelligence**
- **big data**
- **genetics**
- **internet of things**
- **machine learning**
- **radiology**

REFERENCES

1. Copeland, J., (2000). The Turing test. *Minds and Machines, 10*(4), 519–539.
2. French, R. M., (2000). The Turing test: The first 50 years. *Trends in Cognitive Sciences, 4*(3), 115–122.
3. Edwards, S., (2019). World War II at sea: A global history. *The Journal of American History, 106*(1), 237.
4. Turing, A. M., (2009). Computing machinery and intelligence. *Parsing the Turing Test: Philosophical and Methodological Issues in the Quest for the Thinking Computer*, 23 65.
5. Samuel, A. L., (1959). Some studies in machine learning. *IBM Journal of Research and Development, 3*(3), 210–229.
6. Samuel, A. L., (1960). Programming computers to play games. *Advances in Computers, 1*(C), 165–192.
7. Fukushima, K., (1975). Cognitron: A self-organizing multilayered neural network. *Biological Cybernetics, 20*(3, 4), 121–136.
8. Fukushima, K., (1980). Neocognitron: A self-organizing neural network model for a mechanism of pattern recognition unaffected by shift in position. *Biological Cybernetics, 36*(4), 193–202.
9. Huang, K., Mashl, R. J., Wu, Y., Ritter, D. I., Wang, J., Oh, C., et al., (2018). Pathogenic germline variants in 10,389 adult cancers. *Cell*.
10. Silver, D., & Hassabis, D., (2016). *AlphaGo: Mastering the Ancient Game of Go with Machine Learning.* Google Research Blog.
11. Brewka, G., (1996). Artificial intelligence—A modern approach by Stuart Russell and Peter Norvig, prentice hall. Series in artificial intelligence, Englewood Cliffs, NJ. *The Knowledge Engineering Review, 11*(1), 78, 79.
12. Alpaydin, E., (2014). *Introduction to Machine Learning* (Vol. 3, p. 640). London: The MIT Press.
13. Kotsiantis, S. B., (2007). Supervised machine learning: A review of classification techniques. *Informatica (Ljubljana), 31*(3), 249–268.
14. Hastie, T., Tibshirani, R., & Friedman, J., (2009). *The Elements of Statistical Learning - Data Mining, Inference, and Prediction: Springer Series in Statistics* (Vol. 2, No. 1, pp. 93–85). Switzerland: Springer.
15. Ho, T. K., (2002). A data complexity analysis of comparative advantages of decision forest constructors. *Pattern Analysis & Applications, 5*(2), 105–112.
16. Friedman, J. H., (2002). Stochastic gradient boosting. *Computational Statistics and Data Analysis, 38*(4), 367–378.
17. Friedman, J. H., (2001). Greedy function approximation: A gradient boosting machine. *The Annals of Statistics, 29*(5), 1189–1232.
18. Mason, L., Baxter, J., Bartlett, P., & Frean, M., (1999). Boosting algorithms as gradient descent in function space. *NIPS Conference Proceedings*, 512–518.
19. Libbrecht, M. W., & Noble, W. S., (2015). Machine learning applications in genetics and genomics. *Nature Reviews Genetics, 16*(6), 321–332.
20. Längkvist, M., Karlsson, L., & Loutfi, A., (2014). A review of unsupervised feature learning and deep learning for time-series modeling. *Pattern Recognition Letters, 42*(1), 11–24.

21. Lloyd, S. P., (1982). Least squares quantization in PCM. *IEEE Transactions on Information Theory, 28*(2), 129–137.
22. Schmidhuber, J., (2015). Deep learning in neural networks: An overview. *Neural Networks, 61*(1), 85–117.
23. Hazan, E., Klivans, A., & Yuan, Y., (2018). Hyperparameter optimization: A spectral approach. In: *6th International Conference on Learning Representations, ICLR 2018; Conference Track Proceedings*.
24. Bardenet, R., Brendel, M., Kégl, B., & Sebag, M., (2013). Collaborative hyperparameter tuning. In: *30th International Conference on Machine Learning; ICML* (p. 2013).
25. Hutter, F., Hoos, H., & Leyton-Brown, K., (2014). An efficient approach for assessing hyperparameter importance. In: *31st International Conference on Machine Learning; ICML* (p. 2014).
26. Hawkins, D. M., (2004). The problem of overfitting. *Journal of Chemical Information and Computer Sciences, 44*(1), 1–12.
27. Ng, A., (2011). *Regularization—Solving the Problem of Overfitting*. Coursera. Available from: https://www.coursera.org/lecture/machine-learning/welcome-to-machine-learning-iYR2y (accessed on 14 September 2022).
28. Srivastava, N., Hinton, G., Krizhevsky, A., Sutskever, I., & Salakhutdinov, R., (2014). Dropout: A simple way to prevent neural networks from overfitting. *Journal of Machine Learning Research, 15*(56), 1929–1958.
29. Picard, R. R., & Cook, R. D., (1984). Cross- validation of regression models. *Journal of the American Statistical Association, 79*(387), 575–583.
30. Jothi, N., Rashid, N. A., & Husain, W., (2015). Data mining in healthcare—A review. *Procedia Computer Science, 72*(1), 306–313.
31. Koh, H. C., & Tan, G., (2005). Data mining applications in healthcare. *Journal of Healthcare Information Management, 19*(2), 64–72.
32. Cottle, M., Hoover, W., Kanwal, S., Kohn, M., Strome, T., & Treister, N. W., (2013). *Transforming Health Care Through Big Data: Strategies for Leveraging Big Data in the Health Care Industry*. Institute for Health Technology Transformation. New York, iHT2.
33. Wang, Y., Kung, L. A., & Byrd, T. A., (2018). Big data analytics: Understanding its capabilities and potential benefits for healthcare organizations. *Technological Forecasting and Social Change, 126*(1), 3–13.
34. Raghupathi, W., & Raghupathi, V., (2014). Big data analytics in healthcare: Promise and potential. *Health Information Science and Systems, 2*(1), 3.
35. *The Cancer Genome Atlas Program*—National Cancer Institute [Internet]. (2019). Available from: Https://www.cancer.gov/about-nci/organization/ccg/research/structural-genomics/tcga (accessed on 14 September 2022).
36. tcga-infographic-enlarge.__v100169753.png (1400×2580) [Internet]. (2019). Available from: https://www.cancer.gov/PublishedContent/Images/images/nci/organization/tcga/tcga-infographicenlarge__v100169753.png (accessed on 14 September 2022).
37. Peakman, T. C., & Elliott, P., (2008). The UK biobank sample handling and storage validation studies. *International Journal of Epidemiology, 37*(2), 234–244.
38. Sudlow, C., Gallacher, J., Allen, N., Beral, V., Burton, P., Danesh, J., et al., (2015). UK biobank: An open access resource for identifying the causes of a wide range of complex diseases of middle and old age. *PLoS Medicine, 12*(3).

39. *Protocol for the UK Biobank*— Wayback Machine [Internet]. (2019). Available from: https://web.archive.org/web/20060214144838/http://www.ukbiobank.ac.uk/docs/draft_protocol.pdf (accessed on 14 September 2022).
40. Elliott, P., & Peakman, T. C., (2008). The UK biobank sample handling and storage protocol for the collection, processing and archiving of human blood and urine. *International Journal of Epidemiology, 37*(2), 234–244.
41. Gao, G. F., Parker, J. S., Reynolds, S. M., Silva, T. C., Wang, L. B., Zhou, W., et al., (2019). Before and after: Comparison of legacy and harmonized TCGA genomic data commons data. *Cell Systems, 9*(1), 24–34.e10.
42. Bycroft, C., Freeman, C., Petkova, D., Band, G., Elliott, L. T., Sharp, K., et al., (2018). The UK biobank resource with deep phenotyping and genomic data. *Nature, 562*(7726), 203–209.
43. *Background—EPIC* [Internet]. (2019. Available from: http://epic.iarc.fr/about/background.php (accessed on 14 September 2022).
44. Leitsalu, L., Haller, T., Esko, T., Tammesoo, M. L., Alavere, H., Snieder, H., et al., (2015). Cohort profile: Estonian biobank of the Estonian genome center, university of Tartu. *International Journal of Epidemiology, 44*(4), 1137–1147.
45. *Master Decoder: A Profile of Kári Stefánsson* | The Scientist Magazine® [Internet]. 2019. Available from: https://www.the-scientist.com/profile/master-decoder--a-profile-of-kri-stefnsson-65517 (accessed on 14 September 2022).
46. Gulcher, J., & Stefansson, K., (1998). Population genomics: Laying the groundwork for genetic disease modeling and targeting. *Clinical Chemistry and Laboratory Medicine, 36*(8), 523–527.
47. *China Kadoorie Biobank* [Internet]. (2019). Available from: https://www.ckbiobank.org/site/ (accessed on 14 September 2022).
48. Littlejohns, T. J., Sudlow, C., Allen, N. E., & Collins, R., (2017). UK biobank: Opportunities for cardiovascular research. *European Heart Journal, 44*, 1158–1166.
49. Scholtens, S., Smidt, N., Swertz, M. A., Bakker, S. J. L., Dotinga, A., Vonk, J. M., et al., (2015). Cohort profile: LifeLines, a three- generation cohort study and biobank. *International Journal of Epidemiology, 44*(4), 1172–1180.
50. *The Health 202: NIH Wants 1 Million Americans to Contribute to New Pool of Gene Data.* The Washington Post [Internet]. (2019). Available from: https://www.washingtonpost.com/news/powerpost/paloma/the-health-202/2018/01/16/the-health-202-nih-wants-1-million-americans-to-contribute-to-new-pool-of-gene-data/5a5ba45a30fb0469e8840135/ (accessed on 14 September 2022).
51. *FACT SHEET: President Obama's Precision Medicine Initiative,* (2019). whitehouse.gov [Internet]. Available from: https://obamawhitehouse.archives.gov/the-press-office/2015/01/30/fact-sheet-president-obama-s-precision-medicine-initiative (accessed on 14 September 2022).
52. Precision Medicine Initiative (PMI) Working Group, (2015). *The Precision Medicine Initiative Cohort Program—Building a Research Foundation for 21st Century Medicine.* Precision Medicine Initiative Work Group Report to Advisory Committee to Director NIH.
53. *FinnGen, a Global Research Project Focusing on Genome Data of 500,000 Finns, Launched.* EurekAlert! Science News [Internet]. (2019). Available from: https://www.

eurekalert.org/pub_releases/2017-12/uoh-fag121917.php (accessed on 14 September 2022).
54. Le, Q. V., (2013). Building high-level features using large scale unsupervised learning. In: *ICASSP, IEEE International Conference on Acoustics, Speech and Signal Processing—Proceedings*.
55. Greenspan, H., Van, G. B., & Summers, R. M., (2016). Guest editorial deep learning in medical imaging: Overview and future promise of an exciting new technique. *IEEE Transactions on Medical Imaging, 35*(5), 1153–1159.
56. Giger, M. L., (2018). Machine learning in medical imaging. *Journal of the American College of Radiology, 15*(3), 512–520.
57. Foster, K. R., Koprowski, R., & Skufca, J. D., (2014). Machine learning, medical diagnosis, and biomedical engineering research—Commentary. *BioMedical Engineering Online*. Article number: 94.
58. Lundervold, A. S., & Lundervold, A., (2019). An overview of deep learning in medical imaging focusing on MRI. *Zeitschrift fur Medizinische Physik, 29*(2), 102–127.
59. Havaei, M., Davy, A., Warde-Farley, D., Biard, A., Courville, A., Bengio, Y., et al., (2017). Brain tumor segmentation with deep neural networks. *Medical Image Analysis, 35*(1), 18–31.
60. Islam, J., & Zhang, Y., (2018). Brain MRI analysis for Alzheimer's disease diagnosis using an ensemble system of deep convolutional neural networks. *Brain Informatics, 5*(2).
61. Liu, M., Zhang, J., Adeli, E., & Shen, D., (2018). Landmark-based deep multi-instance learning for brain disease diagnosis. *Medical Image Analysis, 43*(1), 157–168.

CHAPTER 6

Heart Disease Prediction Desktop Application Using Supervised Learning

V. PATTABIRAMAN and R. MAHESWARI

Vellore Institute of Technology, Chennai, Tamil Nadu, India,
E-mail: maheswari.r@vit.ac.in (R. Maheswari)

ABSTRACT

Heart disease, often recognized as a cardiovascular disease that, denotes a range of conditions that grief the heart and has become the prominent cause of death worldwide in recent decades. It combines multiple cardiovascular disease risk variables with the requirement for time to produce precise, accurate, and sensitive methods for early identification and treatment of the illness. In the realm of healthcare, data mining is a commonly utilized tool for evaluating large amounts of data. The search employs a variety of data mining and machine-learning approaches to evaluate massive amounts of complex medical data to assist health providers in predicting cardiac issues. This research comprises a model based on supervised learning algorithms such as Support Vector Machine (SVM), K-Nearest Neighbor (KNN), and Random Forest algorithm and several factors linked to heart disease. It makes use of the current Cleveland database, which is made up of persons with heart disease from the heart disease dataset available on Kaggle. The goal of this work is to figure out how likely it is that a patient would acquire heart disease. The outcome of the proposed system reveals that the greatest accuracy score is achieved with the Random Forest algorithm in a desktop application form. After implementing three approaches, the system discovered that the accuracy in the Random Forest was the highest

Computational Health Informatics for Biomedical Applications. Aryan Chaudhary and Sardar M. N. Islam (Naz) (Eds.)

(96%). The research will be enhanced with various machine learning (ML) methods, grouping, and association rules, vector machine assistance, and evolutionary algorithms.

6.1 INTRODUCTION

The coronary heart is the organ that pumps blood, with its lifestyle giving oxygen and nutrients to all of the body's tissues. If the pumping movement of the coronary heart becomes inefficient, essential organs like the mind and kidneys suffer. If the heart stops running altogether, dying takes place within minutes. Heart sickness has been considered one of the most complicated and lifestyles deadliest human illnesses in the world. Life itself is dependent on the green operation of the heart. Symptoms of coronary heart sickness include shortness of breath, a weak spot of the physical body, swollen feet, and fatigue. The coronary heart prognosis and remedy are very complicated, mainly in the growing countries, because of the uncommon availability of diagnostic equipment and different assets, which affect coronary heart patients' right prediction and remedy. This makes coronary heart sickness a prime difficulty to be dealt with.

6.1.1 ROOTS OF CORONARY HEART DISEASE (CHD)

The primary root cause of many CHDs is the sudden blocking of blood supply to the heart of a person or if the coronary arteries may happen to intermittent with the undissolved or more fatty substances. Therefore, the continuous monitoring of the patient's fat level may help to save their lives. But it's far difficult to identify coronary heart sickness due to numerous contributory hazard elements, including diabetes, excessive blood pressure (BP), excessive cholesterol, ordinary pulse rate, and plenty of different elements. The invasive-primarily based strategies to diagnose coronary heart sickness are primarily based on the evaluation of the affected person's clinical history, bodily exam file, and evaluation of involved signs via way of means of health workers. Often there's a postponement in the prognosis because of human errors. Due to such constraints, scientists have grown to become particular towards current methods like Data Mining and ML for predicting sickness.

6.1.2 DATA MINING AND MACHINE LEARNING

Data mining performs a vital function in constructing a sensible version for a clinical gadget to discover coronary heart sickness by investigating the dataset of patients, which involves hazard components related to the illness. Medical practitioners might also additionally offer to assist in its detection. Many software program equipment and various algorithms were proposed via various researchers for growing powerful clinical selection-guided gadgets.

ML predictive models, including SVM, KNN, and random forest, are put to expect whether or not someone is having coronary heart sickness or now no longer. However, clinical records are regularly constricted via means of smaller units of observations than what's usually desired to permit enough education and to try out of fashions constructed the use of machines getting to know algorithms. Heart disorder might also additionally arise because of bad ways of life, smoking, alcohol, and high consumption of fats which might also additionally purpose hypertension. By the WHO's survey, more than 10 million people die due to Heart illnesses. A wholesome way of life and earliest detection are the best approaches to save the person from coronary heart-associated illnesses. The principal venture in today's health department is the allocation of the best high-satisfactory offerings and robust correct verification. Data of an extensive set of scientific statistics created with the aid of using scientific specialists are to be had for observing and removing precious expertise from it. Data mining strategies are the way of eliminating precious and private records from the large number of statistical datasets which to be had. Mostly the scientific database includes discrete records of the dataset.

Hence, selection criteria mostly use discrete statistics datasets that always lead to a complicated and complex task towards prediction and deriving conclusions. ML that's a sub-field of data mining that handles big efficient scale well-formatted data sets. In the scientific field, device studying may be used for identifying, verifying, and forecasting numerous illnesses. The principal intention of this research is to offer a device for doctors to discover coronary heart disorder at an early level. This in flip will assist in providing a powerful remedy to sufferers and keep away from severe results. ML performs a completely vital position to discover the private discrete styles and thereby examine the given statistical data. After this evaluation of statistics data, ML strategies assist in coronary heart disorder forecast and before identification. This work presents an overall evaluation of numerous ML strategies such as KNN, SVM, and Random Forest for forecasting coronary heart disorder at a first level.

The advancement of information technology in the medical sector and related industry is proliferating, which assist medical practitioner in elevating more prediction and decision-making actions. These enhancements aid physicians in handling various medicine-related processes like medications, management of diseases, and discovering patterns and relationships among diagnosis medical data. In particular, the existence and evolvement of ML concepts offer a broader opportunity to progress/ increase the accuracy in prediction, recommendation, etc.

Though many internet of things (IoT) enabled advanced technologies are made available in the medical field, the necessity and role of ML concepts are indeed. For those IoT devices, ML analysis over the medical dataset with appropriate algorithms may help to achieve much accuracy for better treatment of cure. With the practice of the most advanced ML methods, medical data examination is much more efficient. It may help to infer the origin/ source that causes CHD.

6.2 RELATED WORK

Mohammed Abdul Khaleel presented a demonstration for finding frequent diseases using the survey of techniques for data mining on medical data in the local environment [1]. This research focuses on dissecting information mining processes essential for medical data mining, precisely to locate nearby visit ailments like heart disease, lung cancer, and breast disease. Information mining is a method of extracting data to locate the inactiveness of organs; for example, Vembandasamy et al. used it to assess and identify cardiac disease. The Naive Bayes technique was utilized in this case. The Bayes theorem was employed in the Naive Bayes algorithm. As a result, Naive Bayes has a lot of strength when it comes to making assumptions on its own. The data set for this study was gathered from a top diabetes research facility in Chennai, Tamil Nadu, India. The dataset contains the details of more than 500 patients. The tool utilized was Weka, and the categorization was done with a 70% split. Naive Bayes has an accuracy rate of 86.419% in this case.

Research on RHS systems has been made, and the author presented that RHS systems are cost-efficient and efficient in lowering disease. They provide Wanda-CVD, an updated RHM mobile phone-based framework and designed to provide remote teaching and social assistance to members. The research concentrated on social insurance organizations all across the globe, which considers the CVD prevention efforts to be a top priority [2].

Prediction for illness similarities by expending the ID3 algorithm in gadgets like mobile phones and Television (TV) were presented by the researcher [3]. This study explains how to deal with identifying designs that are hidden by cardiac sickness in a planned and hidden manner. The offered framework uses data mining techniques such as the ID3 algorithm. This proposed strategy not only informs people about illnesses but also has the potential to lower the death rate and the number of illness victims.

In their study, the author represented Maximal Frequent Item Set Algorithm (MAFIA) and K-Means clustering techniques. They observed that categorization is crucial for illness prognosis; hence the accuracy of the categorization based on MAFIA and K-means was achieved in a better way [4].

The research proposed an intelligence system using K-Star Algorithm. They present an expectation framework for heart infection in this research. Learning how to quantify vectors computation of the neural system, this framework identifies various 13 clinical features as input. It forecasts the presence or absence of cardiac sickness in the patient and numerous execution measures [5].

The author presented a learning vector quantization-based expectation framework for heart infection in this study. Computation of the neural system is being considered in this work. In this framework, the neural system recognizes 13 clinical features as input and forecasts the presence or absence of cardiac sickness in the patient and numerous execution measures [6].

Avinash et al. has done research on various specific ML algorithm's which will be used for a class of coronary heart sickness. Various analysis has been carried out and thus concluded that a glance at KNN and K-means algorithms may be used for categorizing the data and their accuracy had been in comparison is considered further. This study concludes that accuracy received with the help of victimization turned into most inexperienced aid of using a mixture of specific ways and parameter tuning [7].

Nagamani et al. have papered a machine that deployed chronicle mining methods along with the MapReduce algorithm. The accuracy received in step with this work made remarkable progress in results which is 45 times with the same data set. This result seems to be far better than the accuracy received through the usage of traditional fuzzy artificial neural community. Here, the accuracy of the algorithm used helped to turn the system into a development tool due to its use of dynamic schema and linear scaling [8].

Fahd Saleh Alotaibi has deliberated examination of five specific ML algorithms. The rapid device change resulted in higher accuracy as compared to MATLAB and related devices. During this study, the accuracy of supplying Regression, Random Forest, Naive Thomas Bayes, and SVM class algorithms was compared [9].

Anjan et al. proposed a machine which generates the usage of Naïve Bayesian (NB) ways for closes of dataset and advanced encryption standard (AES) algorithm for steady records switch for prediction of sickness [10].

Sneha et al. recommended an automated system with input video streaming to extract the features and trained the model using SVM. This video streaming system made analysis with DCM disorder, ASD disorder and Normal disorder and summarizes the accuracy of 97.40%, 97.57%, and 98.30%, respectively [11].

Subhashini et al. automated and devised a prediction system with the latest and advanced techniques, including Artificial Neural Network (ANN), SVM, etc., 93% of complete accuracy are achieved with this ensemble model based on Fast Fourier Transform (FFT) [12].

Sabrina et al. has demonstrated their research on two datasets such as American and Italian Dataset. The author applied SVM over American Datasets to predict cardiovascular disease with RBF Kernel Algorithm and improved the outcome using grid search. The proposed model attains 95.25% and 92.15% accuracy using the Italian and American datasets, respectively [13].

Thus many data analytics and ML researchers hastens their research works in many dimensions to design computer or mobile applications to better assist the medical practitioner in predicting and diagnosing heart disease [14].

6.3 METHODOLOGY

This investigation aims to evaluate the risk of heart discomfort as a cause of a computerized cardiac disease prediction desktop application that will support medical professionals and patients. In order to achieve this objective, the authors have studied the usage of a range of ML algorithms for data set analysis. This chapter examines the various related data set analyzes. This research also highlights specific personality attributes that are more crucial for gaining more accuracy than others. This might save money in numerous patient studies since all the features cannot contribute to the result in the same way. In this module, the system is going to compare three different (Random Forest, SVM, and KNN) and plot them into the chart. In this module, the system is going to check prediction on a bulk amount of different user data and check the accuracy of the algorithm. Figure 6.1 represents the proposed application's process flow, which contains the data preprocessing of the heart disease dataset followed by dataset partitioning of training dataset and test dataset. The model is trained and tested using three machine learning (ML) algorithms as KNN, SVM, and Random Forest. The appropriate heart disease prediction with analysis accuracy rate is visualized in the User Interface.

The system utilized data from the heart disease dataset from the ML repository in Kaggle to finish this study. There are many examples of real-world data with 14 various features such as BP, the kind of chest pain (CP), electric cardiograms, etc. In this study, the system used three algorithms to identify cardiovascular causes and build a model that is as exact as feasible. Two categories of details have been considered for the study. Category 1 is the patient's personal details, and Category 2 is their medical details. Under category 1, patients' age and gender were considered. The various medical attributes and its icons that are considered for further analysis are Chest pain (*Cp*), Resting BP (*Trestbps*), Serum Cholesterol (*Chol*), Fasting Blood Sugar (*FBS*), Rest electrocardiograph (*Restecg*), Maximum Heart Rate (*Thalch*), Exercise-induced angina (*Exang*), ST depression (*Oldpeak*), Slope (*Slope*), number of vessels (*Ca*), Thalassemia (*Thal*), Num (class attributes – *Class*) [11].

For chest pain, four types of pains were taken where type 1 represents typical angina, type 2 is atypical angina, type 3 signifies non-angina pain, and type 4 is asymptomatic. The other medical terms referred are resting symbolic BP with units in mm Hg, serum cholesterol in mg/Dl [16]. For FBS, the Boolean parameter true and false have been assigned with values of "zero" and "one" when the measurement is compared, and its value is greater than 120 mg/Dl. Three classes for Rest electrocardiograph are represented in which "0" indicates normal, "1" concludes the wave abnormality, and "2" signifies the left ventricular hypertrophy. Four major vessels colored by fluoroscopy were taken for analysis ranging from "0

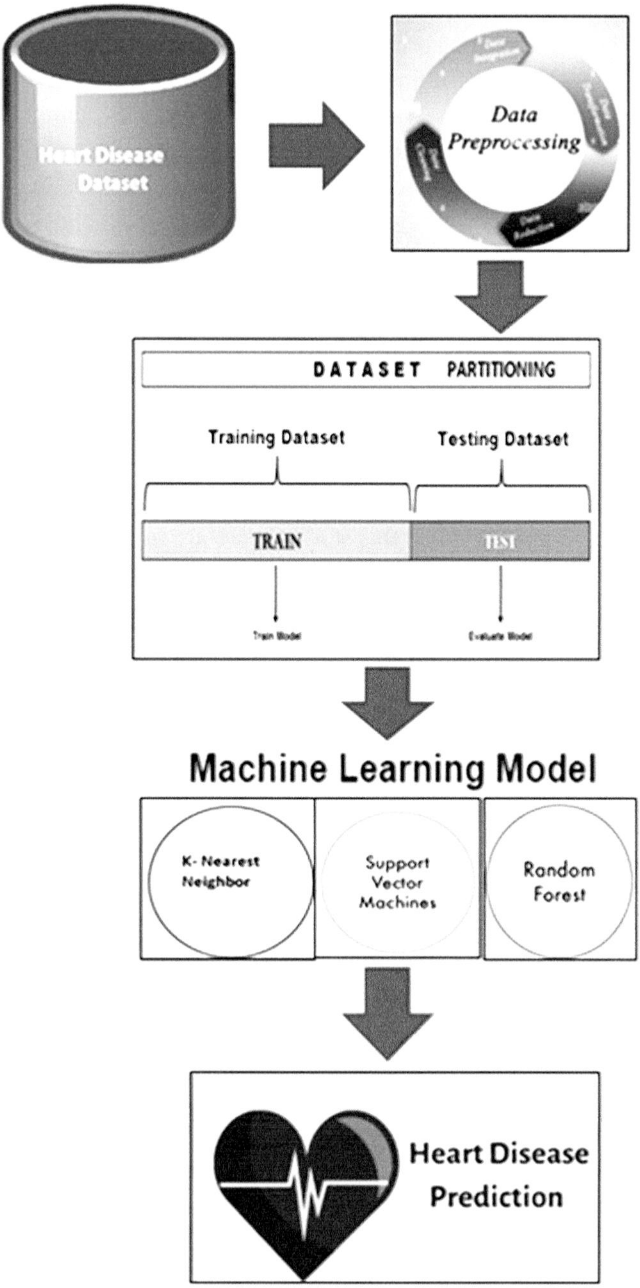

FIGURE 6.1 Process flow diagram of prediction system.

to 3." For Thalassemia, three defect types were observed "3" indicates normal, "6" for fixed defect, and "7" denoting reversible defect. Finally, the diagnosis of heart disease status concludes *"zero" – nil risk, "one" – low risk, "two"-potential risk, "three" – high risk,* and *"four" – very high risk.* Table 6.1 holds the dataset of heart disease with varying attributes such as "Age," "Sex," "Chest Pain," "Rest BP," "Serum Cholesterol," "FBS," "Rest Electrocardiograph," "MaxHeart Rate," "Exercise-induced angina," "ST depression," "Slope," "number of vessels," "Thalassemia," and "Class Attributes." The representative icon of these specified attributes with its detailed description helps indeed for further analysis [17–20].

TABLE 6.1 Heart Disease Dataset and its Attributes

SL. No.	Attribute	Representative Icon	Details
1.	Age	Age	Patients age, in years
2.	Sex	Sex	0 = female; 1 = male
3.	Chest pain	CP	Four types of chest pain ((1) typical angina; (2) atypical angina; (3) non-anginal pain; (4) asymptomatic)
4.	Rest blood pressure	Trestbps	Resting systolic blood pressure (in mm Hg on admission to the hospital)
5.	Serum cholesterol	Chol	Serum cholesterol in mg/dl
6.	Fasting blood sugar	Fbs	Fasting blood sugar >120 mg/dl (0–false; 1–true)
7.	Rest electrocardiograph	Restecg	0–normal; 1–having ST-T wave abnormality; 2–left ventricular hypertrophy
8.	MaxHeart Rate	Thalch	Maximum heart rate achieved
9.	Exercise-induced angina	Exang	Exercise-induced angina (0–no; 1–yes)
10.	ST depression	Oldpeak	ST depression induced by exercise relative to rest
11.	Slope	Slop	Slop of the peak exercise ST segment (1–upsloping; 2–flat; 3–down sloping)
12.	Number of vessels	Ca	Number of major vessels (0–3) colored by fluoroscopy
13.	Thalassemia	Thal	Defect types; 3–normal; 6–fixed defect; 7–reversible defect
14.	Num (class attribute)	Class	Diagnosis of heart disease status (0–nil risk; 1–low risk; 2–potential risk; 3–high risk; 4–very high risk)

6.3.1 K-NEAREST NEIGHBOR (K-NN)

It is a method for supervised classification which classifies the related objects specified in the dataset using their proximity value. This is a kind of learning that's dependent on various examples. The Euclidean distance [11] is used to calculate the distance of an attribute. It takes a collection of named points and represents a new point. The data is sorted based on similarities, and K-NN may be used to fill in the missing values in the data. Several ways for forecasting the data set apply after the missing values have been provided. Improved accuracy may be achieved by combining multiple versions of these algorithms. It is simple to use the K-NN approach without creating a model or making any extra assumptions. The method may be used to categorize, regress, and search data. Although K-NN is the most basic approach, it is influenced by noisy and irrelevant data, which reduces its accuracy [21].

Suppose there are classes, i.e., class A and class B, and the system has a brand new records factor x1 so that this records factor will fall on the range wherein of those related categories. The system wants a K-NN algorithm to clear this sort of problem. With the assist of K-NN, the system can complete the sequence without any problems. These metrics develop an awareness of the class or magnificence of a specific dataset. Consider Figure 6.2, which demonstrates classification outcomes before and after KNN implementation.

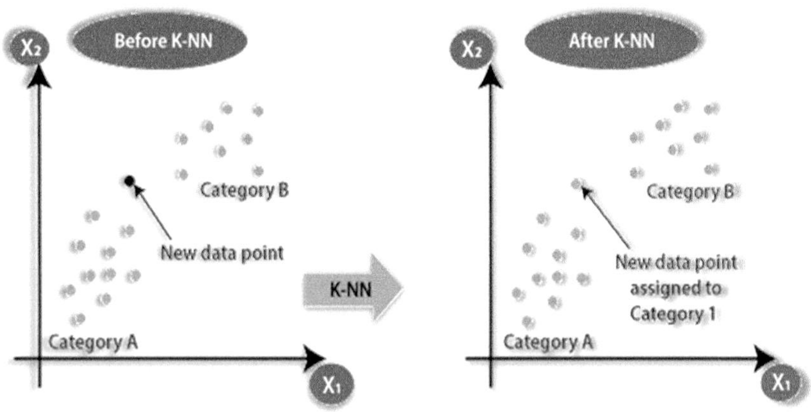

FIGURE 6.2 KNN classification.

6.3.2 SUPPORT VECTOR MACHINE (SVM)

It is a frequently utilized ML algorithm methodology that may have been used mostly as a predictor and a classifier. This locates high energy in the spatial domain, differentiating the categorization classes [18]. The attribute selection is represented by an SVM model as values in the spatial domain, which is prepared in such a way that components in the feature vector representing a variety of categories are separated by a vast range of data values as much as possible. Figure 6.3 indicates the SVM record factor holding various hyperplanes like maximum margin hyperplane, maximum hyperplane, positive hyperplane, negative hyperplane, and support vector.

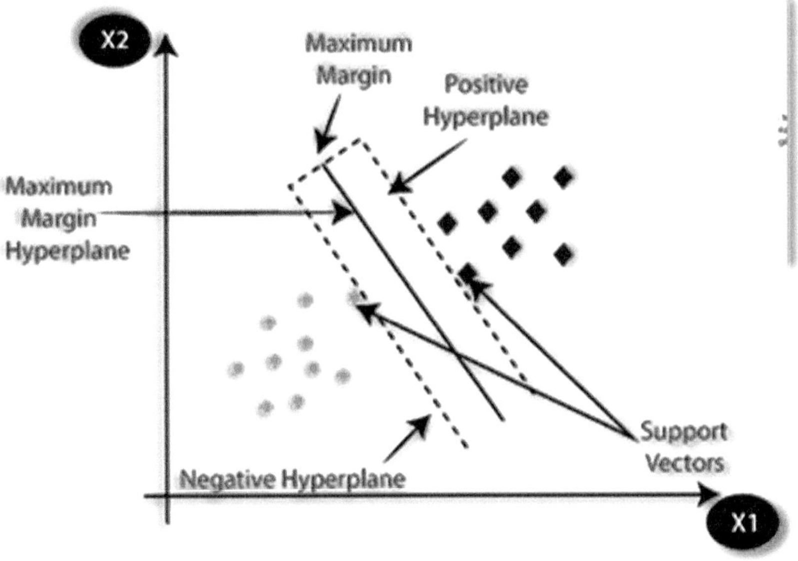

FIGURE 6.3 SVM record factor.

6.3.3 RANDOM FOREST

Random Forest is a multi-decision tree expert system that may be used for analysis and categorization. The findings might include feedback on correctness and parameter significance. A random forest is a filter made up of k tree-structured learners, where k refers to an individually generated factor that combines the cumulative distribution function of random trees.

In this representation, every random tree signifies a piece of judgment for data categorization. Random Forest categorizes and determines the result class in each tree using the power spectrum. The conclusion category inside each tree is merged and determined by the model parameters to obtain the desired classifier [15]. The random forest algorithm works by generating a suitable sample that picks out a data group from the original. In a random forest, each tree may hold a class expectation. The promising class, which evidently accumulates many votes, has been referred to as a model predictor. Figure 6.4 depicts a random forest classification with a large number of trees (preferable).

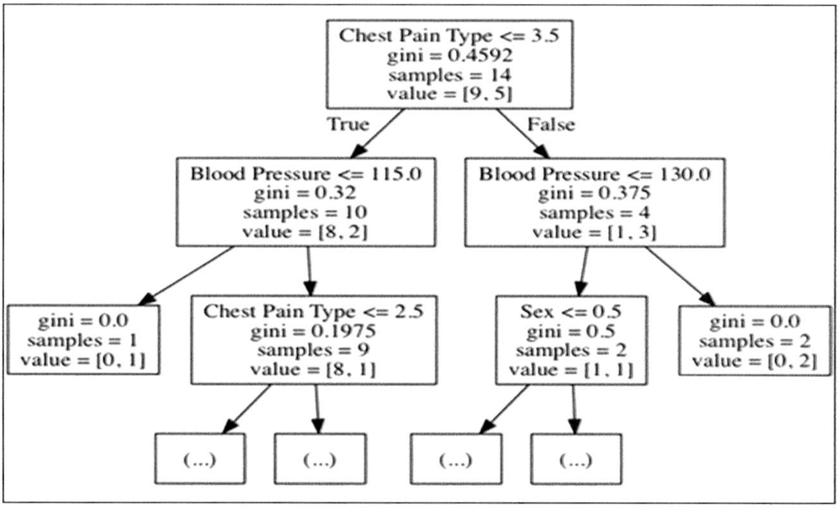

FIGURE 6.4 Random forest schematic.

6.4 EXPERIMENTAL EVALUATION

The work Heart Disease prediction desktop application using supervised learning was created to combat the prevalent disease at an earlier level. ML was developed to battle illness at an earlier stage. In today's competitive world of economic progress, humanity has grown so preoccupied with its well-being that it has forgotten about its own health. According to statistics, 40% of individuals ignore general sickness, eventually developing into dangerous sickness.

6.4.1 ACCURACY COMPARISON MODULE

In this module, the system is going to compare three different (Random Forest, SVM, and KNN) and plot them into the chart. The system is going to check prediction on a bulk amount of different user data and check the accuracy of the algorithm implemented, as shown in Figure 6.5. The accuracy of the Random Forest algorithm was more accurate and good enough when compared to some other algorithms, and it is the best algorithm that fits this kind of problem.

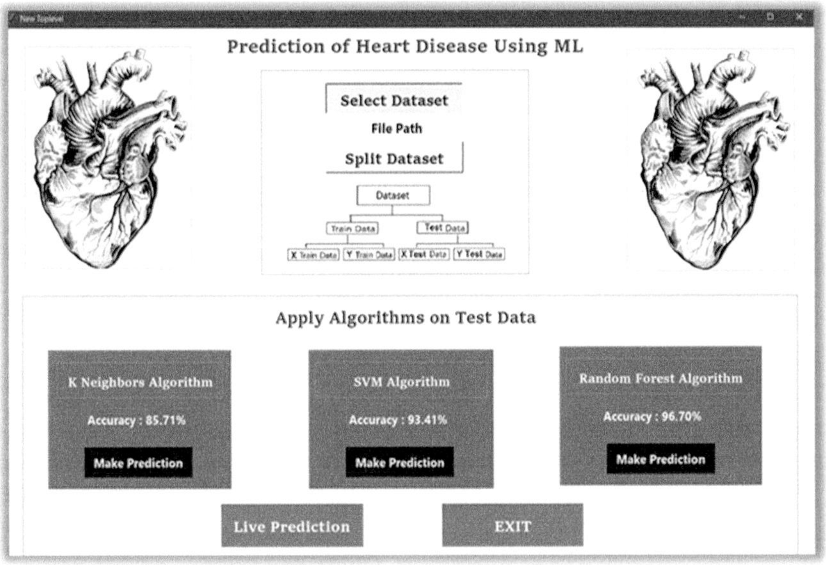

FIGURE 6.5 Prediction of heart disease using ML.

6.4.2 LIVE HEART DISEASE

6.4.2.1 THALASSEMIA

This is a genetic blood disorder (i.e., through genes from parents/grandparents, the children may get affected with this disorder) majorly affected when there is a lack of red blood cells such as hemoglobin. This may affect the passage of oxygen to other body cells leading the person to feel weak, breath shortage and very tired, subsequently causing anemia. Thus persons with thalassemia blood disorder may always suffer from minor or severe anemia. This severe anemia leads to organ damage and may cause the person to death.

6.4.2.2 CHOLESTEROL

Lipid, otherwise body fat, is referred to as Cholesterol. The existence of the total amount of this Cholesterol in the person's blood represents his/her level of serum cholesterol. This serum cholesterol is measured by a unit called milligrams per deciliter (mg/dL). A person with 200 mg/dL or less has been categorized as healthy serum cholesterol, whereas another value signifies the need for a monitor. For healthy Low-density Lipoprotein (LDL), the value is traded as 130 mg/DL or less; for healthy High-density Lipoprotein (HDL), it is 55 mg/DL for women and is 55 mg/Dl for men, and finally, with less than 150 mg/Dl, it is healthy triglycerides. This measure of serum cholesterol plays an essential parameter in recognizing the risk of heart disease.

6.4.3 LIVE HEART DISEASE PREDICTION MODULE

In this module, the user can predict whether he/she has heart disease or not by entering all the attribute values into the interface. Figure 6.6 signifies the user interface design to retrieve patients' personal and medical details. As a part of personal details, Patient Name, Age, Gender has been retrieved. In medical details perspective their Chest Pain type, Major Vessels, Maximum Heart Rate (HR), Thalassemia, Slope of Peak Exercise, Rest Exercise Peak, FBS level, Resting ECG results, Resting Blood Sugar, the status of their regular exercise (Yes/No), Serum Cholesterol level in mg/Dl. All the user inputs have been processed and obtained prediction from the best algorithm that was a random forest where the system achieves 96% accuracy.

6.5 RESULTS AND DISCUSSIONS

After execution, the system summarizes the different results, which are tabulated, and a screenshot of its outputs was presented as figure numbers ranging from 7 to 10. The accuracy of the algorithms SVM, KNN, Random Forest was calculated, and their accuracy outputs are tabulated in Table 6.2.

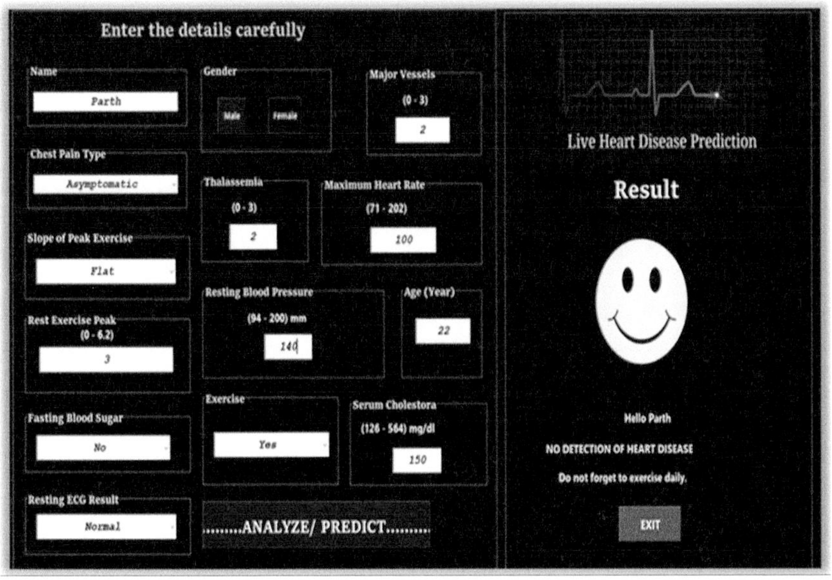

FIGURE 6.6 Live heart disease prediction module.

TABLE 6.2 ML Algorithm Accuracy

Algorithms	Accuracy
Support vector machine	93%
K-nearest neighbor	85%
Random forest	96%

Figure 6.7 represents the analysis of ML algorithm accuracy in percentage, which concludes that the Random Forest algorithm provides the highest accuracy when compared to other ML algorithms such as SVM and KNN.

Figure 6.8 demonstrated the accuracy scores of KNN, which gives the analysis over the accuracy of 85.58% for the training dataset with 85.71% accuracy over the test dataset.

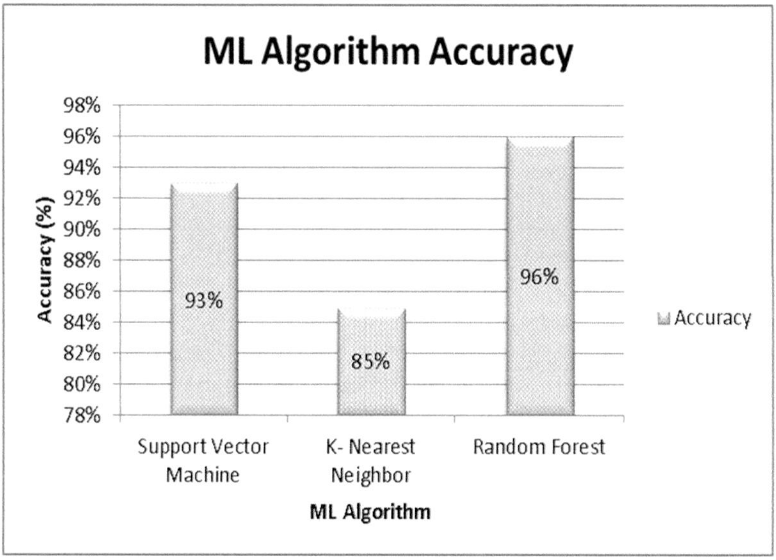

FIGURE 6.7 Analysis of ML algorithm accuracy.

```
 >> Train Result:
********************************

 >> Accuracy : 85.58%
-------------------------------------------------------

 >> CLASSIFICATION REPORT:
                0       1  accuracy  macro avg  weighted avg
precision    0.85    0.86      0.86       0.86          0.86
recall       0.82    0.88      0.86       0.85          0.86
f1-score     0.84    0.87      0.86       0.85          0.86
support    193.00  230.00      0.86     423.00        423.00


 >> Testing Result:
********************************

 >> Accuracy : 85.71%
-------------------------------------------------------

 >> CLASSIFICATION REPORT:

               0       1  accuracy  macro avg  weighted avg
precision   0.88    0.84      0.86       0.86          0.86
recall      0.79    0.91      0.86       0.85          0.86
f1-score    0.83    0.87      0.86       0.85          0.86
support    82.00  100.00      0.86     182.00        182.00
```

FIGURE 6.8 Accuracy scores of KNN.

Figure 6.9 revealed the accuracy scores of SVM, which ensures the improved accuracy of 93.14% for the training dataset with 93.41% accuracy over the test dataset.

```
>> Train Result:
**********************************

>> Accuracy : 93.14%
-------------------------------------------------------------

>> CLASSIFICATION REPORT:
                0        1  accuracy  macro avg  weighted avg
precision    0.94     0.92      0.93       0.93          0.93
recall       0.91     0.95      0.93       0.93          0.93
f1-score     0.92     0.94      0.93       0.93          0.93
support    193.00   230.00      0.93     423.00        423.00


>> Testing Result:
**********************************

>> Accuracy : 93.41%
-------------------------------------------------------------

>> CLASSIFICATION REPORT:

               0        1  accuracy  macro avg  weighted avg
precision   0.94     0.93      0.93       0.93          0.93
recall      0.91     0.95      0.93       0.93          0.93
f1-score    0.93     0.94      0.93       0.93          0.93
support    82.00   100.00      0.93     182.00        182.00
```

FIGURE 6.9 Accuracy score of SVM.

Figure 6.10 publicized the accuracy scores of Random Forest with the achievement of best accuracy of 100% for the training dataset with 96.70% accuracy over the test dataset.

These analysis results conclude the most recommended and promising outcomes of three ML algorithms that may help the researcher take forward further processing with the Random Forest algorithm.

Figure 6.11 depicts the comparison of ML algorithm accuracy pertaining to the train accuracy result versus test accuracy result.

The goal of this research is to build a desktop application to figure out whether or not a patient will acquire heart disease. The utilization of SVM, random forests and K-nearest neighbors (KNNs) in the ML repository was used in this work on supervised classification algorithms for ML. The research was conducted using an Intel Corei7 8th generation processor with a clock speed of up to 2.8 GHz and 16 GB of RAM. To achieve accuracy, data is partitioned (training set and a test set), pre-processed, and controlled classification methods, such as KNN, SVM, and random forests, are used. The accuracy score results of numerous competing training and testing methodologies were recorded using Python programming. The percentage (%) accuracy ratings for several methods are shown in Table 6.3 presents a comparison of the cardiovascular prediction accuracy score in the proposed model with that of others.

```
>> Train Result:
**********************************

>> Accuracy : 100.00%
-------------------------------------------------------

>> CLASSIFICATION REPORT:
                0       1  accuracy  macro avg  weighted avg
precision    1.00    1.00      1.00       1.00          1.00
recall       1.00    1.00      1.00       1.00          1.00
f1-score     1.00    1.00      1.00       1.00          1.00
support    193.00  230.00      1.00     423.00        423.00


>> Testing Result:
**********************************

>> Accuracy : 96.70%
-------------------------------------------------------

>> CLASSIFICATION REPORT:

               0       1  accuracy  macro avg  weighted avg
precision   0.97    0.96      0.97       0.97          0.97
recall      0.95    0.98      0.97       0.97          0.97
f1-score    0.96    0.97      0.97       0.97          0.97
support    82.00  100.00      0.97     182.00        182.00
```

FIGURE 6.10 Accuracy score of random forest.

6.6 CONCLUSION

The ultimate goal is to develop various data mining techniques that would aid in the accurate prediction of heart illness. The system goal is to achieve efficient and reliable prediction desktop applications with fewer characteristics and tests. In this study, the system focuses on 14 essential criteria. KNN, SVM, and random forest classification have been used. The information has been pre-processed and is being used in the model. KNN, SVM, and random forest are the algorithms that provide the greatest results in this scenario. After implementing three approaches, the system discovered that the accuracy in the Random Forest was the highest (96%). Various ML methods, grouping, and association rules, vector machine assistance, and evolutionary algorithms might be used to enhance this research. Because of the limitations of this study, more complex and coupled models are required to improve the accuracy for predicting the occurrence/ existence of early heart disease.

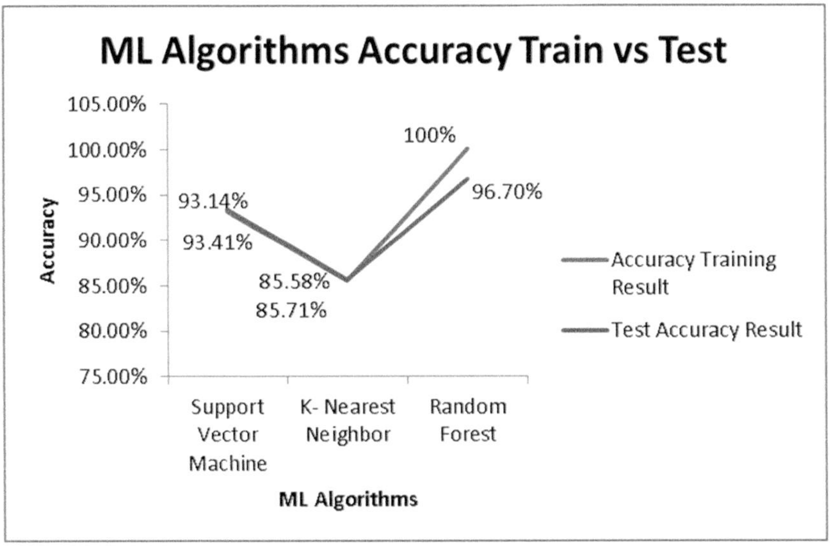

FIGURE 6.11 Analysis of ML algorithm accuracy train vs. test results.

TABLE 6.3 ML Algorithm Accuracy

Algorithms	Accuracy Training Result	Test Accuracy Result
Support vector machine	93.14%	93.41%
K-nearest neighbor	85.58%	85.71%
Random forest	100%	96.70%

6.7 FUTURE WORK

Using the system, getting to know the idea newly educated dataset may be used for an excellent greater correct prediction system. Accounts may be created for every consumer, after which via way of means of referring the beyond desire records of patient's coronary heart circumstance may be monitored to inform if there's any development or if the circumstance has deteriorated. In this research, the system provided a desktop application that's appropriate for real-time coronary heart sicknesses forecast and may be utilized by the patients who've coronary sickness. Unlike many one-of-a-kind systems, it can perform its operation on every screen for prediction. The analysis tool of the system can look ahead to the coronary heart illness through using ML algorithms, and the prediction effects are based absolutely on the coronary heart illness dataset instance. Alternatively, the tool could be very inexpensive; the system used amped pulse sensor and delivered the statistics to the cell through Arduino suite micro-controller. For inspecting the variances and beautifying the alarm if the client's coronary heart price upward thrust than the everyday price of the coronary heart. Nearing display the tool's efficacy were achieved experiments for every monitoring and diagnosis tool. The system ran experiments with some famous algorithms like KNN, SVM, and Random Forest. The check changed into achieved with the holdout check, and the accuracy of the proposed tool changed into 96% achieved with the random forest.

KEYWORDS

- **Arduino suite micro-controller**
- **coronary heart disease**
- **internet of things**
- **machine learning**
- **maximal frequent item set algorithm**
- **naïve Bayesian**

REFERENCES

1. Deeanna, K., (2014). Heart disease: Causes, prevention, and current research. In: *JCCC Honors Journal*.
2. Nabil, A., Costas, S., Mohammad, P., Haik, K., & Majid, S., (2017). Remote health monitoring outcome success prediction using baseline and first month intervention data. In: *IEEE Journal of Biomedical and Health Informatics*.
3. Ponrathi, A., Bradlee, J., & Marcia, J., (2017). Miguel Labrador a mobile health intervention to improve self-care in patients with heart failure: Pilot randomized control trial. In: *JMIR Cardio* (Vol. 1, No. 2).
4. Dhafar, H., Jwan, K. A., Mohamed, I., & Mohammad, B. N., (2017). *The Utilization of Machine Learning Approaches for Medical Data Classification in Annual Conference on New Trends in Information & Communications Technology Applications*.
5. Mai, S., Tim, T., & Rob, S., (2012). Applying k-nearest neighbor in diagnosing heart disease patients. *International Journal of Information and Education Technology, 2*(3).
6. Amudhavel, J., Padmapriya, S., Nandhini, R., Kavipriya, G., Dhavachelvan, P., & Venkatachalapathy, V. S. K., (2016). Recursive ant colony optimization routing in wireless mesh network. *Advances in Intelligent Systems and Computing, 381*, 341–351.
7. Alapatt, B. P., Kavitha, A., & Amudhavel, J., (2017). A novel encryption algorithm for end-to-end secured fiber optic communication. *International Journal of Pure and Applied Mathematics, 117*(19), 269–275.
8. Amudhavel, J., Inbavalli, P., Bhuvaneswari, B., Anandaraj, B., Vengattaraman, T., & Premkumar, K., (2015). An effective analysis on harmony search optimization approaches. *International Journal of Applied Engineering Research, 10*(3), 2035–2038.
9. Amudhavel, J., Kathavate, P., Reddy, L. S. S., & Bhuvaneswari, A. A., (2017). Assessment on authentication mechanisms in a distributed system: A case study. *Journal of Advanced Research in Dynamical and Control Systems, 9*(12), 1437–1448.
10. Sonam, N., & Karandikar, A. M., (2016). Prediction of heart disease using machine learning algorithms. In: *International Journal of Advanced Engineering, Management, and Science (IJAEMS)* (Vol. 2).
11. Sneha, B., & Annadate, M. N., (2018). Supervised machine learning algorithm for detection of cardiac disorders. *Fourth International Conference on Computing Communication Control and Automation (ICCUBEA)*. IEEE.
12. Subhashini, N., & Sathiyamoorthy, E., (2019). A novel recommender system based on FFT with machine learning for predicting and identifying heart diseases. *Neural Computing and Applications* (Vol. 31, pp. 93–102). Springer.
13. Sabrina, M., Claudia, T., Pasquale De, M., Giacomo, F., & Antonio, V., (2019). A machine learning-based approach for predicting the outbreak of cardiovascular diseases in patients on dialysis. *Computer Methods and Programs in Biomedicine* (Vol. 177, pp. 9–15). Elsevier.
14. Keshav, S., & Dilip, K. C., (2020). Heart disease prediction using machine learning and data mining. *International Journal of Recent Technology and Engineering (IJRTE)* (Vol. 9, No.1, pp. 212–219). ISSN: 2277-3878.

15. Tkinter Documentation [online]. URL: https://docs.python.org/3/library/tkinter.html (accessed on 14 September 2022).
16. Saba, B., Zain, S. K., Farhan, H. K., Aitzaz, A., & Khurram, B., (2019). Improving heart disease prediction using feature selection approaches. In: *16th International Bhurban Conference on Applied Sciences and Technology (IBCAST)* (pp. 619–623). IEEE.
17. Martin, G., Monika, S., Anton, G., Ana, P., Matjaz, G., & Gregor, P., (2017). Chronic heart failure detection from heart sounds using a stack of machine learning classifiers. *International Conference on Intelligent Environments (IE)* (pp. 14–19). IEEE.
18. Ashok, K. D., (2018). Performance evaluation of different machine learning techniques for prediction of heart disease. *Neural Computing and Applications* (Vol 29, pp. 685–693). Springer.
19. Qingxue, Z., Dian, Z., & Xuan, Z., (2017). Hear the heart: Daily cardiac health monitoring using ear-ECG and machine learning. In: *8th Annual Ubiquitous Computing, Electronics and Mobile Communication Conference (UEMCON)* (pp. 448–451). IEEE.
20. Samuel, H., Houshang, D., Marina, D. R., Somshubra, M., Fazle, K., Terry, V. H., Kim, E., & Dennis, P. W., (2019). A machine learning based model for out of hospital cardiac arrest outcome classification and sensitivity analysis. *Resuscitation, 138*, 134–140. Elsevier.
21. Chandra, B. G., & Shantharajah, S. P., (2018). An optimized feature selection based on genetic approach and support vector machine for heart disease. *Cluster Computing*, 1–11. Springer.

CHAPTER 7

Coronavirus Outbreak Prediction Analysis and Coronavirus Detection Through X-Ray Using Machine Learning

SUVARNA PAWAR,[1] PRAVIN FUTANE,[1] NILESH UKE,[2] RASIKA BHISE,[1] PRIYANKA MANDAL,[1] TEJAS KHOPADE,[1] and TEJAS RASANE[1]

[1]*School of Computing, MIT ADT University, Pune, E-mail: pawar.suvarna@gmail.com*

[2]*Vishwakarma Institute of Information Technology, Pune, Maharashtra, India*

[3]*Trinity Academy of Engineering, Pune, Maharashtra, India*

ABSTRACT

Corona, generally referred to as COVID-19, is the biggest trouble for society as it hampers human lives in all aspects. This SARS-CoV-2 has caused havoc on people to a large extent all over the world. In real-time, it is hampered in a cumulative manner as it is growing at a faster rate day by day. Machine learning (ML) might be used to track the disease, predict its progress, and design tactics and legislation to control it. Predictive analysis has become a critical component for future prediction as the science of ML has progressed. As the COVID-19 pandemic unfolds, it would be beneficial to forecast the number of positive cases in the future so that stronger measures and control may be implemented. In this work, two supervised learning models experimented to anticipate the future using the COVID-19 time-series dataset during the second pandemic wave from January 20 to

Computational Health Informatics for Biomedical Applications. Aryan Chaudhary and Sardar M. N. Islam (Naz) (Eds.)

March 20. To look at the performance for prediction of COVID-19 disease, the dataset used is completely linear; hence Support vector Regression and Linear regression (LR) are perfectly analyzing and predicting the situations in early times. This leads to the eventual spread of coronavirus disease (COVID-19) to some extent. In the first scenario of COVID-19, we use imaging procedures as X-rays for the chest to know how this disease has damaged the lungs. This makes doctors understand the criticalness grading of COVID-19. This is to employ radiological imaging to emphasize the results of chest X-rays. According to new research, COVID-19 has been discovered in patients with aberrant chest X-ray results. There is a slew of reports on the subject that employ ML techniques such as support vector machine (SVM), LR, etc. Results obtained from SVM and LR are compared with Support Vector Regression for different categories like affected cases. Death cases and recovered cases. While detecting and predicting coronavirus disease, the accuracy achieved is 97.8% using chest X-rays.

7.1 INTRODUCTION

COVID-19 is a highly contagious virus that causes severe chronic respiratory infections (SARSCoV2). Since when it was first discovered in Wuhan, China, in December 2019, it has spread all over the world and caused a pandemic. Cough, fever, malaise, shortness of breath, loss of taste and smell are common symptoms. The World Health Organization (WHO) called the disease a public health emergency of international concern in January and a pandemic in March [2]. COVID-19 has a major impact on economics, education, medical care, logistics, and people's mental health. The pandemic has had an impact on the global economy, with major events being canceled or postponed as a result. World industrial organizations estimate that the outbreak has reduced trade by 13–32%. Many experts assume that it will take 10 years for the economy to return to its original state [3]. According to the WHO, COVID-19 has a significant impact on the health sector of non-communicable diseases such as cancer and Alzheimer's disease. No cures or vaccines for this disease are currently in place, which is a major challenge and a major concern (Figures 7.1–7.4).

ML algorithms aid in the discovery of epidemic tendencies in the context of massive epidemic data, allowing for early intervention to prevent the virus from spreading. By exploiting real-time data, ML models are used in this study to investigate everyday behavior as well as predict the future reachability of the COVID-2019 across the country. COVID-19's prevalence rate is quickly increasing day by day. ML might be used to analyze the disease's growth, predict its progress, and design tactics and legislation to control it. Using a mathematical model, this study analyzes and forecasts the disease's

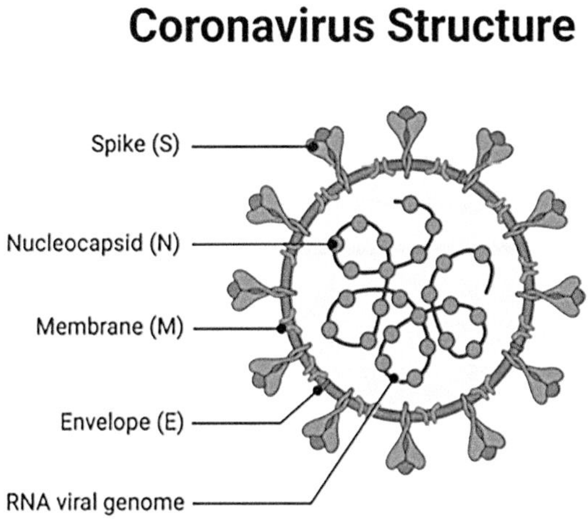

FIGURE 7.1 Image depicts the COVID-19 virus's usual structure. Created with Biorender.com

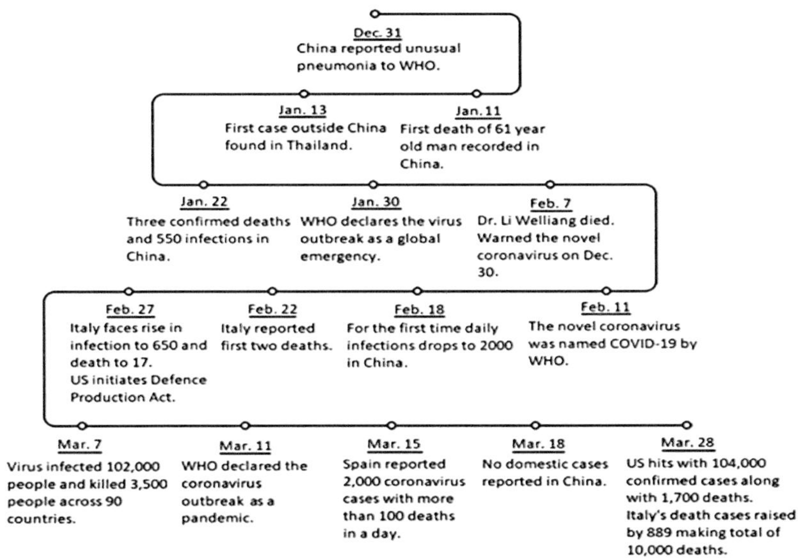

FIGURE 7.2 A timeline of COVID's occurrences around the country COVID-19 epidemic status in countries with the largest number of cases.

Source: Reprinted from Ref. [9]. https://creativecommons.org/licenses/by-nc-nd/4.0/

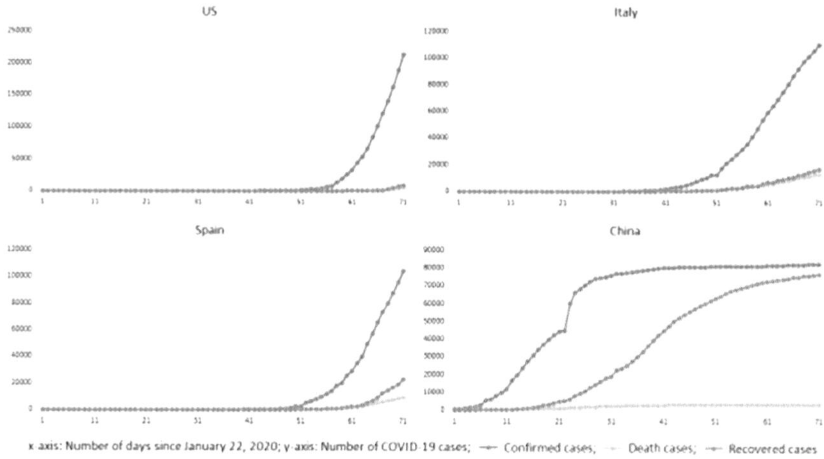

FIGURE 7.3 Since January 22, 2020, shows the COVID-19 epidemic status in countries with the largest number of confirmed cases.

Source: Reprinted from Ref. [9]. https://creativecommons.org/licenses/by-nc-nd/4.0/

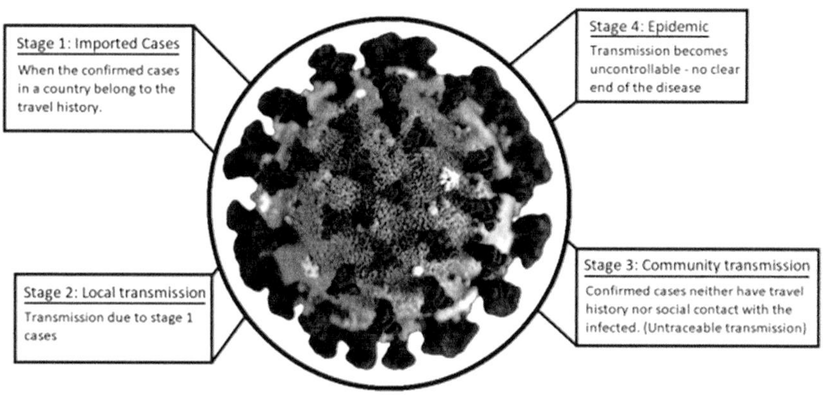

FIGURE 7.4 COVID-19: all stages of transmission.

Source: Reprinted from Ref. [9]. https://creativecommons.org/licenses/by-nc-nd/4.0/

progress. COVID-19's possible danger to countries throughout the world has been forecasted using an improved approach based on ML. We create a model that predicts the virus's spread across different countries and areas. We may use the SVM Algorithm and Linear Regression Model in Python to assess the impact it has had thus far and to analyze the Coronavirus outbreak across various regions using charts and graphs, as well in order to calculate the number of confirmed cases in the future. When enabled, computed tomography (CT) of the thorax has been proved to be a significant supplementary tool in the diagnosis and monitoring of thoracic cancers. COVID-19 was proven to be the most effective when compared to other procedures. Like temperature/breathing symptom monitoring and molecular testing is the current standard method using Nasopharyngeal swab or spit. CT has been adopted as the primary imaging modality in several nations, including China, the Netherlands, Russia, and others, from the initial diagnosis to treatment. Traditional radiographic (x-ray) chest imaging is used heavily in other nations, such as the United States and Denmark, as well as developing countries (Southeast Asia and Africa). We present a deep learning-based technique for identifying COVID-19 infection in chest X-ray pictures in this chapter. For categorizing three types of pneumonia: bacterial, viral, and COVID-19 pneumonia, a deep convolutional neural network (CNN) model is shown. CoroNet will help us in distinguishing between three types of respiratory infections as well as how COVID-19 differs from other infections. Doctors in developing countries would benefit greatly from a model that can diagnose COVID-19 infection from chest radiography images. Positive cases are prioritized, quantified, and tracked. Although this approach does not entirely replace current testing techniques, it can help to minimize the number of situations that require immediate testing or additional expert assessment.

7.2 EXPERIMENTAL METHODS AND MATERIALS

7.2.1 PROPOSED SYSTEM

There are two steps in the proposed Model: assessment and training. The data has been pre-processed before training by deleting null values and inconsequential variables. Throughout the training phase, the Model is trained and tested for the goal of forecasting (Figure 7.5).

7.2.2 DATA PRE-PROCESSING AND METHODS

For transforming the comma-separated values (CSV) file into a Data frame, the Pandas package is used. Filtering data is a process of manipulating and modifying data to match the requirements. Separate data frames, such as anticipated dates in the future and distinct nations, and so on, are created here. The data frame's Lat, Long, and Province/State columns have been eliminated because they have a lower predictive value. For future prediction, we multiply the number of required future dates by the number of actual dates in the data frame. This is accomplished by utilizing the Python date or time library. X-rays were obtained from a variety of sources for the studies. To begin, the GitHub repository was searched for datasets that were linked. Cohen [4] provided a set of X-ray images that were chosen (Figure 7.6).

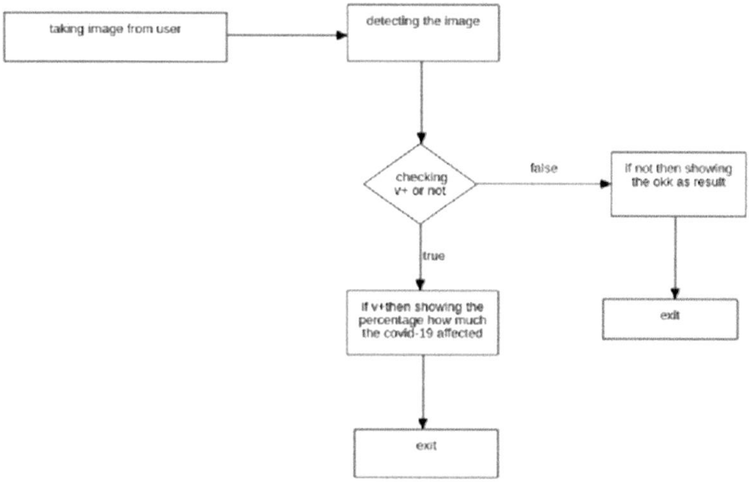

FIGURE 7.5 COVID-19 detection flowchart utilizing X-ray images.

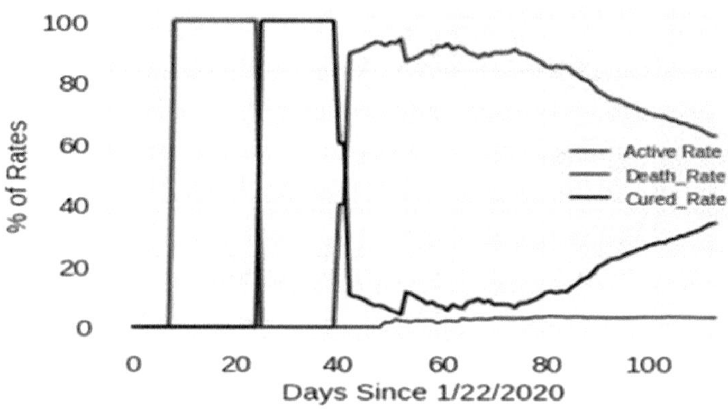

FIGURE 7.6 Active, death, and cured rate.

Source: Reprinted from Ref. [10]. Open access.

7.2.2.1 SUPPORT VECTOR MACHINE (SVM)

The SVM is a supervised learning model that learns by analyzing labeled data. It produces a set of labeled input-output mapping functions as well as extra data. The SVM can be used as a classification or regression approach.

7.2.2.2 SVM CLASSIFICATION (SVC)

The nonlinear classification kernel function creates additional unique input space from the original input space. Hyperplanes with the highest margins are constructed. Only a portion of data within the class borders is used in the final Model.

7.2.2.3 SVM REGRESSION (SVR)

It's a form of SVM that can handle both nonlinear and linear regressions (LRs). Support vector Regression requires X and Y training data that cover the area of interest and are escorted by domain-specific solutions. The Support Vector Regression is an estimate of the function we use to generate the training set in order to reinforce certain data using the Support Vector Method. It's included in the Scikit Learn Python package. The Support Vector Regression model ignores any training data that is too similar to the estimated parameters.

7.2.2.4 LINEAR REGRESSION MODEL

LR is a supervised ML approach. LR is used to anticipate the value of a dependent variable (y) based on the value of an independent variable (x). A linear relationship between x (input) and y (output) is identified as a result of this regression approach (output). Simply said, the LR model determines the best fit line overall input data (labels and features). The LR calculates the smallest distance between all data points to construct a regression line. A combination of independent variables is used to predict a numerical outcome in this sort of predictive analysis. The goal is to discover a connection between two variables. By matching the observed data with a linear equation Importing data using the LR model Python module Sklearn linear models.

7.2.2.5 CONVOLUTIONAL NEURAL NETWORKS

A CNN, sometimes called CNN, is a deep learning (DL) neural network. In a sense, consider CNN to be a ML system capable of taking an input and assigning relevance to different aspects in an image, and distinguishing between them CNN's job is to compress images into a format that is easier to analyze while keeping key qualities for accurate prediction. When we need to scale the approach to huge datasets, this is crucial.

7.2.3 IMPLEMENTATION

We reported the results of the SVM and the LR model in this part. The models are trained on timeseries data from 22 January 2020 to 15 March 2020.

7.2.3.1 SVR PREDICTION

The SVM is a prominent Supervised Learning approach that may be used to resolve classification and regression problems. However, it is largely used in ML to solve classification problems. The goal of the SVM algorithm is to determine the best line or decision boundary for classifying n-dimensional space. By categorizing data points into classes, further data points may be conveniently placed in the correct category in the future. The best choice boundary is referred to as a hyperplane. SVM is used to choose the extreme points/vectors that will help form the hyperplane. To characterize these extreme scenarios, support vectors are utilized, and the method is called a SVM (Figure 7.7).

7.2.3.2 LINEAR REGRESSION

We plot a graph connecting the variables that best suit the given data points in Regression. The ML model can make data-related predictions. "Regression" is defined as "a line or curve that passes through all the data points on a target-predictor graph with the shortest vertical distance between the data points and the regression line." Prediction, forecasting, time series modeling, and identifying the causal-effect link between variables are all common applications (Figure 7.8).

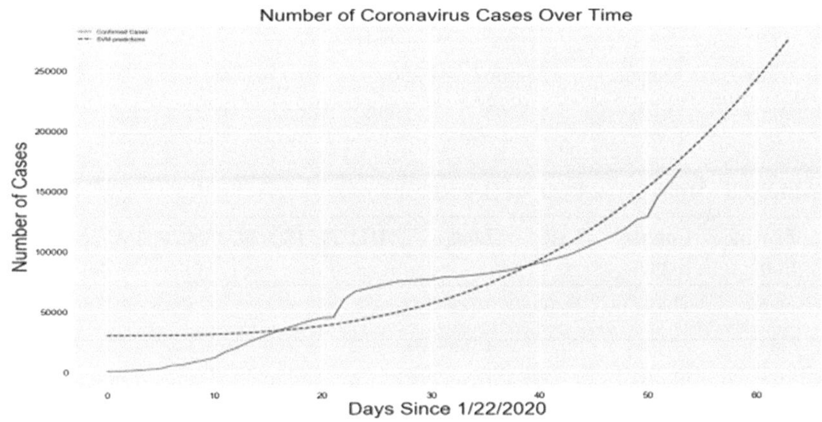

FIGURE 7.7 SVM prediction.

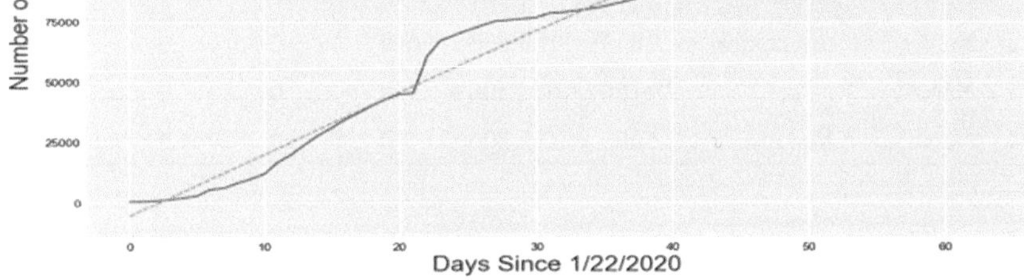

FIGURE 7.8 Linear regression prediction.

7.3 RESULTS AND DISCUSSION

7.3.1 CORONAVIRUS PREDICTION ANALYSIS RESULT (Tables 7.1–7.3; Figures 7.9 and 7.10)

TABLE 7.1 Time Series Data Set Confirmed Cases

	Province/ State	Country/ Region	Lat	Long	1/22/20	1/23/20	1/24/20	1/25/20
0	NaN	Thailand	15.0000	101.0000	0	0	0	0
1	NaN	Japan	36.0000	138.0000	0	0	0	0
2	NaN	Singapore	1.2833	103.8333	0	0	0	0
3	NaN	Nepal	28.1667	84.2500	0	0	0	0
4	NaN	Malaysia	2.5000	112.5000	0	0	0	0

TABLE 7.2 Time Series Data Set Death Cases

	Province/ State	Country/ Region	Lat	Long	1/22/20	1/23/20	1/24/20	1/25/20	1/26/20
0	NaN	Thailand	15.0000	101.0000	2	3	5	7	8
1	NaN	Japan	36.0000	138.0000	2	1	2	2	4
2	NaN	Singapore	1.2833	103.8333	0	1	3	3	4
3	NaN	Nepal	28.1667	84.2500	0	0	0	1	1
4	NaN	Malaysia	2.5000	112.5000	0	0	0	3	4

TABLE 7.3 Time Series Data Set Recovered Cases

	Province/ State	Country/ Region	Lat	Long	1/22/20	1/23/20	1/24/20	1/25/20
0	NaN	Thailand	15.0000	101.0000	0	0	0	0
1	NaN	Japan	36.0000	138.0000	0	0	0	0
2	NaN	Singapore	1.2833	103.8333	0	0	0	0
3	NaN	Nepal	28.1667	84.2500	0	0	0	0
4	NaN	Malaysia	2.5000	112.5000	0	0	0	0

7.3.2 SVR PREDICTION

The Support Vector Regression model was trained and tested using varying ratios of train and test datasets. As noted previously, the Model is trained using different train-to-test dataset proportions, and its performance is measured for each. The Model's performance may be seen by graphing the anticipated and actual numbers (Figure 7.11).

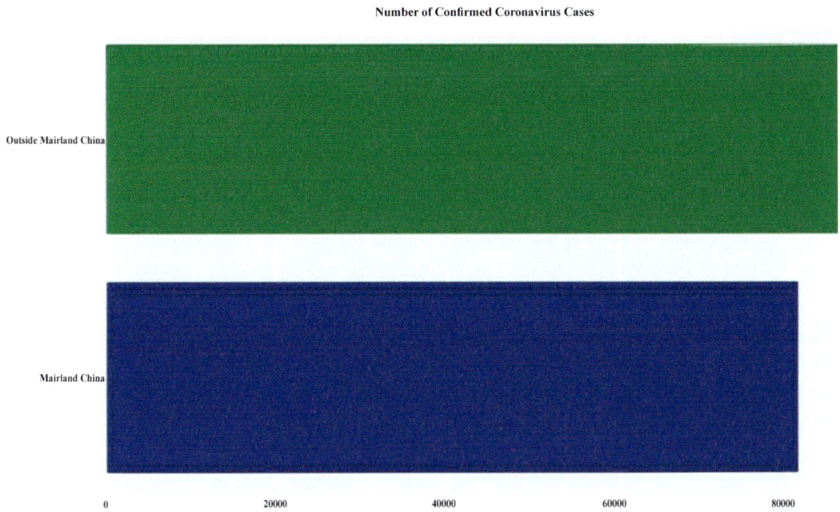

FIGURE 7.9 Number of confirmed coronavirus cases.

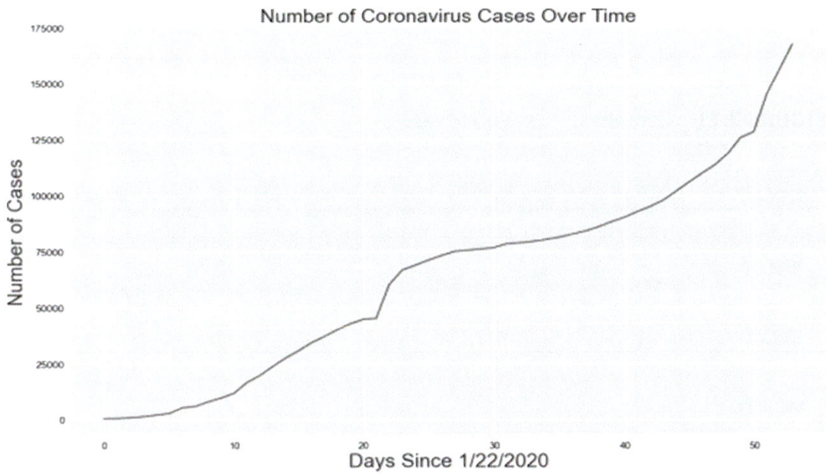

FIGURE 7.10 Total number of coronavirus cases over time.

7.3.3 LINEAR REGRESSION

As previously stated, this Model is trained with various ratios of test datasets, and its performance is assessed for each. By graphing the anticipated and actual numbers, the Model's performance may be visualized.

Visualizing that the LR model's prediction is quite straightforward. Which is why this can be demonstrated by analyzing the Model's performance (Figures 7.12–7.18).

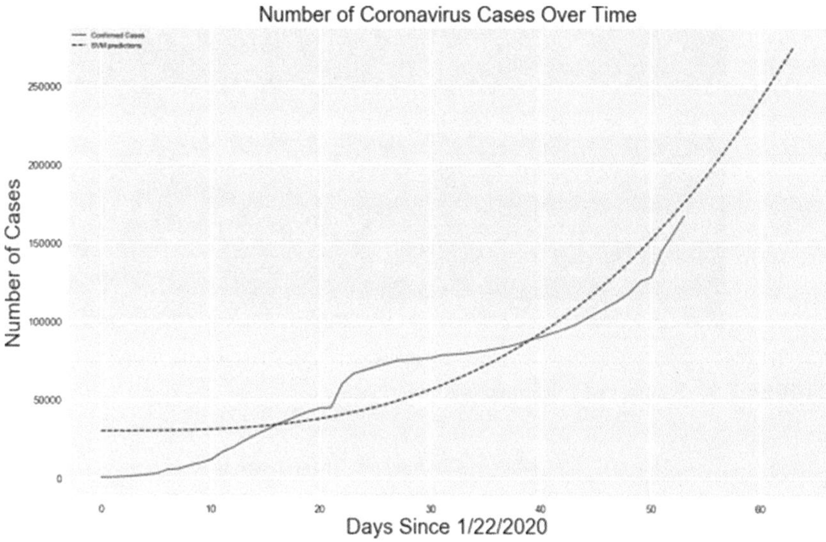

FIGURE 7.11 Confirmed vs. predicted cases.

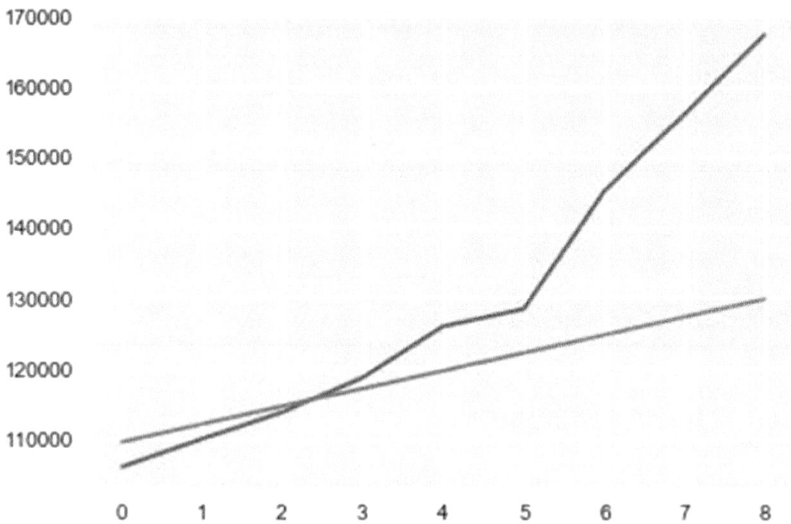

FIGURE 7.12 Using linear regression model to make predictions.

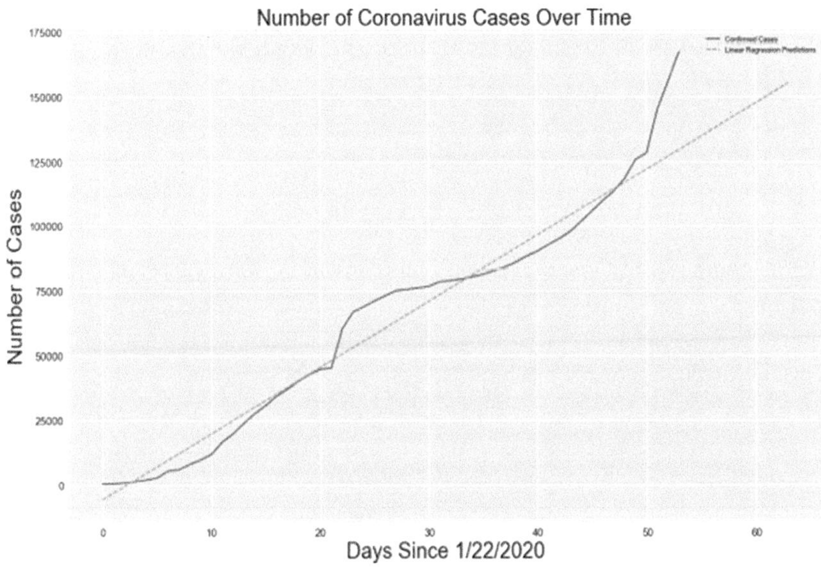

FIGURE 7.13 Number of confirmed cases linear regression prediction.

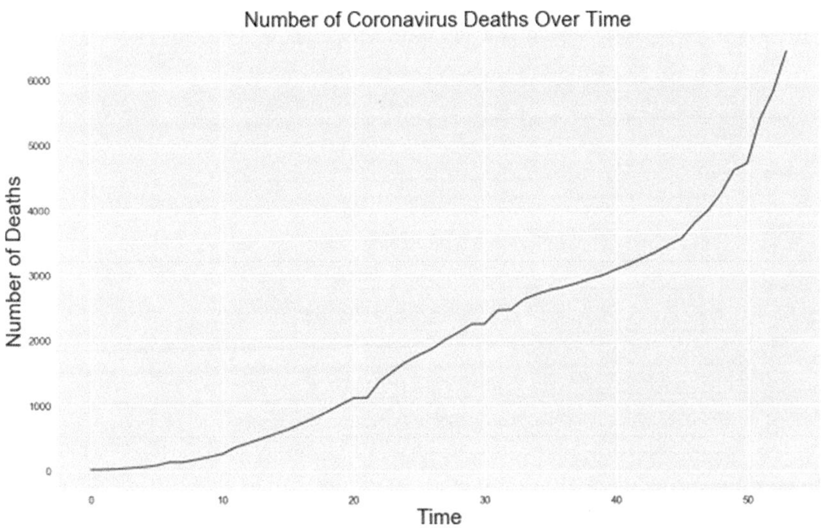

FIGURE 7.14 Total deaths over time.

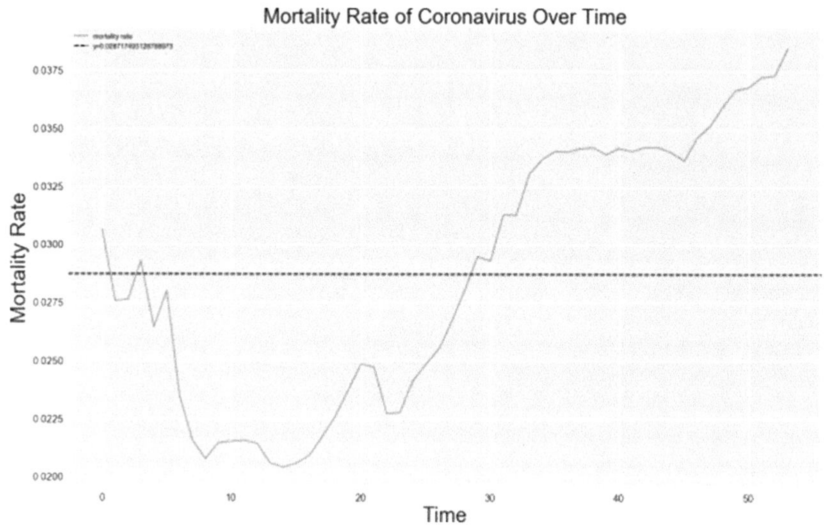

FIGURE 7.15 Mortality rate of coronavirus over time.

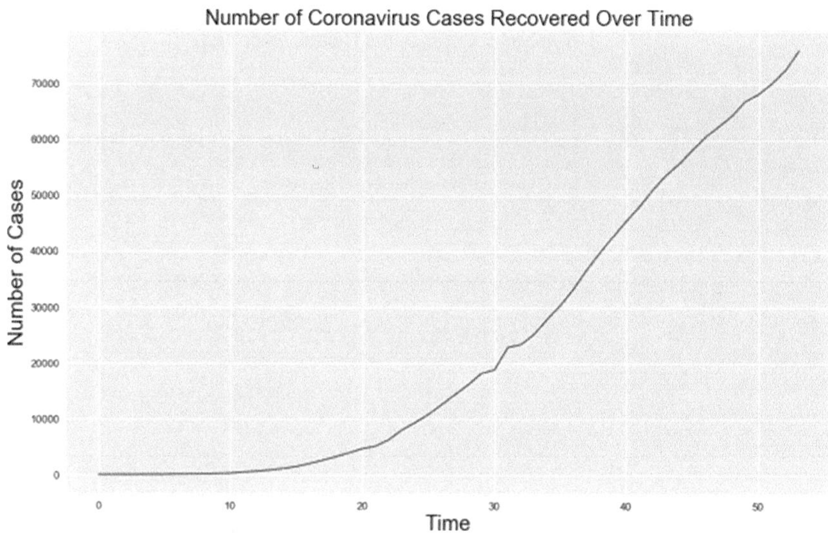

FIGURE 7.16 Coronavirus cases recovered over time.

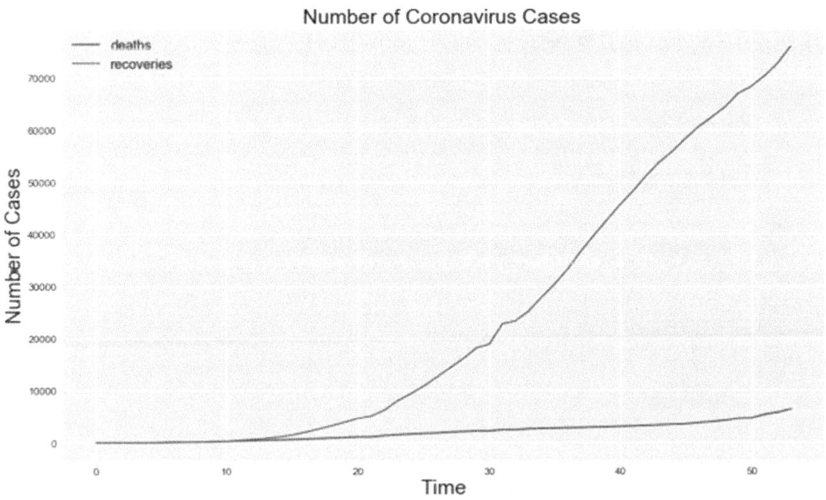

FIGURE 7.17 Number of coronavirus cases recovered vs. the number of deaths.

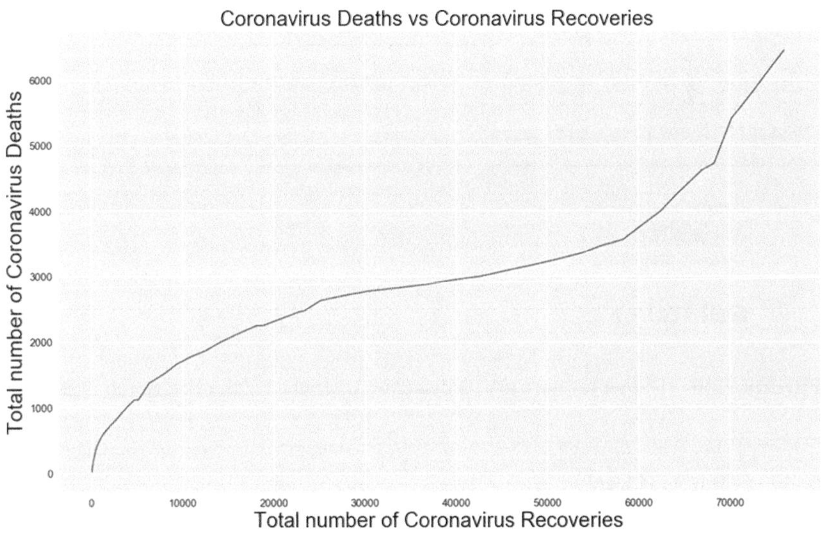

FIGURE 7.18 Coronavirus deaths vs. recoveries.

7.3.4 COVID-19 DETECTION BY USING X-RAY IMAGE RESULTS (Figures 7.19 and 7.20)

FIGURE 7.19 Output for the COVID detection when the result of detection is normal.

FIGURE 7.20 Output for the COVID detection when the result of detection is positive.

7.4 CONCLUSION

In this research chapter, the data of COVID-19, the SVM and LR model were tested for time series analysis of prediction of the expected active, death, and recovered rate. According to current research, the number of COVID-19 pandemic cases is increasing on a daily basis in a linear pattern. This may be seen as a linear growth in the number of incidents, indicating that it will be a major threat in the upcoming years. As a result, we must take all required forethought and measures to prevent the epidemic from spreading further. In this system, we have projected a DL-based Model to detect the COVID-19 cases by using X-ray images.

KEYWORDS

- **comma-separated values**
- **convolutional neural network**
- **COVID-19**
- **linear growth**
- **machine learning**
- **World Health Organization**

REFERENCES

1. Ashish, U. M., Rakshith, A., & Siddesha, S., (2020). *Prediction of COVID-19 Pandemic Based on Regression.*
2. Amit, K. G., Vijander, S., Priya, M., & Travieso-Gonzalez, C. M., (2019). Prediction of COVID-19 pandemic measuring criteria using support vector machine, prophet and linear regression models in indian scenario. *Journal of Interdisciplinary Mathematics, 24*(1), 92–97.
3. Yan, H., Ting, X., Hongping, H., Peng, W., & Yanping, B., (2019). *Prediction and Analysis of Corona Virus Disease, 15*, 3–6.
4. Tulin, O., Muhammed, T., Eylul, A. Y., Ulas, B. B., Ozal, Y., & Rajendra, A., (2019). *Automated Detection of COVID-19 Cases Using Deep Neural Networks with X-ray Images.* 2–9.
5. Ioannis D. A., & Tzani, A. M., (2019). *COVID-19: Automatic Detection from X-ray Images Utilizing Transfer Learning with Convolutional Neural Networks*, 636–637.
6. Asif, I. K., Junaid, L. S., & Mohammad, M. B., (2019). *CoroNet: A Deep Neural Network for Detection and Diagnosis of COVID-19 from Chest X-ray Images.*
7. Hayit, G., Raúl, S. J. E., Wiro, J. N., Eliot, S., & Mads, N., (2019). *Position Paper on COVID-19 Imaging and AI: From the Clinical Needs and Technological Challenges to Initial AI Solutions at the Lab and National Level Towards a New Era for AI in Healthcare*, 3–5.
8. https://www.kaggle.com/sudalairajkumar/novel-corona-virus-2019-dataset (accessed on 14 September 2022).

9. Punn, N. S, Sonbhadra, S. K., Agarwal, S., (2020). *COVID-19 Epidemic Analysis using Machine Learning and Deep Learning Algorithms.* https://www.medrxiv.org/content/10.1101/2020.04.08.20057679v2.full.pdf (accessed on 03 December 2022).
10. Gupta, A. K., Singh, V., Mathur, P., & Travieso-Gonzalez, C. M., (2021) Prediction of COVID-19 pandemic measuring criteria using support vector machine, prophet and linear regression models in Indian scenario. *Journal of Interdisciplinary Mathematics,* 24(1), 89–108. doi: 10.1080/09720502.2020.1833458

CHAPTER 8

Numerical Analysis of Bioheat Transfer in Thermal Medicine

N. MANJUNATH

Department of Mechanical Engineering, College of Engineering Trivandrum, Kerala, India, E-mail: iammanjunathn@gmail.com

ABSTRACT

Computerized simulations of various treatment methods are beginning to demonstrate their true potential in clinical situations. This is occurring in all of the traditional healthcare and treatment modalities. The mathematical models are known to be generally useful in invoking mathematical rigor and precision in the formulation and evaluation of problems. Numerical simulations provide a theoretical framework that can be used to substantiate, evaluate, and interpret experimental results. In this chapter, a detailed study on numerical simulations of the different procedures in the field of 'thermal medicine' is analyzed and compared.

Thermal medicine is the method of application or alteration of human body temperature in order to cure certain ailments or to condition the body for other treatments. The different treatment methods in thermal medicine include; hyperthermia, thermal ablation, cryosurgery, and heat-activated drug delivery. All of these methods are actively applied in cancer treatment. Even though these methods have been in use for decades, the scope of optimization of these methods utilizing computer-aided designing/ computer-aided engineering (CAD/CAE) tools are now being explored by various researchers. Such research has already begun to show their

Computational Health Informatics for Biomedical Applications. Aryan Chaudhary and Sardar M. N. Islam (Naz) (Eds.)

importance in the prediction of the outcome of experiments or treatments (in terms of temperature distributions for the applications).

Heat transfer is a major area of study under mechanical engineering and also the one with the widest range of tools and techniques for the purpose of analysis as well as experimentation. The mathematical models of thermal medicine are founded on *bio-heat transfer* models, which can exploit most of the existing techniques in heat transfer for computerized simulations.

This work aims at verifying the reliability of the application of various bioheat equations for the analysis of heat transfer problems occurring in the biological tissues during cancer treatment methods. Direct experiments in humans are against our ethical concerns, so in such a condition, the numerical analysis of treatment procedures can be a promising technique to find out the apt and effective treatment for a specific type of cancer.

8.1 INTRODUCTION

Thermal medicine is a discipline of the study of thermal treatments and their biological, pharmacological, and therapeutic effects. The importance and benefits of the thermal cures known and practiced from the ancient times, the ancient men discovered the importance heat and water in the earlier times itself which created the major civilization in the comparatively hotter areas and near to water bodies.

Thermalism is believed to born in ancient Greece, but the Romans were the ones who practiced and spread its science. The aqueducts and Thermae of ancient Rome propounded the advantages of thermalism which reached far up to Asia. The sanus per aqua (SPA) we see now is short for "SPA," developed out of the Roman Thermae. But as humankind modernized, they learned the merits of internal thermal therapies as well as external thermal therapies. This chapter is about the discovery and different numerical simulations utilized to improve the different kinds of thermal therapies now exist [1].

8.2 EVOLUTION OF THERMAL MEDICINE

The major application, and therefore, the hotbed of research of hyperthermia was in cancer treatment, that too specifically in tumorous cancers.

Oncological hyperthermia can be believed to begin last decades of the 18th century. Even though it cannot be compared to hyperthermia as we know it today, the first evidence of thermal effects on cancer cure was discovered when someone found out fever aids in the treatment. In 1779, Kizowitz found that high fever due to malaria can retard or even inhibit tumor growth. It started in France, but soon many other medical practitioners in the neighboring European countries began to ponder the scope of induced fever for cancer therapy.

Then the major events that happened in the second half of the 19th century by Busch (1866) recorded a complete reduction of face sarcoma with two subsequent years of disease-free survival. In 1887, Bruns confirmed complete diminution of melanoma of a patient after treating with a temperature of 40°C and disease-free survival of eight-years. Till the beginning of the second half of the 20th century, this treatment was called *febrile therapy*. Hyperthermia was only considered as a body reaction and not as a separate treatment modality [2].

The simultaneous discovery of X-rays and radium respectively gave newer and interesting tools to cancer research. Brachytherapy and radiotherapy diminished the interest in febrile therapy. Even William B Coley, who developed a mixed bacterial vaccine known as *Coley toxin* to induce fever for the febrile therapy, didn't approve the increased temperature as the important mechanism in tumor remission [3].

Now hyperthermia (increased temperatures) is used along with radiation and chemotherapies. The different modalities of thermal Medicine now exist discussed in the next section.

8.3 DIFFERENT PRACTICES IN THERMAL MEDICINE

Thermal medicine, in a broad sense, is limited to two categories viz; hyperthermia and cryobiology. Hyperthermia is a set of wide and highly diverse treatment modalities but is mostly sidelined as adjuvant therapy in most of its applications. However, hyperthermia means an increased temperature beyond the thresholds of thermoregulation. Hyperthermia can occur due to many reasons such as heat stress, heat fatigue, heat syncope, heat exhaustion, etc. But in this chapter, by hyperthermia, we mean therapeutic hyperthermia. While cryobiology is a less diverse but highly used branch in medicine. Also, there is another category of thermal medicine, hypothermia, that involves decreasing the local or whole-body temperature.

8.3.1 HYPERTHERMIA THERAPY

Hyperthermia therapy (HT), also known as thermal therapy or thermotherapy, is considered as next in line with surgery, chemotherapy, and radiation therapy in the treatment of cancer. Hyperthermia is used in either combination with chemo or radiotherapy or as a standalone procedure. HT is the method of naturally or artificially elevating the temperature of the whole body or a particular region above its normal limits set by the thermoregulatory mechanism [4, 5]. In oncology, hyperthermia finds a significant role since the cancerous cells or tumors can be destroyed, or its growth can be prevented. Hyperthermia temperature is mostly considered as the temperature between 40 to 48°C, the temperatures above 50°C causes coagulation of proteins, thermal ablation occurs in the range of 60 to 90°C and above 120°C charring of the tissues happens [6].

Many researches have shown that elevated temperatures can destroy the cancer cells and can make them susceptible to other therapies modalities (radiation, chemo-, and immunotherapies) [7, 9]. Hyperthermia encourages reoxygenation of cells, improves the delivery and efficacy of chemotherapeutics [13] and enhances the immune reactions against thermotolerance developed by heat shock proteins (HSPs) [14]. Thermotolerance is the phenomenon of buildup resistance to the additional heat stress, up to 72 hours of the inducement of thermal shock in tissues [10]. Thus, further heating during this period will be ineffective, but thermotolerance will not affect the radiosensitization [15].

All the hyperthermia procedures come under the categorization of minimally invasive or non-invasive treatments requiring minimal hospitalization and bed rest which makes it more attractive in terms of lesser expenses [16] The hyperthermia therapies are grouped into three; local, regional, and whole-body hyperthermia (WBH) according to the location, depth, and stage of the tumor [14].

The local hyperthermia is of superficial, intracavital, intraluminal, and interstitial types and is applied for local tumors of sizes of 3 cm up to 6 cm. Superficial applications refer to those applications where is the location of the tumors are on the skin or just below it. Antennas or electrodes emitting the selected type of energy are made to come in contact with an area of interest. Needles or tiny tubes are inserted into the tissues through with thermometers or thermocouples for temperature measurements are positioned [17].

The intracavital and intraluminal types of hyperthermia utilize the natural openings of the human body to reach the locally advanced or deep-seated tumors. The interstitial type local hyperthermia is applied for the class of tumors which is fit to be treated with brachytherapy. Interstitial hyperthermia is similar to brachytherapy involves an array of applicators or electrodes inserted into tissues to patients under anesthesia. Interstitial hyperthermia is best suited for lesion size < 5 cm. It uses microwaves, radiofrequency, or ultrasound (US) waves focused on the tissue volume under treatment.

Regional hyperthermia is also known as part-body hyperthermia and is concerned with heating large parts of the body. It is utilized in advanced tumors treatments involving larger body parts such as the abdomen (prostate, cervical, bladder, ovarian, etc.), major pelvis or thighs (soft tissue sarcomas). The three major methods in regional hyperthermia involve heating the deep-seated tumors with external applicators, thermal perfusion for cancers in limbs and for larger portions continuous hyperthermic peritoneal perfusion (CHPP). In regional hyperthermia also, arrays of antenna or electrodes are placed in the area of interest. The antennas then release microwave or radiofrequency, or US energy on the lesion. The temperature rise developed is limited to the range of 41–42°C. The specific absorption rate (SAR) of the tissues causes the absorbed electromagnetic energy to be converted to thermal energy [4, 11]. Regional hyperthermia is also combined with heat-activated drug delivery, and for this case, the temperature range should be slightly lower.

WBH is the method of inducing a slightly higher temperature than that of the thermoregulation temperature to the whole body. The patients suffering from soft tissue sarcoma, melanoma, ovarian cancer, etc. (metastatic diseases) are chosen for WBH. Distributed cancerous cells are sensitized to drugs or destroyed with an elevated temperature of the whole body. The temperature ranges of 41.8–42.2°C are induced using thermal chambers, hot water blankets or infrared radiators. The WBH is coupled with general anesthesia for extreme WBH and moderate WBH. But the complications involved in WBH are the chances of thermal stress to the heart, lungs, brain, etc. [15]. The WBH is found effective for neurological diseases or even for psychological disorders like chronic depression.

And according to another classification based on the temperature achieved due to a particular hyperthermia modality in the tissue, the methods can be classified into; adjuvant *hyperthermia, real hyperthermia,* and *ablative hyperthermia.*

Adjuvant hyperthermia is said to be achieved when the temperatures of the tissues are raised to the range of 38–41°C, mostly coupled with radiation and chemotherapy therapies to yield better results and accelerate the process at the same time without side effects. In 'real hyperthermia,' the temperature of the local tissues must reach 43–46°C. Irreversible cellular damage is attained at this range without necrosis of the tissue. The 'ablative hyperthermia' treatment is carried out at the temperatures range of 50–100°C and should be achieved in less time so as to cause the complete necrosis of targeted tissues. Thermo-ablative techniques are performed using typical forms of electromagnetic energy, such as radiofrequency (Radio Frequency Ablation (RFA)), microwaves (Microwave Ablation, (MWA)), acoustic energy and laser energy. By acoustic energy here, the frequency range fall mostly in the US region and High-Intensity Focused Ultrasound (HIFU) is the most successful method employed.

8.3.2 HYPOTHERMIA

Hypothermia is the lowering of the local or whole-body temperature beneath the thresholds of the normal body temperature. Since all the bodily functions are biochemical reactions, it requires a specified range of temperature for the reactions to take place. Hyperthermia can be classified into three; *primary, secondary,* and *therapeutic* hypothermia. The primary hypothermia is the chilling caused due by the lowering of body temperature below 36°C; perioperative Hypothermia is an example when some patients start to shiver when the anesthesia effects wear off after a surgery [18]. Secondary Hypothermia is the decrease in body temperature due to some underlying medical conditions such as hypopituitarism, hypoglycemia, vasodilatation, hypoadrenalism, erythrodermas, hypothyroidism, Parkinson's disease, Wernicke's disease. Therapeutic hypothermia ranges from the application of cold packs for burns or blunt traumas up to post-resuscitation therapy.

8.3.3 CRYOBIOLOGY

Cryobiology deals with extremely low temperatures, which mean those causing phase transformations. The major areas of cryobiology in Medicine are *cryopreservation* and *cryosurgery*. Cryopreservation is the method of preserving biological samples under low temperatures for longer time periods without getting damaged [19].

Cryosurgery cryoablation or cryotherapy is a minimally invasive surgical method utilizing extremely low temperatures to destroy cancerous or tumorous tissues in the body. And is attained by the absorption or removal of heat from targeted tissue by introducing cold surfaces into them. The low temperatures are achieved by delivering different cryogens through metal or ceramic probe tips having millimeters in diameter, where the expansion or vaporization of the liquid cryogens takes place, absorbing heat from the surrounding tissues. An ice ball is formed around the probe tip, and the tumor tissues cooled beyond its freezing point are destroyed permanently. The clinical threshold to achieve complete cell death is −40°C, and it is easy to detect that temperature while the cryosurgery process using suitable thermometers. During the freezing process, the cells will try to sustain themselves by some cellular, vascular, and immunological mechanisms [21–24]. The major reason for cell death is the physical occurrence of ice both the inside as well as outside of the cell wall, which will, in turn, check the osmotic balance of the cell. With the advances in medical imaging and cryoprobe technologies, cryosurgery can now be performed in deep tissue diseases easier with minimal invasion.

8.4 MATHEMATICAL MODELS AND NUMERICAL SIMULATIONS

Bioheat transfer is unique to living systems and deals with the transference of heat or thermal energy inside the living systems. It is known that all biological process involves some kind of chemical reaction. The temperature dependency of all the biochemical processes makes the study of heat transfer occurring inside living systems important and interesting. Blood circulation is the key mechanism of thermoregulation, biochemical, injury, and therapeutic processes in all living systems with a developed circulatory system. Blood perfusion can either accelerate or decelerate the effectiveness of thermal medicine, depending on the modality. The heat transfer in local tissues is influenced by blood flow rates, its heat capacities, and the local blood vessel geometries. The correlation between blood perfusion and temperature regulation is not linear hence making it highly complicated to understand and model.

Researchers have shown that all the physiological functions get seriously affected by a temperature difference of 6–10°C from the normal range of body temperature. This phenomenon is exploited in the therapeutic modalities and diagnostic processes, which involve heat addition or absorption from the body tissues. The change in the local tissue temperature will hinder the normal biochemical processes, thereby either destroying the tissues or making them susceptible to other treatment modalities (chemo or radiotherapy). Thus, heat transfer can either be the sole treatment, or it can be a facilitator for other therapies. The assessment of normal functioning of body tissues under different heat transfer modes helped us to classify the processes into three viz, *hyperthermia, hypothermia,* and *cryobiology*. Diller et al. have extensively revised the hyperthermia, where an increased temperature is induced locally, Hypothermia which is decreasing the local temperature, and cryobiology, where the temperatures are lowered to the subfreezing temperatures [24].

Henry Penne's [25] in 1948, presented the milestone work providing the mathematical model of heat transfer within the human body and the effects of blood perfusion on it. He conducted experiments involving nine human subjects whose forearms were pierced with thermocouple coupled needles. The human subjects were unanesthetized to prevent the errors that may occur due to the anesthetic effects on blood perfusion. His studies showed a temperature increase of 3–8°C from the skin to the interiors of the human forearm and substantially proved that it is caused by the metabolic heat generation as well as the heat transfer due to blood perfusion. The classic Pennes' Bioheat equation can be written as:

$$\rho c\frac{\partial T}{\partial t} = \nabla \cdot k\nabla T + \rho_{bl}c_{bl}\omega(T_a - T) + Q_{met} \quad (1)$$

The model proposed by Pennes combined the thermal diffusion and metabolic heat generation with the standard Fourier heat transfer equation. The metabolic heat generation Q_{met} is considered uniform over the tissue under study. This model assumes blood as homogeneous and isotropic to reduce the complexities relating to the non-Newtonian nature of blood. The term (Ta-T) assigns the temperature difference between arterial blood temperature and that of the local tissues. The terms r_{bl}, c_{bl}, and ω are the density, specific heat, and perfusion rate of the blood, respectively. The Pennes' model equates the transient temperature change in the tissue as the summation of conductive heat transfer among the tissues, convective heat transfer due to blood perfusion and metabolic heat generated.

However, the oversimplified model of Pennes received a lot of criticisms in the following years. The assumptions of uniform metabolic heat generation, constant values for blood properties and the neglection of geometry, location of the vasculature, etc., were the weakest points in his model. Many researchers reinvestigated and modified the Pennes' model and its underlying assumptions, but it still the foundation of all other models came later.

The local temperature difference is very much affected by the blood perfusion, and it may be a significant term that can hardly be neglected. The pathophysiological processes like inflammation, therapeutic processes like local heating or cooling and environmental factors like heat stress all these processes can cause a substantial difference between the blood and the tissue temperature. The convective heat transfer developed due to this temperature difference may tend to nullify itself by dispersion of thermal energy. However, the convective heat transport developed not only the function of mere temperature difference but of different factors such as blood perfusion rate, local vascular geometry (anatomy), type of tissue, and the cause of temperature gradient. Diller et al. [28] compiled and published somewhat comprehensive information on blood perfusion data of different tissues as well as for different species.

Wulf, Chen, and Holmes, Weinbaum, and Jiji, etc., were some of the researchers who immediately revolted against the Pennes' classical model and presented their versions shortly after the first [26, 29]. If we try to divide the progression of bioheat models into three phases in the timeline then, the researchers in the first phase were obsessed with investigations of blood perfusion and local vessel anatomy along with the directional effects of blood flow in them. The investigators of the second phase were keen to inquire about the possibilities of statistical models in bioheat models. Whereas the third phase, researchers began to employ applied mechanics and sophisticated heat transfer models [27–29, 31].

8.4.1 MATHEMATICAL MODELS ON THERMAL DAMAGE

The scope of mathematical modeling and numerical analysis becomes void if there is not a criterion of threshold temperature value to ascertain the cell death in hyperthermia processes. So various investigators conducted studies in the temperature and time-at-temperature to cell death and tissue damage. The thresholds of the thermal damage are majorly based on two concepts which are the cell survival rate at a given temperature, and the effective temperature causing the permanent cell damage. The two significant thermal damage models are the *Arrhenius model* and *cumulative effective minutes (CEM)* model.

Svante' Arrhenius, in 1889 [32], conducted experiments that led to the formulation of thermal dose units. His *in-vitro* studies demonstrated that the dependency of duration and degree of temperature has an effect on the rate of cell destruction and is exponential in nature. The Arrhenius analysis is carried out by a number of investigators to determine the heat at which the inactivation of cells occurs. For the analysis, the rate of cell killing is plotted against the reciprocal of temperature at which it occurs. And the heat of inactivation is determined using the equation.

$$K = Ae^{(-E/RT)} \tag{2}$$

where; E is the heat of inactivation in kcal/mole; A is Arrhenius constant; R is the molar gas constant in Kcal/mole-K; and T is the absolute temperature in K.

Saporito and Dewey [33] developed a method for converting one time-temperature combination to another. This method is comparatively simple and is known as 'thermal is an effective dose.' In this method, the time-temperature data are transformed into an equivalent number of minutes at 43°C. There is not much explanation of the criterion behind the selection of this particular temperature beyond the fact that 43°C is believed to be the effective temperature of cell death by the clinical practitioners who conducted early hyperthermia studies. The equation used for the conversion is:

$$CEM43°C = t\, R^{(43\text{-}T)} \tag{3}$$

where; CEM 43°C is the cumulative number of equivalent minutes at 43°C; t is the time interval in minutes; T is the average temperature during time interval t. R is the number of minutes needed to compensate for a 1°C temperature. The CEM43°C value gives the entire history of the exposure.

8.5 CONCLUSION AND FUTURE SCOPE

Now we are at the junction of robust knowledge of constitutive biological mechanisms, which can empower researchers and engineers involved in deeper analysis of diseases and to design procedures and devices for specific purposes. If suitably exploited, the knowledge of constitutive properties of tissues, cells, and molecules can be used to model devices and methods. The challenge that arises by the biological systems is its severe complexity along with the number of genetic and biochemical variables, nonlinear relationships between them, energies involved and laws that govern the processes. These processes are extremely difficult to model without incorporating patient and disease-specific data. Given are some of the future scopes in bioheat transfer study following.

- The development of proper molecular thresholds for thermal injury to different kinds of tissues combining the knowledge of kinetic processes and the limits of reversibility.
- Devising a better blood perfusion response model for the thermal stress in both normal and cancerous tissues.
- Developing instruments for real-time thermal monitoring along with suitable feedback control methods will help to achieve targeted results.
- Devise more noninvasive and minimally invasive methods and instruments for thermometry that work well with the clinically useful resolution of temperature and time.
- Conduct studies to understand the molecular thermodynamics of tissues and pathological structures in the context of environmental stress factors.
- Establish proper methods for the determination and planning of targeted organ or tissue-specific optimized doses for thermal.

Each biological problem, however small it may be, has a significant role and can result in some major revelations if properly investigated. In the coming future, if a proper study on the events to be analyzed is carried out with the existing state of the art knowledge and technologies, then the future can be reshaped in a way that mankind never ever dreamed before.

KEYWORDS

- **computer-aided designing**
- **computer-aided engineering**
- **continuous hyperthermic peritoneal perfusion**
- **heat shock proteins**
- **hyperthermia therapy**
- **real-time thermal monitoring**

REFERENCES

1. Melillo, L., (1995). Thermalism in ancient world. *Med. Secoli., 7*, 461–483. PMid: 11623482.
2. Roussakow, S. V., (2012). Critical analysis of electromagnetic hyperthermia randomized trials: Dubious effect and multiple biases. In: *Proceedings of 31st Meeting of CHS.*
3. Coley, W. B., (1910). The treatment of inoperable sarcoma by bacterial toxins: (the mixed toxins of the Streptococcus erysipelas and the Bacillus prodigious). *Proceedings of the Royal Society of Medicine, 3,* 1–48.
4. Dewhirst, M. W., Gibbs, F. A. Jr., Roemer, R. B., & Samulski, T. V., (2000). Hyperthermia. In: Gunderson, L. L., & Tepper, J. E., (eds.). *Clinical Radiation Oncology* (1st edn., pp. 256–282). Chapter 14. New York, NY: Churchill Livingstone.
5. Glossary, (2003). *Int. J. Hyperthermia, 19*(3), 385–390.
6. Stauffer, P. R., & Goldberg, S. N., (2004). Introduction: Thermal ablation therapy. *Int. J. Hyperthermia, 20*(7), 671–677.
7. Overgaard, J., & Horsman, M. R., (1997). Hyperthermia. In: Steel, G. G., (ed.) *Basic Clinical Radiobiology* (2nd edn., pp. 212–221). Edward Arnold.
8. Van, D. Z. J., (2002). Heating the patient: A promising approach? *Ann. Oncol.,13*, 1173–1184.
9. Jones, E. L., Samulski, T. V., Vujaskovic, Z., et al., (2004). Hyperthermia. In: Perez, C. A., Brady, L. W., Halperin, E. C., & Schmidt-Ullrich, R. K., (eds.), *Principles and Practise of Radiation Oncology* (4th edn., pp. 699–735). Chapter 24 Lippincott Williams & Wilkins.
10. Kapp, D. S., Hahn, G. M., & Carlson, R. W., (2000). Principles of hyperthermia. In: Bast, R. C. Jr., Kufe, D. W., Pollock, R. E., et al., (eds.), *Cancer Medicine e.5* (5th edn.). Hamilton, Ontario: B.C. Decker Inc.
11. Kim, J. H., Hahn, E. W., & Ahmed, S. A., (1982). Combination hyperthermia and radiation therapy for malignant melanoma. *Cancer, 50*(3), 478–482.
12. Asea, A., Kraeft, S. K., Kurt-Jones, E. A., et al., (2000). HSP70 stimulates cytokine production through a CD 14-dependent pathway, demonstrating its dual role as chaperone and cytokine. *Nat. Med., 6*, 435–442.
13. Repasky, E., & Issels, R., (2002). Introduction: Physiological consequences of hyperthermia: Heat, heat shock proteins and the immune response. *Int. J. Hyperthermia, 18*(6), 486–489.
14. Li, G. C., Mivechi, N. F., & Wietzel, G., (1995). Heat shock proteins, thermotolerance and their relevance to clinical hyperthermia. *Int. J. Hyperthermia, 11*, 459–480.
15. Myerson, R. J., Roti, R. J. L., Moros, E. G., et al., (2004). Modeling heat-induced radio sensitization: Clinical implications. *Int. J. Hyperthermia, 20*(2), 201–211.
16. Furusawa, H., Namba, K., Thomsen, S., Akiyama, F., Bendet, A., Tanaka, C., Yasuda, Y., & Nakahara, H., (2006). Magnetic resonance-guided focused ultrasound surgery of breast cancer: Reliability and effectiveness. *J. Am. Coll. Surg., 203*, 54–63.
17. Peek, M. C., Ahmed, M., Napoli, A., Usiskin, S., Baker, R., & Douek, M., (2016). Minimally invasive ablative techniques in the treatment of breast cancer: A systematic review and meta-analysis. *Int. J. Hyperther., 33*,1–12.

18. Kutz, M., (2002). *Standard Hand Book of Biomedical Engineering and Design.* McGraw-Hill Professional, New York.
19. Hoffmann, N. E., & Bischof, J. C., (2002). The cryobiology of cryosurgical injury. *Urology, 60*, 40–49.
20. Kuflik, E. G., (2004). Re: Evidence-based review of the use of cryosurgery in treatment of basal cell carcinoma. *Dermatol. Surg., 30–33*, 478.
21. Cooper, I. S., (1964). Cryobiology as viewed by the surgeon. *Cryobiology, 51*, 44–51.
22. Cooper, I. S., (1965). Cryogenic surgery for cancer. *Fed. Proc., 24*, S237–S240.
23. Gonder, M. J., Soanes, W. A., & Smith, V., (1965). Chemical and morphologic changes in the prostate following extreme cooling. *Ann. N. Y. Acad. Sci., 125*(2), 716 729.
24. Cahan, W. G., (1972). Cryosurgery for cancer. *Ca-Cancer J. Clin., 22*(6), 338–343.
25. Pennes, H. H., (1948). Analysis of tissue and arterial blood temperature in the resting human forearm. *J. Appl. Phys.,1*, 93–102.
26. Lagendijk, J., (1982). The influence of blood flow in large vessels on the temperature distribution in hyperthermia. *Phys. Med. Biol., 27*, 17–23.
27. Lagendijk, J., Schellekens, M., Schipper, J., & Linden, P., (1984). A three-dimensional description of heating patterns in vascularized tissues during hyperthermic treatment. *Phys. Med. Biol., 29,* 495–507.
28. Diller, K. R., (1992). Modeling of bioheat transfer processes at high and low temperatures. *Adv. Heat Trans., 22*, 157–357.
29. Lemons, D. E., Chien, S., Crawshaw, L. I., Weinbaum, S., & Jiji, L. M., (1987). Significance of vessel size and type in vascular heat transfer. *Am. J. Physiol., 253*, R128–R135.
30. Lindahl, T., (1967). Irreversible heat inactivation of transfer ribonucleic acids. *J. Biol. Chem. 242*, 1970–2173.
31. Diller, K. R., Valvano, J. W., & Pearce, J. A., (2005). Bioheat transfer. In: Kneith, F., (ed.), *CRC Handbook of Heat Transfer* (2nd edn.).
32. Arrhenius, S., (1889). On the reaction rate in the inversion of cane sugar by acids. *Phys. Chem., 4*, 226–248.
33. Sapareto, S. A., & Dewey, W. C., (1984). Thermal dose determination in cancer therapy. *Int. J. Radiat. Oncol. Biol. Phys., 10*, 787–800.
34. Kong, G., & Dewhirst, M. W., (1999). Hyperthermia and liposomes. *Int. J. Hyperthermia, 15*, 345–370.

CHAPTER 9

Evolution of Artificial Intelligence and Deep Learning in Healthcare

KESHAV KAUSHIK

School of Computer Science, University of Petroleum and Energy Studies, Dehradun, Uttarakhand, India, E-mail: officialkeshavkaushik@gmail.com

ABSTRACT

The imitation of human intelligence by computers, primarily computer systems, is known as artificial intelligence (AI). Sophisticated computers can simulate cognitive abilities such as memory, remembering, problem-solving, and decision-making that are linked with the human brain. Deep learning (DL) is a more specialized extension of AI that allows systems to group data and make forecasts with greater accuracy while requiring less human intervention. A neural network with three or even more layers is used in DL. These neural networks aim to imitate the human brain's cognitive skills better, allowing it to process and learn from bigger volumes of data. While we are still a long way from replicating the human brain, DL is a positive step forward. Take, for instance, a self-driving vehicle. Consequently, AI approaches have made big ripples in the healthcare field, creating a heated discussion on whether AI physicians may eventually replace human doctors. Human doctors, we think, will not be substituted by automation in the near future, but AI will surely aid physicians in making better clinical judgments, and in some areas of healthcare, AI may entirely replace judgment. The widespread availability of healthcare data and the rapid development of big data analysis techniques have made current practical applications of healthcare analytics possible. When

Computational Health Informatics for Biomedical Applications. Aryan Chaudhary and Sardar M. N. Islam (Naz) (Eds.)

driven by clinical research queries, powerful AI algorithms may disclose clinically essential information buried in enormous volumes of data, which can improve clinical decision-making. The chapter highlights the role of AI in healthcare. The chapter will also enlighten the readers about the research scope of DL in *Healthcare*. The chapter will be helpful for AI and DL enthusiasts, healthcare professionals, PhD scholars, researchers, and students.

9.1 ROLE OF ARTIFICIAL INTELLIGENCE IN HEALTHCARE

AI in healthcare refers to the application of machine learning (ML) technologies and approaches to mimic human cognition in the evaluation, presentation, and understanding of challenging medical and healthcare information. AI is defined as the ability of computational methods to make intelligent decisions based only on incoming data. AI technology is distinguished from traditional medical solutions by the process of collecting data, processing it, and providing a well-defined product to the end-user. AI to do this requires ML techniques, including DL. Sophisticated algorithms are competent in spotting patterns in human nature and establishing their own logic. Throughout the last decades, AI, notably ML and, more importantly, DL, has made amazing progress. Substantial resources are increasingly being committed to medical concerns, but this has the tendency to worsen the digital gap by disregarding impoverished communities and their specific context. DL is, in essence, a difficult approach that needs a high level of technical skill.

AI is a relatively new concept in the field of healthcare. AI aids in the prediction of illness patients for medical procedures. The use of AI in healthcare is vast [1], and it is used by not just physicians but also patients, the pharmaceutical industry, healthcare services, insurance companies, and medical institutes. Dermatology, echocardiography, neurology screening, retinal care, diagnostics, surgery, angiography, and other fields benefit from AI. It offers easy access to both doctors and patients, as well as input to the medical community for study. AI aids in monitoring patients, diagnosis, and clinical and medical investigations. AI and automation are hot themes in conversations about the sustainability of professional employment, social transformation, and economic performance throughout the world. The essential principles behind AI and Big Data, as well as their importance to public health, are described in this study [2]. The authors outline the possible consequences and problems for medical practitioners and diagnosticians by highlighting the issues at hand. In light of previously published research, the potential advantages of sophisticated ML and AI are discussed.

Although AI has enormous potential to enhance people's health and well-being, its use in medical practice is presently restricted. One of the primary challenges to deployment is a lack of transparency since physicians must be satisfied that the intelligent framework can be recognized. Comprehensible AI has the ability to solve this problem and is a step toward AI that is trustworthy. The authors [3] examined current literature in this study to give direction to practitioners and academics on the architecture of explainable AI technologies for the healthcare area, as well as to contribute to the formalization of the subject of explainable AI. Several experts predict that AI will ultimately be likely to substitute radiologists as a result of the rising focus on AI in radiology. This research aims to evaluate this invention and the ways in which it is altering healthcare in order to assess its impact on physicians. This research investigates the role of AI-based solutions in performing medical activities in specialties such as radiography, histopathology, optometry, and cardiology [4]. It is concluded that AI-based technology will supplement rather than change the original physician-patient relationship. Because IoT-based healthcare apps vary from those that operate on conventional wireless networks, the demand for security is significantly greater. As a result [5], smart gateways are competent in dealing with a wide range of difficulties that ubiquitous health services encounter, including as dependability, scalability, and safety. The healthcare market is critical to people's well-being all across the globe. The need for a strong healthcare business has grown in tandem with the rapid rise of the worldwide population. This study [6] aims to emphasize the relevance of wearable in IoT healthcare, as well as the underlying technologies of the current wearable technologies and different IoT communication protocols. Furthermore, the features of the various wearables employed in IoT healthcare are highlighted in this study. AI has a wide variety of applications in medical and healthcare. Some of them are highlighted in Figure 9.1.

9.1.1 FRAUD DETECTION IN HEALTHCARE

Healthcare is an important part of people's lives, particularly as the older population grows, and it must be inexpensive. One of these healthcare programs is Medicare. Claims fraud is a key contribution to rising healthcare expenses, although it may be mitigated by detecting it early. Numerous ML algorithms for detecting Medicare fraud are compared in this research. The authors [7] compare four performance indicators as well as class imbalance reduction through interpolation and an 80–20 under sampling strategy with supervised, unsupervised, and blended ML techniques. In this chapter [8], multiple ML techniques are used to provide a comparative study on healthcare fraud detection approaches. Especially contrasted to the other techniques, it is evident that the Multilayer Perceptron Algorithm gives a substantial performance. DL algorithms examine criminal activities and health information from a range of sources, comprising claims history, hospital-related information, and clinical manifestations, to identify health coverage fraud claims.

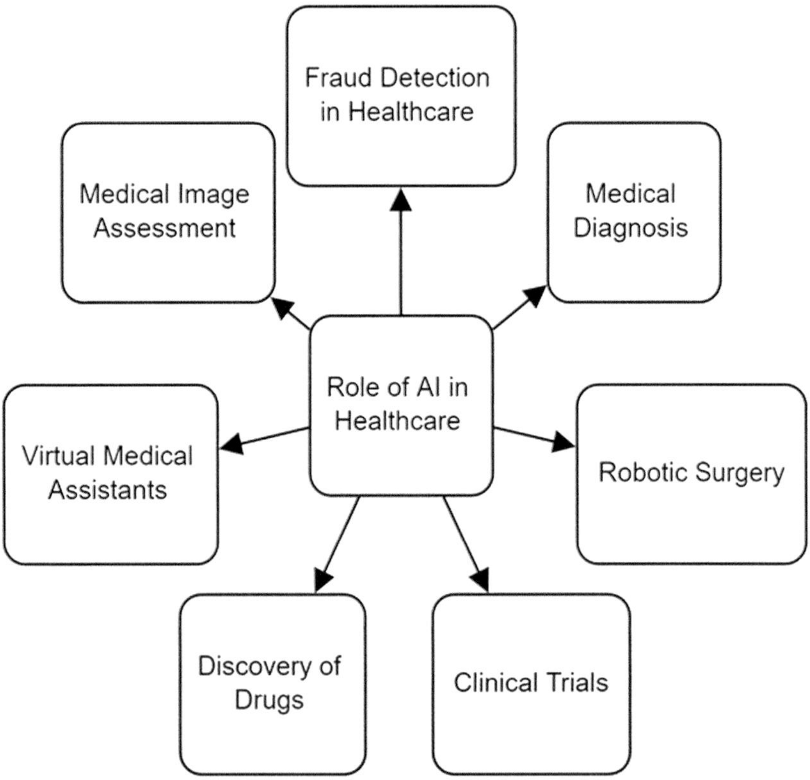

FIGURE 9.1 Role of AI in healthcare.

9.1.2 MEDICAL DIAGNOSIS

In medical diagnostics [9] and treatment, AI gives dependable assistance to overburdened healthcare professionals and institutions, reducing workload pressure, and increasing practitioner productivity. ML in diagnostic purposes benefits medical decision-making, management, technology, government, and procedures. It may be used to detect cancer, triage important medical imaging findings, highlight acute anomalies, help radiologists prioritize life-threatening patients, identify cardiac arrhythmias, predict stroke outcomes, and help patients with chronic disorders.

9.1.3 ROBOTIC SURGERY

Robotic limbs for prosthetic devices, micro-robots that cure injuries from the inside, and robo-assistants in operations are just a few of the applications in the medical industry that are now in use. Telepresence robots that evaluate patients to spare up time for medical staff may be available shortly. AI is improving its ability to perform what humans do, but more efficiently, quickly, and for less price. In the field of medicine, both AI and robotics have a lot of potentials. AI and robotics are increasingly becoming a part of our healthcare system, just because they are in human everyday lives.

9.1.4 CLINICAL TRIALS

AI is being increasingly frequently applied in a variety of fields, including the pharmaceutical industry. In this overview [10], the application of AI in many pharmaceutical sectors is emphasized, including pharmaceutical research, therapeutic goals, boosting pharmaceutical effectiveness, and clinical trials; such usage minimizes human effort while also reaching objectives in a short period of time.

9.1.5 DISCOVERY OF DRUGS

Despite the fact that the use of computational techniques in the pharmaceutical business is widely recognized, new ways that may enhance and optimize drug discovery and continued development are urgently needed. Despite the fact that there are no one-size-fits-all solutions to this demand for innovation, AI [11] has lately sparked a lot of attention. In reality, not only has the research establishment made significant contributions in this area but there's also been a developing relationship between the pharmaceutical business and AI firms.

9.1.6 VIRTUAL MEDICAL ASSISTANTS

From a basic calculator to complicated ML gadgets, AI is being employed in a variety of devices, applications, and protocols. Voice-activated virtual assistant technology is a relatively easy and effective AI-based application that is rapidly being deployed, notably in the consumer-grade and residential usage areas. Voice commands [12], according to some technology enthusiasts, will one day enable us to manage a variety of digital assistants, including office managers and librarians.

9.1.7 MEDICAL IMAGE ASSESSMENT

The use of AI in radiography has sparked a lot of excitement in recent years. Although some of this excitement may have been inspired by erroneous expectations that the technology will be able to do better than physicians in some activities, there is a large evidences that reveals its limitations in medical imaging. The true potential [13] of the technology is most probably somewhere within the middle, and AI will ultimately play a significant role in medical imaging. AI is a perfect contender for providing the uniformity, uniformity, and reliability required to assist radiologists in their quest to offer outstanding patient care because of computers' boundless potential. The growth of this sector in medical imaging is now hampered by significant hurdles.

9.2 DEEP LEARNING TECHNIQUES AND ALGORITHMS FOR HEALTHCARE

DL models' computing power has allowed rapid, accurate, and efficient healthcare procedures. DL networks are revolutionizing patient care, and they play a critical role in clinical practice for health systems. DL approaches in healthcare include image processing techniques, NLP, and reinforcement learning. DL may be a powerful tool for identifying patterns of certain illnesses that arise in our bodies much faster than a doctor. AI has been combining astounding breakthroughs in practically every field, and healthcare is no exception. In the healthcare business, there is an unprecedented quantity of unlabeled and incomplete information that may be examined to derive important insights using AI. Virtual treatment and perhaps surgery are also possible applications of the technology. DL uses a mix of data inputs, weights, and biases to simulate the human brain's functioning process. DL's core method continues to be clustering data and creating very accurate predictions.

DL has been proved to be a potential technology for pattern identification in medical systems. This chapter's [14] objective is to study DL methodologies used in medical systems by evaluating advanced network topologies, applications, and industry developments, which is inspired by this concern. The purpose of this research is to give an in-depth view into the use of DL techniques in medical applications in order to bridge the gap between DL methods and human medical representativeness. During the content evaluation for this chapter, the authors [15] realized that DL outperforms traditional analysis and machine categorization approaches for large and diverse datasets. DL algorithms attempt to construct a model by incorporating all available data. This review study shows how different DL methods have been employed in the past, but they will be used in much more healthcare domains in the future to increase diagnostic quality. This chapter [16] discusses all DL approaches and their experimental examination, as well as their benefits and drawbacks. With actual variation case examples, the autoencoder, restricted Boltzmann machine, deep belief network, recurrent neural network (RNN), convolutional neural network (CNN), and feedforward neural network are all included in this study. The authors [17] explored how end-to-end systems may be built and how these computational approaches can influence a few critical areas of medicine. Our study of computer vision is mostly focused on medical imaging, and the authors illustrate how NLP may be applied to areas like electronic health record (EHR) data. Likewise, reinforcement learning in the framework of robotic-assisted surgery is examined, as well as extended deep-learning approaches for genomics. There are various DL technologies that are used in the healthcare. Some of the prominent ones are mentioned in Figure 9.2.

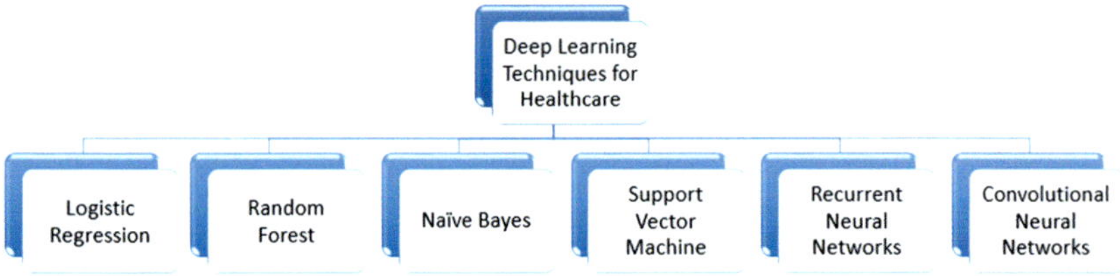

FIGURE 9.2 Deep learning techniques for healthcare.

9.2.1 LOGISTIC REGRESSION

The main purpose of logistic regression is to predict the likelihood of a binary event happening. Predictive models created with this method can have a positive impact on your company or organization. You can enhance decision-making by using these models to analyze linkages and forecast consequences. Healthcare epidemiologists should be familiar with large-scale analytical techniques, especially ML, in order to fully exploit the possibilities of big generalized data in medical healthcare and to effectively interact effectively with information systems professionals and data analysts. This review [18] focuses on ML as a whole, as well as its first implementations in the rapidly developing field of healthcare technology epidemiology.

9.2.2 RANDOM FOREST

Random Forest is a ML technique that may be used for both regression and classification. For a more accurate prediction, Random Forest tends to grow multiple decision trees that are then merged together. The Random Forest model is based on the idea that numerous uncorrelated models perform significantly better as a cluster than they do individually. The researchers [19] of this work used a variety of ML algorithms as well as public healthcare information stored in the cloud to create a system that allows for real-time and remote health surveillance using IoT infrastructure and cloud services.

9.2.3 NAÏVE BAYES

The Bayes' Theorem is used to produce the Naive Bayes classifiers, which are a collection of classification algorithms based on the Bayes' Principle.

It's a collection of strategies that all have the same premise: each pair of classified features is distinct from the others. To accomplish classification tasks, a Naive Bayes classifier, a stochastic DL approach, is used. The Bayes' Theorem and the greatest posteriori hypothesis are used to create this classifier. To save time, the naïve assumption of conditional probability independence is often used. Medical professionals provide data on health conditions, which is collected and kept on the server end of the partition. After merging descriptions of the sickness and symptoms, diagnostics, projections, and prescriptions, the system's software provides answers to problems for auditing and treating patients. This program will be very beneficial to individuals in order to prevent them from complications. In this suggested system [20], the Voicebot is utilized to address the issues that people have while they are sick, which have an impact on their health.

9.2.4 SUPPORT VECTOR MACHINE

Support Vector Machines (SVMs) are a quick and trustworthy classification system that works well with small amounts of data. An SVM is a classification-based supervised ML technique for two-group classification tasks. After being provided collections of labeled data for each category, SVM models may categorize new text. They offer two distinct advantages over newer approaches like as neural networks: quicker processing and higher performance with smaller samples. This makes the approach particularly well suited to text classification issues, where it's usually only to have access to a few thousand labeled samples. This study [21] proposes an excellent solution based on voice recognition for providing a simple monitoring system to elderly, ill, and handicapped individuals. The objective is to create a low-cost platform based on voice recognition that allows IoT devices deployed in smart hospitals and nursing homes to be accessed without the need for a centralized master station.

9.2.5 RECURRENT NEURAL NETWORKS

The RNN operates on the premise of preserving a layer's result and sending it back into the input to anticipate the layer's output. An RNN is a feedforward neural network that can accommodate variable-length sequence input and is an expansion of a traditional feedforward neural network. RNN can manage time series because it contains a recurrent hidden state whose stimulation is dependent on the activation of the prior time. The authors introduced a body sensor-based behavior identification system based on deep RNN, a potential DL technique based on consecutive layers, in this chapter [22].

9.2.6 CONVOLUTIONAL NEURAL NETWORKS

Convolutional Neural Networks (CNNs) are DL neural networks that are meant to analyze organized arrays of input such as photographs. CNNs are extensively employed in machine vision and have become the state-of-the-art for a variety of visual image processing applications categorization, as well as in natural language processing (NLP) for text categorization. This research [23] describes a sensor fusion system based on CNNs that use a depth lens and a wearable inertial sensor to track six transition motions as well as falls in smart healthcare.

9.3 CHALLENGES OF AI AND DEEP LEARNING MODELS

AI is the computer emulation of intelligent human behavior. This conduct encompasses a wide range of activities that need human intellect. Language translations, decision-making, and visual pattern recognition are only a few examples of these activities. DL, on the other side, uses ANN and is inspired by an understanding of human brain biology. As a result, methods for addressing categorization issues have emerged that are exceedingly effective. DL is a kind of AI that uses large amounts of accurate data and complicated neural networks to teach robots to learn in the same manner that people do. Such cutting-edge solutions undoubtedly encounter a number of difficulties. The challenges related to AI and DL models are highlighted in Figure 9.3.

9.3.1 SECURITY

DL networks have a lot of potential in the area of enhancing cyber security. It's important to remember that DL models might modify their output as a result of a little change in input data. This might expose you to malicious assaults. DL models are currently used in part by automobiles. If someone gained access to the model and changed the data input, the car may act strangely, which might be harmful. DL's broad range of applications in a variety of disciplines raises a number of security concerns, including securing devices and systems, preventing assaults in IoT networks, and controlling resource-constrained IoT networks. Numerous security mechanisms [24] for IoT have been suggested to solve scalability and resource-constrained security challenges, such as web application intrusion detection and prevention.

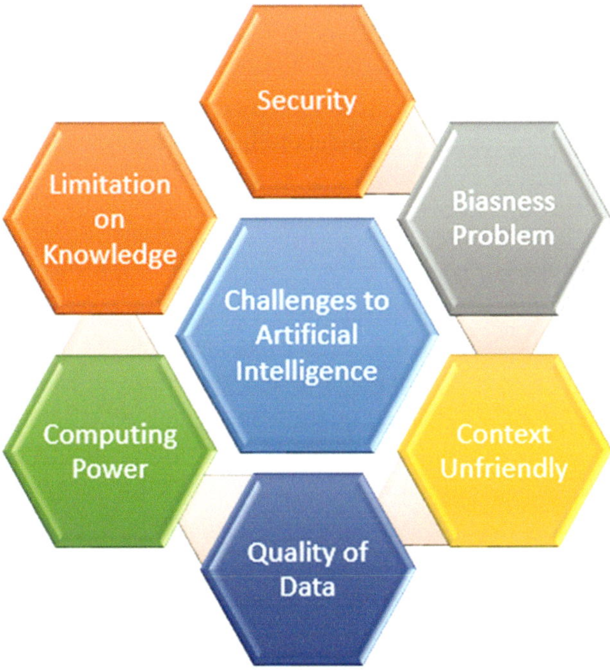

FIGURE 9.3 Challenges to artificial intelligence.

9.3.2 BIASNESS PROBLEM

The quantity of data used to train an AI technology determines whether it is excellent or terrible. As a result, it goes without saying that the number and quality of data accessible to train AI systems are critical. However, the majority of data acquired by institutions is of poor quality and importance.

They are, first, and foremost, prejudiced, substantially defining the characteristics and character of a limited population with preferences such as age, religion, and other similar factors.

9.3.3 CONTEXT UNFRIENDLY

The word 'deep' in DL refers to the architecture rather than the amount of knowledge that the algorithms can achieve. Let's look at a video game as an example. A DL system can be taught to play Mortal Kombat very effectively, and once fully trained, it will be capable of defeating humans. If you switch to Tekken, you'll have to retrain the neural network. This is due to the fact that it is oblivious to the surrounding circumstances. With the rise of IoT devices and the importance of actual analysis, the time required to swiftly retrain DL models will be insufficient to keep up with the input of data.

9.3.4 QUALITY OF DATA

DL works best when it has access to a large amount of high-quality data. The effectiveness of the deep teaching method improves as the amount of data available increases. When quality data is not given into the system, a DL system might fail catastrophically. Researchers deceived Google's DL algorithms last year. They changed the existing evidence in the sense that they introduced 'noise' to the data that was previously there. It wasn't a case of high-alert mistakes. In the image recognition techniques, the system misidentified weapons for turtles, resulting in inaccuracies.

9.3.5 COMPUTING POWER

The amount of computer resources these resource-intensive methods need turn off most developers. ML and pattern recognition are the foundations of AI technology, and they need a growing amount of computing power and GPUs to perform well. Asteroid monitoring, healthcare distribution, tracking of cosmic entities, and many more areas are examples of where we have concepts and expertise to use DL architectures.

9.3.6 LIMITATION ON KNOWLEDGE

Nevertheless, AI has the potential to be a better replacement to traditional approaches in a number of industries. AI knowledge is the real problem. Only a tiny fraction of the populace is conscious of AI's potential, outside of computer enthusiasts, college students, and academics. Numerous Small and Medium Businesses, for example, may organize their work or learn new skills to increase productivity, use things, sell, and manage products online, analyze, and understand client behavior, and react to the market swiftly and effectively. They are also unfamiliar with computer service providers such as Google Cloud, Amazon Web Services, and others.

9.4 RESEARCH SCOPE OF ARTIFICIAL INTELLIGENCE IN HEALTHCARE

AI has a lot of promise for revolutionizing healthcare delivery, resulting in better patient outcomes and more effective care. Considering these benefits, nevertheless, AI integration in healthcare has lagged behind technological improvements. According to a previous study [25], it is critical to comprehend the many organizational elements that influence the incorporation of new technology in healthcare. AI reporting standards that give the possible origin of bias in studies employing AI treatments have the opportunity to enhance the performance of AI research by improving their content and implementation, as well as the accuracy and transparency of their documentation. With a plethora of AI study guidance documents coming from various professional organizations, this review chapter [26] offers researchers some fundamental concepts for choosing the most suitable reporting standards for a study, including an AI intervention.

Medicine has always been inundated with large amounts of complicated data arriving at breakneck speed. Data is generated by a variety of companies in the healthcare industry, including hospital accreditation, health insurance, hospital instruments, life sciences, and health research. With technology improvements, there is a strong prospect of transforming healthcare with this data. Big data analytics, DL, and machine intelligence allow for the discovery of trends and connections, leading to actionable insights for improved healthcare delivery. Despite considerable contributions to the literature on this topic, we still lack a complete picture of the current state of research and application. This chapter [27] examines the existing research in order to offer academics with evidence that will help them support continued development in this field. The authors covered the most recent advances in AI applications in biomedicine, covering disease diagnoses, living aid, biomedical data processing, and biomedical research, in this review [28]. The goal of this review is to maintain track of current scientific achievements, comprehend technology availability, recognize the immense possibilities in biomedicine, and provide inspiration to researchers in related fields. It's fair to say that, like AI itself, AI's implementation in medicine is still in its infancy. With tremendous developments projected in the future years, new discoveries and improvements will continue to push the AI boundaries and widen the range of its capabilities.

KEYWORDS

- **artificial intelligence**
- **deep learning**
- **internet of things**
- **machine learning**
- **recurrent neural network**
- **support vector machines**

REFERENCES

1. Wisetsri, W., (2021). Rise of artificial intelligence in healthcare startups in India. *Advances in Management, 14*(1). https://www.researchgate.net/publication/349604103 (accessed on 14 September 2022).
2. Benke, K., & Benke, G., (2018). Artificial intelligence and big data in public health. *International Journal of Environmental Research and Public Health, 15*(12), 2796. https://doi.org/10.3390/IJERPH15122796.
3. Markus, A. F., Kors, J. A., & Rijnbeek, P. R., (2021). The role of explainability in creating trustworthy artificial intelligence for health care: A comprehensive survey of the terminology, design choices, and evaluation strategies. *Journal of Biomedical Informatics, 113*, 103655. https://doi.org/10.1016/J.JBI.2020.103655.
4. Ahuja, A. S., & Schmidt, C. E., (2019). The impact of artificial intelligence in medicine on the future role of the physician. *PeerJ, 7*(10), e7702. https://doi.org/10.7717/PEERJ.7702.

5. Kaushik, K., Dahiya, S., & Sharma, R., (2021). Internet of things advancements in healthcare. *Internet of Things*, 19–32. https://doi.org/10.1201/9781003140443-2.
6. Singh, K., Kaushik, K., Ahatsham, & Shahare, V., (2020). Role and impact of wearables in IoT healthcare. *Advances in Intelligent Systems and Computing, 1090*, 735–742. https://doi.org/10.1007/978-981-15-1480-7_67.
7. Bauder, R. A., & Khoshgoftaar, T. M., (2017). Medicare fraud detection using machine learning methods. *Proceedings-16th IEEE International Conference on Machine Learning and Applications, ICMLA 2017*, 858–865. https://doi.org/10.1109/ICMLA.2017.00-48.
8. Lavanya, S., Manojkumar, S., & Kumar, P. M., (2021). Machine learning based approaches for healthcare fraud detection: A comparative analysis. *Annals of the Romanian Society for Cell Biology, 25*, 8644–8654. https://www.annalsofrscb.ro/index.php/journal/article/view/2409 (accessed on 14 September 2022).
9. Kaur, S., Singla, J., Nkenyereye, L., Jha, S., Prashar, D., Joshi, G. P., El-Sappagh, S., et al., (2020). Medical diagnostic systems using artificial intelligence (AI) algorithms: Principles and perspectives. *IEEE Access, 8*, 228049–228069. https://doi.org/10.1109/ACCESS.2020.3042273.
10. Paul, D., Sanap, G., Shenoy, S., Kalyane, D., Kalia, K., & Tekade, R. K., (2021). Artificial intelligence in drug discovery and development. *Drug Discovery Today, 26*(1), 80. https://doi.org/10.1016/J.DRUDIS.2020.10.010.
11. Cavasotto, C. N., & Di Filippo, J. I., (2021). Artificial intelligence in the early stages of drug discovery. *Archives of Biochemistry and Biophysics, 698*, 108730. https://doi.org/10.1016/J.ABB.2020.108730.
12. Muttanahally, K. S., Vyas, R., Mago, J., & Tadinada, A., (2021). Usefulness of artificial intelligence-based virtual assistants in oral and maxillofacial radiology report writing. *World Journal of Dentistry, 12*(2), 97–102. https://doi.org/10.5005/JP-JOURNALS-10015-1807.
13. Prevedello, L. M., Halabi, S. S., Shih, G., Wu, C. C., Kohli, M. D., Chokshi, F. H., Erickson, B. J., et al., (2019). Challenges related to artificial intelligence research in medical imaging and the importance of image analysis competitions. *Radiology: Artificial Intelligence, 1*(1), e180031. https://doi.org/10.1148/RYAI.2019180031/ASSET/IMAGES/LARGE/RYAI.2019180031.FIG1.JPEG.
14. Shamshirband, S., Fathi, M., Dehzangi, A., Chronopoulos, A. T., & Alinejad-Rokny, H., (2021). A review on deep learning approaches in healthcare systems: Taxonomies, challenges, and open issues. *Journal of Biomedical Informatics, 113*, 103627. https://doi.org/10.1016/J.JBI.2020.103627.
15. Faust, O., Hagiwara, Y., Hong, T. J., Lih, O. S., & Acharya, U. R., (2018). Deep learning for healthcare applications based on physiological signals: A review. *Computer Methods and Programs in Biomedicine, 161*, 1–13. https://doi.org/10.1016/J.CMPB.2018.04.005.
16. Pandey, S. K., & Janghel, R. R., (2019). Recent deep learning techniques, challenges and its applications for medical healthcare system: A review. *Neural Processing Letters, 50*(2), 1907–1935. https://doi.org/10.1007/S11063-018-09976-2.
17. Esteva, A., Robicquet, A., Ramsundar, B., Kuleshov, V., DePristo, M., Chou, K., Cui, C., et al., (2019). A guide to deep learning in healthcare. *Nature Medicine, 25*(1), 24–29. https://doi.org/10.1038/s41591-018-0316-z.

18. Roth, J. A., Battegay, M., Juchler, F., Vogt, J. E., & Widmer, A. F., (2018). Introduction to machine learning in digital healthcare epidemiology. *Infection Control & Hospital Epidemiology*, *39*(12), 1457–1462. https://doi.org/10.1017/ICE.2018.265.
19. Kaur, P., Kumar, R., & Kumar, M., (2019). A healthcare monitoring system using random forest and internet of things (IoT). *Multimedia Tools and Applications, 78*(14), 19905–19916. https://doi.org/10.1007/S11042-019-7327-8.
20. Revathy, S., Niranjani, R., & Roslin, K. J., (2020). *Health Care Counselling Via Voicebot Using Multinomial Naive Bayes Algorithm*, 1063–1067. https://doi.org/10.1109/ICCES48766.2020.9137948.
21. Ismail, A., Abdlerazek, S., & El-Henawy, I. M., (2020). Development of smart healthcare system based on speech recognition using support vector machine and dynamic time warping. *Sustainability, 12*(6), 2403. https://doi.org/10.3390/SU12062403.
22. Uddin, M. Z., Hassan, M. M., Alsanad, A., & Savaglio, C., (2020). A body sensor data fusion and deep recurrent neural network-based behavior recognition approach for robust healthcare. *Information Fusion, 55*, 105–115. https://doi.org/10.1016/J.INFFUS.2019.08.004.
23. Dawar, N., & Kehtarnavaz, N., (2018). A convolutional neural network-based sensor fusion system for monitoring transition movements in healthcare applications. *IEEE International Conference on Control and Automation, ICCA*, 482–485. https://doi.org/10.1109/ICCA.2018.8444326.
24. Thakkar, A., & Lohiya, R., (2020). A review on machine learning and deep learning perspectives of IDS for IoT: Recent updates, security issues, and challenges. *Archives of Computational Methods in Engineering, 28*(4), 3211–3243. https://doi.org/10.1007/S11831-020-09496-0.
25. Lebcir, R., Hill, T., Atun, R., & Cubric, M., (2021). Stakeholders' views on the organizational factors affecting application of artificial intelligence in healthcare: A scoping review protocol. *BMJ Open, 11*(3), e044074. https://doi.org/10.1136/BMJOPEN-2020-044074.
26. Ibrahim, H., Liu, X., & Denniston, A. K., (2021). Reporting guidelines for artificial intelligence in healthcare research. *Clinical & Experimental Ophthalmology, 49*(5), 470–476. https://doi.org/10.1111/CEO.13943.
27. Mehta, N., Pandit, A., & Shukla, S., (2019). Transforming healthcare with big data analytics and artificial intelligence: A systematic mapping study. *Journal of Biomedical Informatics*, *100*, 103311. https://doi.org/10.1016/J.JBI.2019.103311.
28. Rong, G., Mendez, A., Bou, A. E., Zhao, B., & Sawan, M., (2020). Artificial intelligence in healthcare: Review and prediction case studies. *Engineering*, *6*(3), 291–301. https://doi.org/10.1016/J.ENG.2019.08.015.

CHAPTER 10

Medication Extender Drone Using CoppeliaSim

R. FAERIE MATTINS,[1] PAVITRA VASUDEVAN,[1] S. SRIVARSHAN,[1] R. MAHESWARI,[1] and VENUSAMY KANAGRAJ[2]

[1]*School of Computer Science and Engineering, Vellore Institute of Technology, Chennai, Tamil Nadu, India, E-mail: maheswari.r@vit.ac.in (R. Maheswari)*

[2]*Engineering Department, University of Technology and Applied Sciences-Al Mussanah, Sultanate of Oman*

ABSTRACT

The use of unmanned aerial vehicles (UAV) in the field of transportation has helped to save a lot of human power. Drones can deliver payloads faster and more efficiently because they are small and lightweight. For this reason, drones in health centers have endless applications. Every healthcare facility has a centralized drug distribution unit, where drugs are made available to the patient from a central drug warehouse, such as a pharmacy. In-patients are administered medication upon orders from the physician, and the necessary medications are delivered to the patient's room by a nurse or are collected by the attendant. During a pandemic situation like COVID-19, it is best to avoid human intervention and encourage social distancing. Therefore, deploying a drone to deliver medicines will ensure that manpower is not needed to distribute the medications that the patients need. The Medication Extender Drone (ME-Drone) under consideration will collect and dispense prescribed drugs to patients

Computational Health Informatics for Biomedical Applications. Aryan Chaudhary and Sardar M. N. Islam (Naz) (Eds.)

within the hospital premises. The hospital layout is predefined, following delivery authentication. Furthermore, it will optimize the path of travel by taking the shortest possible route to the patient's room. For this purpose, it follows Dijkstra's algorithm. It will ensure that any unused drugs are returned to the Pharmacy, and the Drone will return to its docking station once the process is completed. In addition, this study talks about simulating an interactive web application for the control center to track the Drone, as well as a mobile application for the user to order the prescribed medications. The mobile application will be the Point of contact between the user (attendant) and the Drone as it is dispatched, as soon as an order is placed, to the Pharmacy. On the other hand, the web application will aid in monitoring the Drone's location from the control center. Also, the drug delivery ensures security by using a One-Time Password (OTP) for authentication.

10.1 INTRODUCTION

Drones are UAV, which means they don't have a human pilot or crew on board. Drone use in real life has been progressively expanding in recent years. The fundamental reason for this is that drones may be controlled remotely or autonomously. There are many remote-controlled drones in use. However, autonomous drones are utilized far less frequently. The concept of autonomous drones has made human lives easier because no further human labor is required after the Drone has been programmed to perform its required mission. The current main usage of autonomous drones is shipping and delivery, aerial photography, weather forecasting, precision agriculture, etc.

The usage of autonomous drones in the field of shipment and delivery has helped in saving a lot of manpower. Since drones are small and lightweight, it is easier for them to deliver packages faster and more efficiently. For example, they can avoid any kind of unwanted traffic since they travel by air. Shipment companies are gradually adapting to the autonomous drone way of life by using drones to deliver packages. However, despite this massive technological advancement, these inventions are not integrated in our daily life. It is critical that these technologies, namely autonomous drones, make humanity's daily lives easier and more comfortable. As an initial step towards such a goal, an autonomous simulation drone is introduced in this work, which is called a ME-drone, to deliver medicines from pharmacies to patients, and if any pharmaceuticals are left over, it is programmed to return them to the pharmacy.

10.1.1 COPPELIASIM

For simulating the autonomous drone, a simulator called CoppeliaSim is used. CoppeliaSim is a robot simulator that was previously known as V-REP and is used in education, industry, and research. It's based on a distributed control architecture that uses Lua scripts or C/C++ plug-ins to operate as synchronous controllers. Various middleware solutions with programming languages such as Python, C/C++, MATLAB, and Java can execute additional asynchronous controllers in another thread, process, or machine. For forward and inverse kinematics calculations, CoppeliaSim employs a kinematics engine, as well as many physics simulation libraries for rigid body simulation. Models and scenarios are created by putting together a hierarchical framework of diverse items. Motion planning, synthetic vision and imaging processing, collision detection, minimum distance calculation, unique graphical user interfaces, and data visualization are some of the additional features supplied by plug-ins. In this work, Lua script has been implemented to simulate the Drone. Also, multiple models and scenarios from CoppeliaSim are used to build the overall hospital architecture.

10.1.2 JUSTINMIND

For any kind of application, user interface/ user experience (UI/UX) plays the most important role. For the user to access any application, they need a user interface via mobile app, website, etc., to utilize them. There are many products in the market which are built effectively, but due to the lack of a good UI, the products aren't being used much in the market. Similarly, there is a need for a user-friendly UI/UX even for the usage of autonomous drones. In this work, software called JustinMind has been used for UI/UX creation.

JustinMind is a tool for wireframing and prototyping high-fidelity internet and mobile app prototypes. It's known for its ability to produce realistic renderings of finished goods, as well as collaboration, interactivity, and design features. Overall, it's become one of the most used prototyping tools in the industry. Prototypes and wireframes for JustinMind software can be shared in the cloud and simulated on mobile devices. The prototyping tool also generates HyperText markup language (HTML) for entire prototypes. JustinMind's major capabilities include UI design, web interaction design, mobile app gesture design, form design, data visualization in grids and tables, developer handoff, and user testing. One is the mobile user app, and the other is the web interface for the controller monitoring the Drone from the control center. Some of the features that were not implementable in the free version of JustinMind were implemented by extending it using some HTML and JavaScript.

10.1.3 SECURITY AND AUTHENTICATION

It is critical for the system to check for validation in any UI/UX for every user input. A user must, for example, input the necessary credentials on a login page. The system must throw an error if the user enters invalid credentials. Only the JustinMind pro edition had access to these capabilities. As a result, HTML, and JavaScript were used to accomplish these functionalities in this work. For the completion of the user interface, JustinMind was used to construct the UI/UX, and JavaScript was used to validate it. For any successful application, it is very crucial to have proper security. Security in apps prevents them from being misused by hackers. Hence, in ME-drone, an OTP feature has been added for security and authentication. When a user places an order using the ME-drone app, they receive an OTP that must be entered on the Drone in order for the drugs to be delivered. This guarantees that the medicine has not been misplaced and that it has been administered to the user correctly. The JavaScript function is used to implement this feature.

10.2 RELATED WORKS

Judy et al. propose two novel designs of drone healthcare delivery networks. Both models discuss the use of a strategy that involves a combination of transport by land, a collection of warehouses and a drone flight system to establish an effective and efficient delivery network. The systems use the monetary cost as a constraint while providing the most efficient mode of delivery for the healthcare services [1]. Niels et al. discuss the emerging trend of using UAV for the transport and delivery of goods. The chapter attempts to solve the Traveling Salesman Problem (TSP) in association with UAVs. The authors present the problem as an integer program and provide various cluster and heuristic-based algorithms to solve the program. The chapter also demonstrates the efficiency of their algorithms by providing example situations and use cases [2].

Mesar et al. propose a method of delivery of medical supplies in difficult to traverse or inaccessible locations by the use of aerial drones. The author's work entails the development of a drone that is able to carry a maximum load of 4.5 kgs and capable of automated flight. The flight path was programmed into the Drone via a grid of coordinates. Finally, the system was compared to existing methods of delivery to sensitive areas, and the performance was measured. The Drone completed a delivery that would have taken about 5 hours on foot in under 21 minutes [3]. Anshul et al. proposed a Drone-based healthcare delivery system called "drone ambulance." The Drone proposed in the chapter was an aerial vehicle with six rotors for a stable flight with a dropping mechanism attached that would enable it to drop the container at the designated location. The container may hold emergency healthcare equipment like medicine, defibrillator, etc. Two kilograms was the maximum load that the Drone was able to carry. The authors claim that the model would be of immense use in COVID-19 infected areas due to the ease with which the drones are able to mobilize and tend to emergencies [4].

Maheswari et al. propose a system in their paper that aims to replace hospital attendants fetching medicines for the patients with a drone system. This system works on the principle that hospitals generally have a central location/warehouse that keeps stock of all the medicines. It takes time and energy for an attendant to go to the location and fetch the medication for the patients. The presence of the COVID-19 pandemic also does not help matters. Having a drone fetch these medicines will not only be more efficient but will also minimize human contact. The Drone contained a lightweight, secured container that opens only on OTP verification to ensure that the medicines are delivered properly. The authors have also proposed a system where the drones utilize multiple sensors for obstacle detection and mapping out the shortest path to travel to the patient [5]. Angurala et al. introduce a system called DBCMS in their paper. DBCMS, which stands for Drone based COVID-19 medical service, is a system that employs drones for the collection and analysis of blood samples from

COVID-19 infected patients, thereby reducing the risk to the medical staff in the frontline. The system is divided into multiple modules. The first module is the collection of blood samples from patients, where OTP authentication is used to authorize the patient who is due for testing. The second module is concerned with critically unwell patients. Here drones cannot be used; hence a call for an ambulance is automatically sent. The third module deals with unprecedented situations where the control is passed to a senior doctor. The area of coverage by the Drone is divided into various clusters where K-means clustering is performed to classify the areas [6].

Rangel presented a paper which details the development of a low-cost drone system for usage in pandemic scenarios. The author claims that the system would be able to overcome the problems due to the restriction in physical contact placed due to the pandemic. The system termed as "medical drone system" (MDS) by the author works in three phases. The first phase was called the Cargo Drone. This Drone was used to transport medical equipment and medicine from one place to another. The second phase was the Aerial Eye Drone. This Drone was used to remotely check the temperatures of the patient aerially using thermal cameras without having the need to come in contact with them. The third phase was the Sterilizer Drone. This Drone was used to spray disinfectant in an infected area in an attempt to sterilize it. These three phases came together to create the MDS. The author claims that all three drones were created using a 3D printer [7]. Alsamhi et al. propose a system where blockchain technology is used in conjunction with decentralized drones to help in curbing the COVID-19 pandemic. The drones in this system are capable of delivering medicine and goods in real-time. The authors also explain the capabilities of a drone equipped with a thermal camera in helping to monitor the body temperatures of multiple people at the same time. The chapter explains the usage of blockchain as a way to control and coordinate multiple drones effectively and efficiently. Additionally, the blockchains would be able to secure the drones from malicious entities seeking a backdoor into the system, aiming to disrupt the operations or use it for personal gain/benefit. "DroneChain" is the name of the blockchain that the authors have used in the system [8].

Khaled Awad Ballous et al. have proposed an autonomous drone that is capable of being agile and fast whilst carrying heavy portable medical equipment. Additionally, their chapter also gives a clear insight on the estimated possible failure of their proposed Drone using SolidWorks analysis [9]. Sara Imran Khan et al. have proposed four-vehicle routing algorithms for UAVs. They are the Genetic, Ant Colony Optimization, Particle Swarm Optimization and Captivated vehicle routing problem (CVRP) algorithm. All these algorithms were tested on a UAV, and the results state that CVRP gave the best results with a runtime of 0.06 seconds [10].

Sanjana et al. have designed a drone system that facilitates rapid delivery of a first aid kit to a location of emergency whenever the ambulance assigned is stuck in traffic. The Drone makes use of multiple sensors like gyroscope, accelerometer, and GPRS technology to enable swift response to the call. They have also designed a web application that is used to make the call for the drone [11]. J. Sanfridsson et al. show how drones can be used to help in out-of-hospital cardiac arrests by delivering defrillibators. The Drone delivered the defibrillators and instructed the bystanders on how to use them so that they could treat the patient suffering from cardiac arrest [12].

The research put forth by Pradeep Abeygunawaradana et al. deals with the delivery of medicines using an automated drone [13]. The E-medic system developed consists of 4 major parts: Autonomous Drone, obstacle detection and avoidance, healthcare platform and an ETA calculation system. The healthcare platform has 3 components: doctor's portal, patient's application, and pharmacy portal. The doctor's portal mainly allows the doctor to upload the prescriptions and consists of an OCR recognition system for written prescription recognition. A mobile application is provided to the patients to place an order, track the Drone, and get the estimated arrival time of the Drone. Order and delivery management is done by the pharmacy portal, and in crucial situations, can take control of the Drone. The Drone uses GPS to determine the shortest path to its destination, and with the help of a vision sensor attached to its front, it can detect and avoid obstacles. However, it will not be able to detect obstacles to its side or back. The REST API is employed to establish communication between the healthcare platform and the Drone. Implemented using flask, this can track and set the location of the Drone while assigning tasks to the Drone. Though there are several similar systems, the novelty of this work is highlighted in the interconnection of the healthcare platform and the Drone.

The use of drones to deliver COVID-19 viral tests to potentially sick people was highlighted by the study presented by Maximilian Kunovjanek et al. [14]. A powerful disaster response tool was developed by leveraging the current drone infrastructure. Hence, drones operated by public and private entities can be used to distribute essential goods in the case of an emergency, which proves to be a good backup transport system. The Drone enabled testing process begins with a base station operator placing a self-test kit onto the Drone, and the entire route is monitored and controlled by the drone operator. Drone disinfection is required, after every delivery, to reduce the possibility of patient cross-contamination and to prepare for the subsequent dispatch of self-tests. The total distance traveled by the Drone is consequently estimated by summing the distances of a return trip from the docking station to each patient location, and hence is not cost-efficient. A feasibility study to determine the practical context of the approach was done by retrofitting a drone originally made for other purposes to deliver COVID-19 test kits, which proved to be successful. For added security and verification of customer identity, quick response code (QR code) scanning is done through the mobile application. As a result, when it comes to the time it takes to deploy tests, drones remain a viable alternative to vehicle-based test distribution approaches in all cases.

The existing systems all entail the usage of a drone to supply medical supplies and healthcare equipment to inaccessible areas. Drones are being used only when the places that require these items are impossible or very hard to reach by foot or vehicle. These drones are primed for remote flight with the capability for autonomous flight in case of emergencies or signal loss. The systems in use do not have exclusive use of drone technologies. Drones are used in conjunction with other services like a chain of warehouses, delivery vehicles, and a person to conduct the delivery in most cases. An algorithm that calculates the most efficient use of all these services has been put into place to improve the effectiveness of the system.

Some studies have researched the usage of the TSP in order to solve the constraint of finding the shortest route for the Drone to travel in order to increase effectiveness and decrease energy consumption. Most of the drone systems equipped to fight against the COVID-19 pandemic usually consist of a drone fitted with a thermal camera to monitor the temperature of people without coming into contact with them. Some works employed the use of drones to collect and analyze blood samples from potential COVID-19 victims. The trend followed explains how drones are being used to do the tasks of humans so that minimal contact between humans is achieved so as to stop the spread of the virus. There are various algorithms in use to improve the efficiency of the drones, such as calculating the shortest path. The TSP solution seen above is one such algorithm. Other algorithms include splitting the flight area into small clusters and using clustering algorithms such as k-means clustering to segregate the clusters and tend to the entire area cluster by cluster. The most common method for verification that the supplies are reaching the correct person or destination seems to be verified using OTP. However, some systems report the usage of blockchain technology in order to ensure the safety, security, and reliability of the system.

10.3 MEDICATION EXTENDER DRONE

This work consists of a system where a drone is used to deliver the medication from the hospital's Pharmacy to a patient staying in the hospital room. A mobile application will take in the patient details and map them with the hospital database. It will retrieve the prescription of the patient, and it allows the user (attendant) to order any medicine, if not all, from the prescribed drugs listed in the application. This order is sent to the control center, which will be the Drone's docking station. When the control center receives an order via the mobile app, it will wire the Drone such that it reaches the patient's room after collecting the list of prescribed and ordered drugs. Any certified operator or professional in the control center will be able to track/monitor the Drone via a web application.

After the Drone reaches the patient's room, the OTP displayed on the user's mobile application should be entered in the Drone, which will then allow the user to take the medicines in the container carried by the Drone. After authenticating the delivery, the Drone will then return to the Pharmacy to get sanitized and returning the unused medicines, if any. After completing the process, the Drone will return to its docking station in the control center. The Drone will take the optimized path (shortest route) to and from the patient's room on any floor of the hospital. The Drone achieves this by following Dijkstra's Algorithm, coded in the Lua programming language. Any changes to the hospital layout must be hard-coded onto the Drone's path so that the Drone will have a predetermined path for drug distribution at all times.

The Drone will perform obstacle detection to identify any obstacles during its flight. Ultrasonic proximity sensors have been used for this purpose, which helps detect objects in proximity of the Drone without the need for physical contact. The tool used for simulating this ME-drone is CoppeliaSim, where a fabricated hospital layout is used. The detailed workflow is given in Figure 10.1.

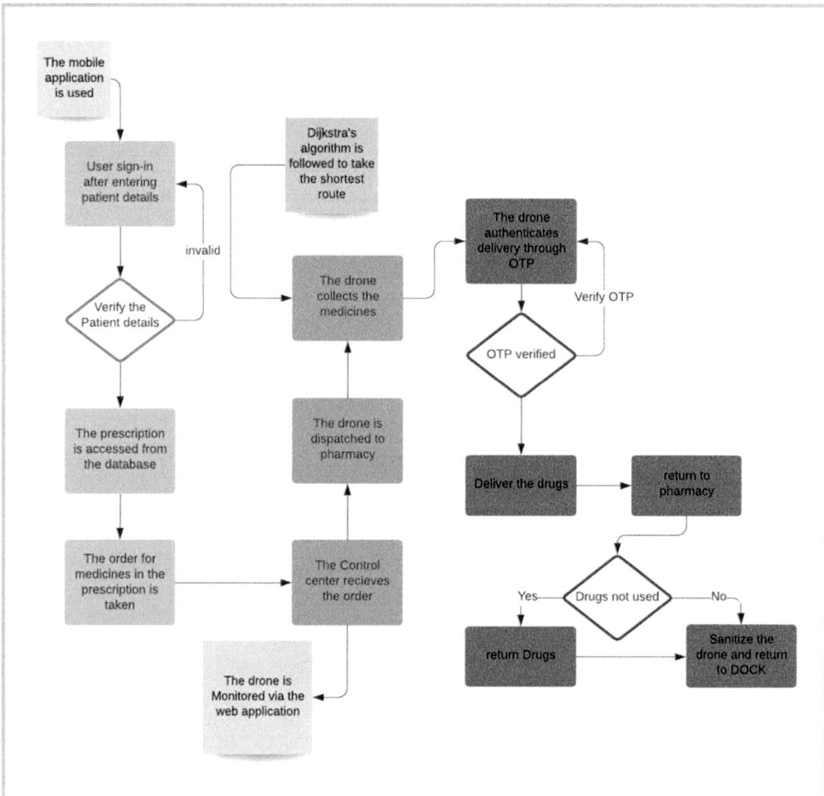

FIGURE 10.1 Flow chart of the ME-drone.

10.4 METHODOLOGY

In this work, the entire workflow is divided into five working modules. These modules are the foundation and the core of the entire system. The first module focuses on developing the hospital layout. This is the most fundamental module as all further simulations happen in this layout. The next module works on building the path for the Drone to follow. Since there are multiple rooms in the hospital layout, the path must be constructed properly for every room. Following this is module three, which focuses on building the user interface. The user interface consists of two parts, the mobile app for the user and the web interface for the controller in the control center. Further implementation of security and authentication via OTP is implemented in this module. The fourth module focuses on the Drone's most significant characteristics. The shortest path algorithm and the obstacle detection algorithm utilized in the Drone is explained in this module. The final module concentrates on the overall simulation's final steps. This module shows how the Drone returns to the control center after returning the unused drugs to the Pharmacy. The detailed module-wise flowchart is given in Figure 10.2.

- **Module 1:** Developing the hospital layout:
  - o **Floor 1:** Consist of the waiting area, control center and pharmacy.
  - o **Floor 2:** Consists of the patient rooms and waiting area.
- **Module 2:** Setting the drone path for flight:
  - o Making drone flight in a predefined path.
- **Module 3:** Development of ME-drone application software:
  - o Development of interactive user interface for both mobile app and web interface;
  - o Validating the user credentials;
  - o Display of prescription from the already provided user information;
  - o Implementing authentication to confirm the order;
  - o Display the drone's path and view via the web interface.
- **Module 4:** Shortest flight path findings to reach the destination:
  - o Designing an algorithm to find the shortest path;
  - o Analyzing the obstacles along the path.
- **Module 5:** Drone return to the dock:
  - o Unused drugs returned to pharmacy;
  - o Drone returning to the control center.

10.4.1 DEVELOPING THE HOSPITAL LAYOUT

This is the first and the most important module of the entire system. This module consists of the entire hospital layout plan. The hospital consists of two floors. The ground floor, also called floor one consists of a control center, Pharmacy, and a few chairs for the patients to sit and wait for their appointment. Figure 10.3 shows the side view of floor 1 and its features.

Floor 1 consists of 3 chairs for patients to wait and 3 plants for decoration. There are two rooms on floor 1. The first one is the control center located on the bottom right of the entrance. In this room, there is a man sitting in front of a laptop to control the Drone. This point is the Drone's starting point. The next room is the Pharmacy. It is located on the top left side from the entrance. The Pharmacy consists of two shelves that have the medicines. There is a man sitting in the Pharmacy to take the order from the user and send it via Drone. On the top left side, there are stairs that connect floors 1 and 2.

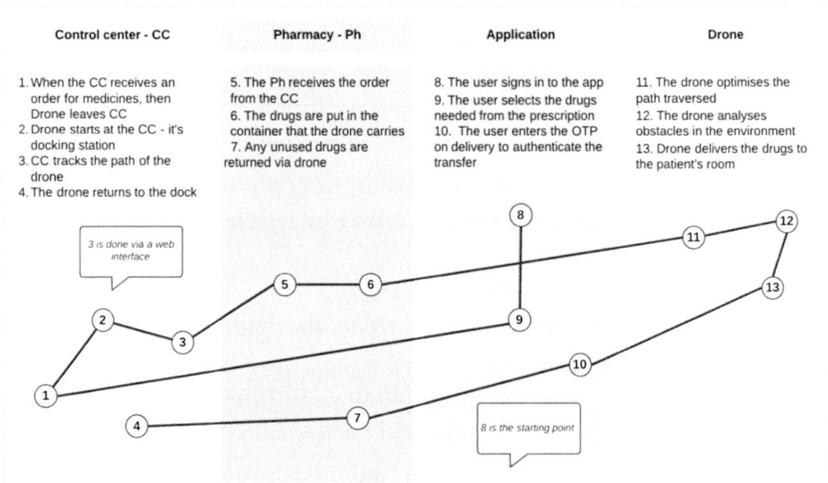

FIGURE 10.2 Module-wise workflow of ME-drone.

FIGURE 10.3 Side view of first floor – the control center and pharmacy.

The next floor is the first floor, also known as floor 2. Floor 2 consists of 5 patient rooms. Each patient room consists of a bed, chair, and a patient.

This floor also consists of two windows for ventilation. There are also 3 chairs for people to sit on and 2 plants for decoration. Figure 10.4 gives the top view of floor 2. The Drone travels via the stairs to reach floor 2 and then follows the path to deliver the drugs to the respective patient's room.

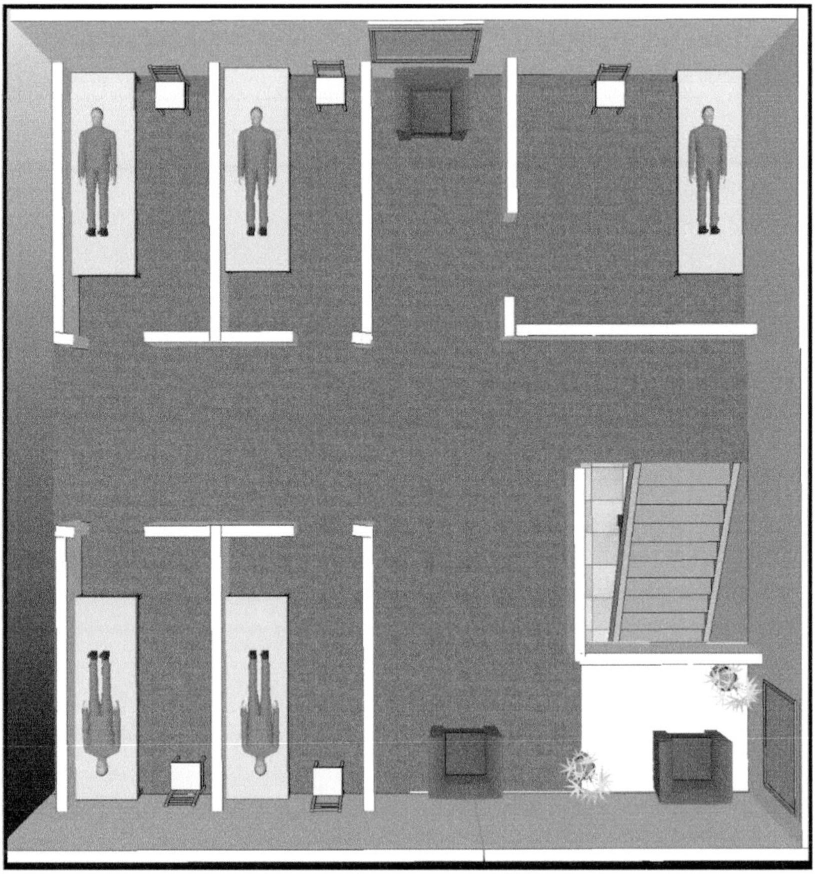

FIGURE 10.4 Top view of the second floor – patient rooms.

10.4.2 SETTING THE DRONE PATH

After a neat fabrication of the hospital layout, the Drone is set up for a flight. The Drone is imported from CoppeliaSim mobile robot repository. This is shown in Figure 10.5. The Drone has been elevated, by changing the position in the Z-axis, to enable safe flight, preventing unwanted traffic. The Quadcopter follows a target ball as it moves through the hospital layout. The sim. FollowPath in CoppeliaSim moves any object along the specified path. Using this function, Lua parameters such as velocity and acceleration can also be configured. The path handle parameter, which is given to the target ball, is configured after following the Dijkstra's algorithm. The Quadcopter can follow only predefined paths, which proves to be a disadvantage.

FIGURE 10.5 The drone elevated in a plain environment.

10.4.3 DEVELOPMENT OF ME-DRONE APPLICATION SOFTWARE

This module focuses on the user interface of the work. The user interface is of two parts: mobile app and web interface. The first part is the mobile app is made for the user to order the drugs. The UI/UX design was made using the software tool JustinMind. The first opening page is the home page. The home page consists of the ME-drone, and its motive was given, along with a button "get started," which redirects the user to the login page. On the login page, the user needs to enter the credentials provided by the hospital to log in to the system. While visiting the hospital for the first time, the patient enters all their details, and this information is added to the hospital database. The hospital further provides the patient with a passcode. For accessing this app, the patient must enter their name, contact number and the hospital-provided passcode. This is illustrated in Figure 10.6. If the user enters the invalid credentials, then the app will throw an error stating, "The credentials do not match!" If the user enters the correct credentials, then the app will show a text box stating "login success," and the user will get redirected to the prescription page. The validation of the credentials is invoked using the JavaScript functions. After the patient's visit from the doctor, the prescription is updated in the hospital database. Hence, upon entering the valid credentials, the user can view the list of medicines prescribed by the doctor. The list is given using checkboxes, where the user can select multiple medicines and order them. Upon clicking the order button, the user gets redirected to the OTP page, where the JavaScript function generates a random OTP, which is unique for every user. This OTP needs to be entered on the Drone to receive the drug package. This is shown in Figure 10.7.

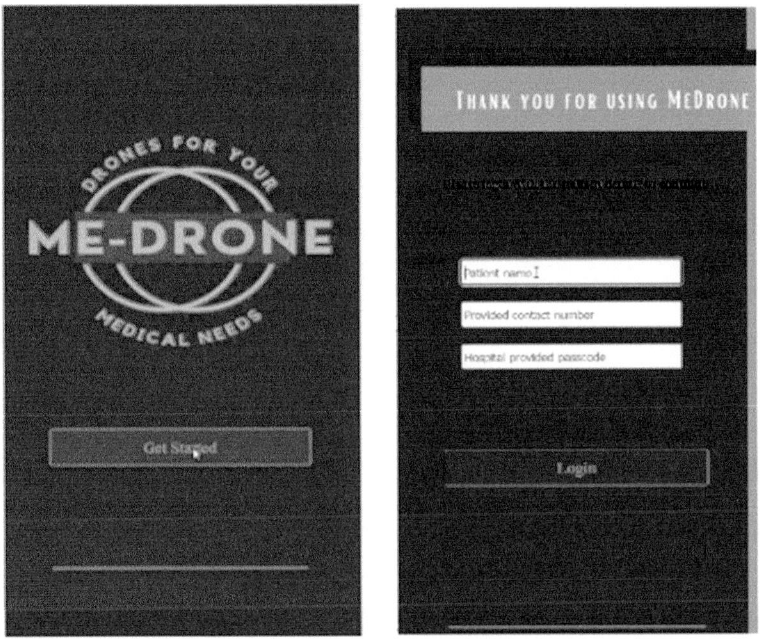

FIGURE 10.6 ME-drone app – home page and login page.

The second part is the web interface made for the control center. The controller of the Drone can monitor the Drone from the control center. The web interface had the home page, which consisted of the ME-drone, and its motive was given, along with the "Start tracking the Drone" button. This is shown in Figure 10.8. Upon clicking the button, the user gets redirected to the tracking page, where a live video feed is shown. For simulation purposes, this work shows only a predefined video. There are 3 videos on the screen. The first one is the top view of the hospital. This video shows the path at which the Drone is traveling. The next two videos are the ones taken from the Drone. They are the floor view and the Drone's front view. This camera is attached to the Drone. Hence the controller can view all the angels of the Drone from the control center. This is illustrated in Figure 10.9.

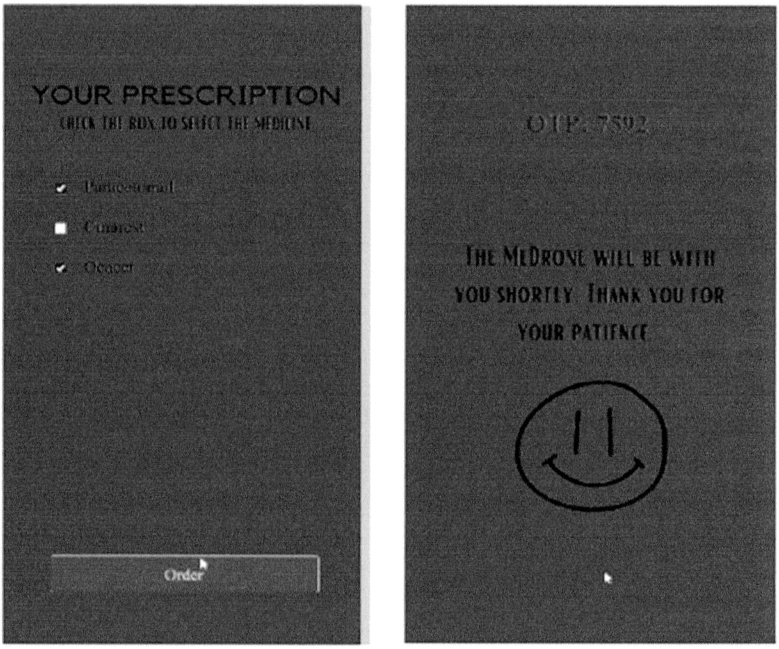

FIGURE 10.7 ME-drone app – list of medicines and OTP page.

10.4.4 SHORTEST FLIGHT PATH

The next step was to create an algorithm to find the shortest path from the control center to the Pharmacy, then to the patient's room and then back again. It also consists of the usage of sensors to detect the presence of obstacles on the way. The problem of finding the shortest path presents itself as a TSP. One of the more popular and effective ways to solve this is

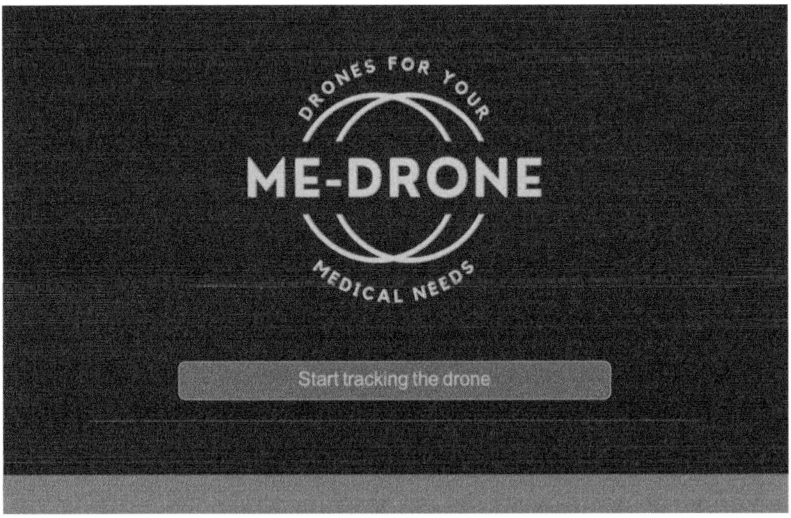

FIGURE 10.8 Home page of the web interface.

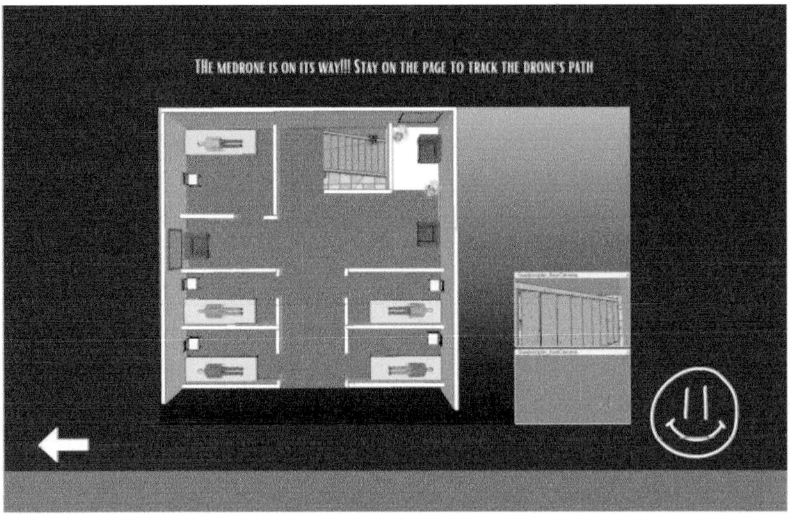

FIGURE 10.9 Monitoring the drone using camera feeds via the web interface.

to use Dijkstra's Algorithm [15]. This algorithm was conceived by Edsger W. Dijkstra in the year 1956 and was presented three years later in his work titled "A note on two problems in connection with graphs." The technique calculates the shortest path between a specified source node in the network and every other node. It may also be used to find the shortest pathways from a single node to a single destination node, with the method halting once the shortest path to the target node has been identified.

- **Algorithm 10.1** can be presented in the following steps:
  1. Set all points as unvisited.
  2. A list containing all the unvisited points is initialized.
  3. Each Point is assigned a predetermined value; the starting point is set to 0, and the rest are set to infinity, signifying that the shortest path to these Points is unknown.
  4. Take the current point, the starting point in the initial state, and visit all of its unvisited neighboring points.
  5. Calculate the distance to each of the neighboring points from the current point.
  6. Compare the new distance with the initially assigned value and assign the new value to the smaller of the two.
  7. After visiting all the neighboring points, mark the current point as visited.
  8. Remove the current point from the unvisited points list.
  9. Choose the point with the least value from the unvisited points list.
  10. Travel to that point and mark it as the current point.
  11. If all the points in the unvisited list have infinite value or all the points in the list are marked as visited, then the algorithm reaches its end.

Algorithm 10.1 has been implemented in the CoppeliaSim simulator using the Lua scripting language. The simulator allows child scripts to be attached to objects in the scene. A script defining the path that should be taken along with the algorithm has been assigned to the Drone object in the scene. For the detection of obstacles on the path, CoppeliaSim provides various sensors that are able to detect them and give a signal when they enter a particular area of influence. Six such obstacle detection sensors are added to the Drone object. Figure 10.10 shows how individual Paths come together to form a graph that the Drone can traverse. The area that these sensors cover comes together to form a spherical region. Figure 10.11 provides an illustration of the Obstacle Detection Sensors attached to the Drone Object. Whenever an obstacle enters this spherical region, the sensors give out a signal and a sound to announce the detection of an obstacle. These signals have been captured and visualized in the form of a graph.

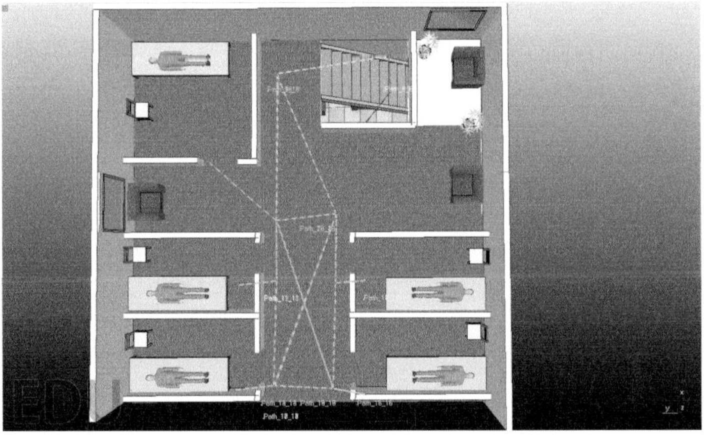

FIGURE 10.10 A graph depicting the drone's flight route in floor 2.

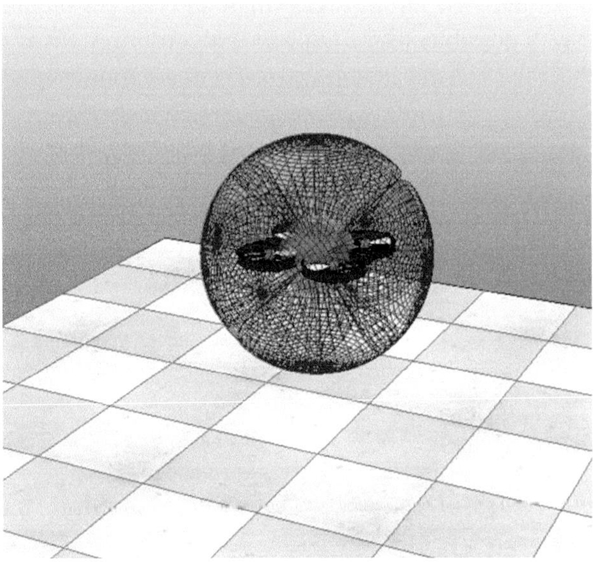

FIGURE 10.11 Illustrating the proximity sensors used by a drone.

10.4.5 DRONE RETURN TO THE DOCK

The drone dispenses the drugs and returns them to the Pharmacy. Here, it gets sanitized and returns unused drugs, if any. Then the Drone reaches the docking station, which is the control center. The return path taken is also the optimal path, the shortest route, determined by Dijkstra's algorithm.

10.5 RESULTS AND DISCUSSION

For the obstacle detection part, six proximity sensors have been used. Each proximity sensor is cone-shaped and senses any obstacle in front of it for the given radius. Since six cones are used to cover the Drone, it gives complete coverage and makes sure any obstacle can be sensed. Figure 10.12 shows the time-series graph of the six proximity sensors. From Data to Data4 indicates the proximity sensors from 1 to 6. This is a Boolean graph where if there is an obstacle detected, the line graph moves to 0, and if there is no obstacle in the path, the line graph stays on 1. The time-series graph depicts the output of the proximity sensor over the time of flight of the Drone. Using Dijkstra's algorithm, the Drone takes flight and reaches the destination in the shortest way possible. Since Dijkstra's algorithm is always optimal, it is evident that the Drone is guaranteed to fly in the shortest path possible.

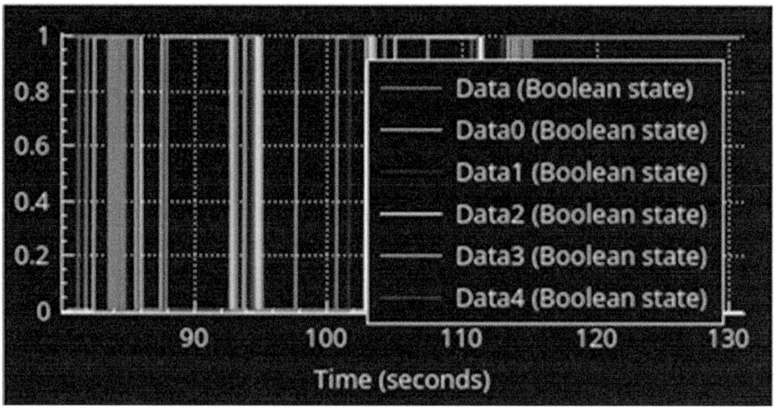

FIGURE 10.12 Time plot to depict object detection by proximity sensors.

10.6 CONCLUSION AND FUTURE SCOPE

This chapter focuses on simulating a ME-drone, using the tool CoppeliaSim that is used for drug distribution within hospital premises. The main objective of bringing about this Drone is to avoid human intervention in times like the COVID-19 pandemic, where social distancing is the need of the hour. The ME-drone takes the shortest route to the destination (patient's room), for which Dijkstra's Algorithm is used and coded in Lua. It collects medicine from the Pharmacy, delivers medicines after authentication, gets sanitized after every delivery, returns any used drugs to the Pharmacy, and returns to the docking station only to prepare to fly for another order. The mobile application and web application shown here are just simulations done using HTML, JavaScript, cascading style sheets (CSS), and tools such as Canva and JustinMind and after integrating with a hospital database, will be developed taking into consideration the key principles of UI design.

In this work, only obstacle detection has been implemented. In future, obstacle avoidance can be implemented. Also, there is no use of the database in this work since this was focused on being only a simulation. In future, this work can be expanded and made into a proper App and website which is connected to the database. Also, this simulation can be implemented using a real drone and be used in hospitals in real-time. An even bigger and more complicated architecture of the hospital can be constructed to give a better idea of the hospital in real-time. In this system, the Drone carries the drugs only to one patient at a time. This can be extended and sent to multiple patients at the same time. More security and authentication can be improved. For example, RFID authentication can be implemented in the hospital rooms or along with OTP. A QR code can be given for the Drone to scan. All this work can be extended in the future and used in real-life scenarios.

KEYWORDS

- **drone**
- **medical drone system**
- **medication extender drone**
- **one-time password**
- **travelling salesman problem**
- **unmanned aerial vehicles**

REFERENCES

1. Scott, J., & Scott, C., (2017). Drone delivery models for healthcare. In: *Proceedings of the 50th Hawaii International Conference on System Sciences*. doi: 10.24251/HICSS.2017.399.
2. Agatz, N., Bouman, P., & Schmidt, M., (2018). Optimization approaches for the traveling salesman problem with drone. *Transportation Science, 52*(4), 965–981.
3. Mesar, T., Lessig, A., & King, D. R., (2018). Use of drone technology for delivery of medical supplies during prolonged field care. *Journal of Special Operations Medicine: A Peer Reviewed Journal for SOF Medical Professionals, 18*(4), 34–35.
4. Singh, A., Kumar, P., Pachauri, K., & Singh, K., (2020). Drone ambulance. In: *2020 2nd International Conference on Advances in Computing, Communication Control and Networking (ICACCCN).*, 705–708.
5. Maheswari, R., Ganesan, R., & Venusamy, K., (2021). MeDrone-a smart drone to distribute drugs to avoid human intervention and social distancing to defeat COVID-19 pandemic for Indian hospital. In: *Journal of Physics: Conference Series, 1964*(6), 062112.
6. Angurala, M., Bala, M., Bamber, S. S., Kaur, R., & Singh, P., (2020). An internet of things assisted drone based approach to reduce rapid spread of COVID-19. *Journal of Safety Science and Resilience, 1*(1), 31–35.
7. Rangel, R. K., (2021). Development of low cost medical drone, using COTS equipment. In: *2021 IEEE Aerospace Conference, (50100)*, 1–12.
8. Alsamhi, S. H., Lee, B., Guizani, M., Kumar, N., Qiao, Y., & Liu, X., (2021). *Blockchain for Decentralized Multi-Drone to Combat COVID-19.* CoRR. abs/2102.00969.
9. Khaled, A. B., Ahmed, N. K., Ahmad, A., Al-Shabi, M., & Mamdouh El, H. A., (2020). Medical kit: Emergency drone. *SPIE, Unmanned Systems Technology XXII, 114250,* 248–253. doi: 10.1117/12.2566115.
10. Sara, I. K., Zakria, Q., Hafiz, S. M., Soumya, R. N., Anil, K. B., Verma, K. D., & Deo, P., (2021). UAVs path planning architecture for effective medical emergency response in future networks. *Physical Communication, 47,* 101337.
11. Parvathi, S., & Prathilothamai, M., (2020). Drone design for first aid kit delivery in emergency situation. In: *2020 6th International Conference on Advanced Computing and Communication Systems (ICACCS)* (pp. 215–220).
12. Sanfridsson, J., Sparrevik, J., Hollenberg, J., Nordberg, P., Djärv, T., Ringh, M., Svensson, L., et al., (2019). Drone delivery of an automated external defibrillator – a mixed method simulation study of bystander experience. *Scandinavian Journal of Trauma, Resuscitation and Emergency Medicine, 27*(1), 1–9.
13. Abeygunawaradana, P., Gamage, N., De Alwis, L., Ashan, S., Nilanka, C., & Godamune, P., (2021). E-medic – autonomous drone for healthcare system. In: *2021 International Conference on Computing, Communication, and Intelligent Systems (ICCCIS)* (pp. 994–999). doi: 10.1109/ICCCIS51004.2021.9397104.
14. Maximilian, K., & Christian, W., (2021). Containing the COVID-19 pandemic with drones - Feasibility of a drone-enabled backup transport system, *Transport Policy, 106*, 141–152. ISSN 0967–070X, doi: 10.1016/j.tranpol.2021.03.015.
15. Dijkstra, E. W., (1959). A note on two problems in connexion with graphs. *Numerische Mathematik, 1*(1), 269–271.

CHAPTER 11

Big Data and Visualization-Oriented Latency-Aware Smart Health Architecture

M. S. GURU PRASAD,[1] PRABHDEEP SINGH,[1] and MOHIT ANGURALA[2]

[1]*Graphic Era (Deemed to be University), Dehradun, Uttarakhand, India, E-mail: ssingh.prabhdeep@gmail.com (P. Singh)*

[2]*Khalsa College of Engineering and Technology, Amritsar, Punjab*

ABSTRACT

Internet of things (IoT) main aim is to build a smart, connected city that is comprised of pervasive environmental and human sensing combined with low-capacity computing dispersed over a large number of devices. Citizens in various smart surroundings can obtain a great deal of information by using this method of data collection. Users' data from linked wearable devices can be combined with ubiquitous environmental sensing and versatile actuation to improve smart health applications. This information is made available by the smart city infrastructure, which can be used to develop smart health apps as well. The current state-of-the-art in smart health applications is comprised of black-box, end-to-end solutions that are neither intended for usage with heterogeneous data nor customizable to a changing set of sensing and actuation parameters and conditions. In this chapter, we proposed latency-aware smart health architecture to perform intelligent health concerns, enabling the capability to scale with a large amount of data available, to employ general-purpose machine

Computational Health Informatics for Biomedical Applications. Aryan Chaudhary and Sardar M. N. Islam (Naz) (Eds.)

learning (ML), and to reduce computational redundancy and complexity [41]. This improves response times for important circumstances in smart health, as well as other benefits. In a smart city setting, the identification of health-related symptoms and subsequent actuation are made more efficient.

11.1 SMART HEALTHCARE

Health service systems that use technology such as wearable devices, the internet of things (IoT), and mobile Internet to dynamically access information while also connecting people, materials, and institutions associated with healthcare are referred to as smart healthcare systems. Smart healthcare systems then actively manage and respond to the needs of the medical ecosystem intelligently [1]. All partners in the healthcare industry should be able to communicate with each other, ensuring that participants get the services they need and assist them in making educated choices. It's capable of supporting a decent deployment of resources. Patients, doctors, hospitals, and research institutions are all part of the healthcare system. The IoT, cloud computing, mobile Internet (5G), artificial intelligence (AI), and big data, as well as for biotechnology, represent the cornerstone of smart healthcare [37]. The healthcare system as a whole benefit from the use of these new technologies in smart healthcare [2]. For patients who are constantly monitoring their health, wearable devices may be used to give medical care through virtual assistants, and distant locations can be used to provide services; a range of complex clinical judgments are used by physicians to support systems that enhance diagnosis. A doctor's form manages integrated information platforms, which include tools such as picture archiving, communication systems, and laboratory information management systems, in addition to the electronic medical record. Medical treatments might benefit from the use of surgical robots and mixed reality technology.

11.2 SMART HEALTHCARE

Smart healthcare has as its service aims clinical and scientific research institutions, regional health decision-making organizations, and individuals or families. The subcategories of smart healthcare are discussed further.

11.2.1 ASSISTING DIAGNOSIS AND TREATMENT

AI, surgical robots, and mixed reality have improved the identification and treatment of illnesses. In the creation of clinical decision support systems, AI has contributed to several achievements, such as hepatitis, lung cancer, and skin cancer diagnosis [3]. AI diagnostic findings are more accurate than those of human physicians. Most frequently, particularly in pathology and imaging, ML-based algorithms outperform even the most skilled clinicians. When it comes to clinical decision support systems, IBM Watson, an intelligent cognitive system that analyzes all relevant clinical data and literature, is the most outstanding and excellent product [4]. To diagnose diabetes and cancer, the program has a significant impact. Clinical decision support systems, which employ expert advice based on algorithms, may improve diagnostic accuracy, reduce the number of missed diagnoses and misdiagnoses, and enable patients to get timely and appropriate medical treatment. An accurate assessment of the patient's health and illness state helps to design a specific treatment plan, and the program has been endorsed by medical professionals. Doctors can fine-tune the radiation treatment, monitor illness progression, and eliminate the risk of manual operation [5]. As far as surgery goes, surgical robots have raised the bar considerably. Surgeons and patients will both benefit from improved outcomes and quicker recovery times, as well as enhanced equipment versatility and compatibility. As an added benefit, implementing remote surgery will be simpler. Mixed reality technology aids in surgical planning and execution by streamlining these processes. It is possible to build an interactive information loop between the virtual and the real world by modeling the target and comparing it to its real-world counterpart. Subversive changes will be brought about by the development of this technology in medical education and research as well as in clinical care.

11.2.2 HEALTH MANAGEMENT

Chronic illnesses have steadily risen to the top of the human disease spectrum in the early 21st century, and they have now become an epidemic in and of itself. When it comes to long-term conditions like diabetes and other chronic illnesses, prevention, and treatment are more critical than ever. As a result, it seems that the conventional hospital- and doctor-centered health management approach is ineffective in coping with the rising number of patients and illnesses [6]. Smart healthcare's new health management paradigm emphasizes patient self-management. Advanced sensors, microprocessors, and wireless modules are combined in third-generation wearable/implantable devices to allow continuous intelligent monitoring of a patient's physiological indications. These devices use less energy and are more comfortable for the patient, and they allow the data to be combined with health information from other sources. With this strategy, the focus shifts from event monitoring to ongoing awareness and comprehensive treatment. Medical institutions can better track the disease's prognosis because of the reduction in related hazards. New technology, such as smartphones, smartwatches, and other devices, makes this type of monitoring possible. Biosensors have been attempted to be integrated into cellphones. In addition to enhancing mobility, a high-performance smartphone allows users to more easily keep tabs on their surroundings and their health.

Smart houses aid the elderly and the handicapped in their own homes. Sensors and actuators are incorporated into the residential infrastructure of smart homes, which monitor the physical and environmental conditions of its occupants. Smart houses may also be used to enhance the quality of life for the occupants. The medical company's adoption of smart homes is heavily influenced by both automation and health monitoring [7]. Individuals in need of care may be able to reduce their reliance on medical experts and enhance the quality of life at home with the help of these technologies, which provide basic services and gather health data.

There are several applications and a health information platform that patients may use to monitor their health. Wearable medical sensors in the Stress Detection and Alleviation system, for example, continually monitor a person's blood pressure (BP) level and automatically assist the body in reducing stress [8]. A hierarchical health decision support system may also be built using data from several portable devices, which can then be used to accurately diagnose illness most efficiently feasible. The cloud calculator and big data can help identify patient risks and provide advice ahead of time while supporting clinical decision-making. An alternative approach is to build an open mHealth platform that facilitates the exchange of information between healthcare providers, patients, and researchers. Using telemedicine, patients may receive information and services quickly, and physicians can keep an eye on their patient's health.

11.2.3 DISEASE PREVENTION AND RISK MONITORING

Patients' data is traditionally gathered by health authorities, who then cross-reference it with guidelines issued by authorized organizations before making their illness risk prediction model publicly available. Inaccurate counsel is given to people because of a temporal lag caused by this practice [9]. Disease risk prediction is dynamic and individualized in the context of smart healthcare. With the help of this system, patients, and clinicians may engage, keep tabs on their own illness risk, and take action based on their data. Predicting the likelihood of developing an illness may now be done in real-time by using wearable devices and smart applications to gather data, upload it to a cloud, and then run the findings using big data algorithms. It has been proved that these methods are successful [10]. At any time, patients, and healthcare professionals alike may utilize them to change the healthcare habits of the people, and they can also contribute to the establishment of regional health programs that seek to reduce sickness risk.

11.2.4 VIRTUAL ASSISTANTS

It is an algorithm, not a living being, that performs the functions of a virtual assistant. User preferences or wants are taken into account when calculating responses from virtual assistants, who converse with users using methods like voice recognition or big data. Smart healthcare relies heavily on virtual assistants to facilitate communication between physicians, patients, and healthcare providers. They make it easier for people to get medical care. The smart device's virtual assistant can readily translate regular, daily English into one that uses medical terminology so that people may more properly locate the medical service they need. Patients' basic information is used to build the virtual assistant's responses, making it easier for physicians to keep track of their patients and plan medical operations more efficiently [11]. In medical facilities, the use of virtual assistants may save a lot of resources, both human and material, while also better meeting the demands of all parties involved. Medical service participants' experience may be substantially improved by using nuance technology to communicate with diverse virtual assistance, particularly between general assistants and highly specialist assistants. The use of virtual assistants to promote human mental health, which may alleviate the shortage of human psychotherapists and provide spiritual well-being for more patients, can also be employed to help cure illnesses.

11.2.5 SMART HOSPITALS

Regional, hospital, and family healthcare are all key parts of smart healthcare. Smart hospitals rely on ICT-based settings, especially those based on IoT optimization and automated operations, to improve existing patient care processes and introduce new features [12]. "Smart hospital" services include everything from medical personnel to patient care to administrative functions. Hospital management choices must take into account the needs of these patients. The IoT-based information platform in hospital management unites digital gadgets, smart buildings, and hospital staff [38]. Additionally, this technology may be used to identify patients in hospitals, manage medical personnel, and track tools and biological specimens in a hospital environment. For example, in pharmaceutical manufacturing and distribution, inventory control, anti-counterfeiting, and other procedures make use of smart healthcare. It is possible to supply each patient with their own RFID tag and save their data in a database that can be easily monitored and accessed through mobile devices in order to guarantee reliable, stable, and efficient distribution of hospital supplies. Resource allocation, quality assessments, and performance evaluations may all be carried out via an integrated management platform, which can also help hospitals save costs, improve resource efficiency, and aid in expansion decisions [13]. Patients may make use of a variety of services, including physical examination systems, online appointments, and contacts between patients and doctors. Using these automated technologies streamlines the patient's medical care. Patients have to wait less time and are treated kindlier as a result. All in all, integration, refinement, and automation will shape the destiny of smart hospitals in the future (see Figure 11.1).

11.2.6 AI IN HEALTHCARE

The healthcare industry's most critical tool is new technology. AI's potential success in the healthcare industry has been bolstered by the increased availability of health information and the quick development of massive data diagnostic tools [14]. AI can disengage information that should be incorporated in healthcare choices by recognizing important medical conditions. People may now live healthier and longer lives thanks to contemporary healthcare technology, which is now available to a broad range of creative entrepreneurs throughout the globe. First, the beginnings of accessibility and software have decided the development, allowing the health sector to digitize part of the pen-and-paper-based system activities now maintained by service release. Using AI in the healthcare business may save costs, enhance innovation, and improve customer service. The healthcare system is dealing with rising costs and needs more money to provide the services it provides. Identifying key relationships in a sample and using AI to assess and anticipate healthcare outcomes are two of the most prevalent uses of AI [15]. There are several medical applications for this technology, including cancer detection and therapy, chronic disease management, and more.

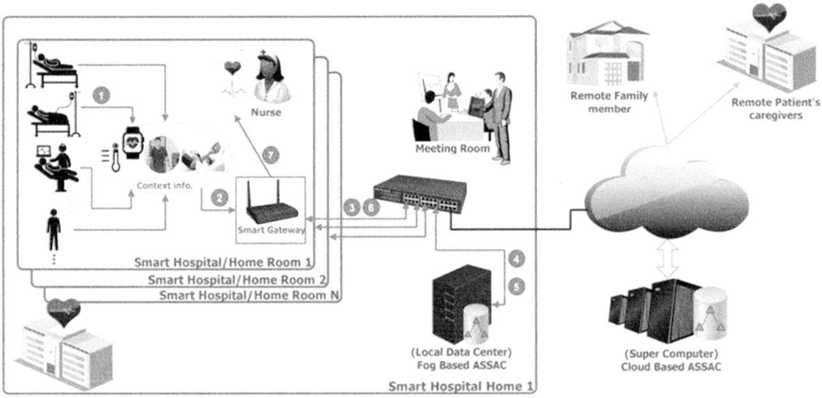

FIGURE 11.1 Smart fog computing for efficient situations management in smart health environments.

Source: Reprinted from Ref. [33]. https://creativecommons.org/licenses/by/4.0/

11.2.7 ARTIFICIAL INTELLIGENCE AND MACHINE LEARNING

AI and ML are having an increasing impact on the healthcare industry. The lives of millions of patients each year may be transformed by AI. As of 2025, the healthcare sector will be a $200 billion business, and AI will have a profound impact on it. To enhance and monitor their health, it is projected that customers would purchase the most recent healthcare applications and AI technologies available. There are now clinical AI robots that can give cognitive behavioral therapy to patients in their homes and places of employment, as well as assistive robots and diagnostic robots that can diagnose and treat patients wherever they are. Detection or exclusion of health conditions/diseases may be made using this pricey imaging approach. The medical image-assistive AI developed by Google DeepMind may help healthcare providers diagnose and cure a variety of vision-threatening disorders before they become more serious.

11.2.8 AUTOMATED ROBOTIC PROCESS CONTROL

As AI is being used in healthcare to provide robotic process automation, the way patients are treated is also being transformed by ML. It is expected that ML technologies will be employed to their full potential in healthcare settings.

As a result, imaging diagnostics are becoming more and more important, lowering patient wait times, allowing early and accurate diagnosis, and reducing medical staff burdens. The following resources might help you learn more about the positive effects AI and ML are having on the medical industry and the people they serve.

11.2.9 COMPUTER VISION

Computer vision saves a lot of time since medical professionals use their imagination and creativity to arrive at a precise diagnosis and decrease incorrect predictions via the use of computer vision. Predicting disease in real-time is one of the many advantages of computer vision. Detection of cancer, for example, may be done considerably earlier and so spare individuals from unnecessary suffering. Computer vision may also be used for health monitoring, not only as a fitness tracker, but for other uses such as tracking the quantity of blood loss after operations, assessing the body's fat percentage, and more.

11.2.10 WEARABLE TECHNOLOGY

Activity trackers and smartwatches like the Apple Watch, Fitbit, and Garmin are already commonplace. There are no indications of these wearable gadgets slowing down in society, which is a good thing. There is a growing market for wearable health technology as insurers, doctors, and firms marketing these devices see the value in monitoring patients' and users' physical activity and heart rate (HR).

11.2.11 AUGMENTED AND VIRTUAL REALITY

Virtual and augmented reality has already started to disrupt several industries, and the healthcare industry is no exception. Patients with long-term health conditions will be able to monitor their health from the comfort of their own homes. If a patient requires immediate medication delivery, a doctor may instruct the patient how to give the drug through a virtual reality (VR) tour. Doctors will be able to guide patients through forthcoming procedures and teach them about the problems or hazards they may face, as well as how they may overcome them, as the breadth of VR continues to expand. Physicians can see organs in 3D or high-definition graphics with this new technology, and they may zoom in on specific problem regions to make better treatment options. By 2025, the healthcare VR and AR industry is expected to be worth $11.14 billion.

VR and augmented reality will be used to do surgeries, and surgeons will be able to build models and practice the procedure before they perform it in the operating room. Surgeons can practice and see what could happen before they do the procedure. Even while this technology is still in its infancy and may not be able to cover every aspect of surgery, the potential of AR and VR aiding surgery is fascinating.

11.2.12 ELECTRONIC HEALTH RECORDS

Information about a patient's diagnosis, therapy, and progress notes, test findings, and operations are all included in medical records. Recordings are often made on paper, but they may also be heard or seen as audio or video files [16]. Their electronic storage allows for legal archiving and retrieval. The EHR software is better than paper records since it doesn't take up space and doesn't decay over time. Professionals in the medical field now have much simpler access to patient data because of electronic records. To keep crucial patient information from falling into the hands of the wrong people, electronic records have been given additional levels of protection. In addition to insurance eligibility verification, refill requests, direct messaging, referral exchange, and appointment reminders by text or e-mail, the patient experience is maintained easily and free of problems.

11.2.13 5G TECHNOLOGIES

When it comes to healthcare IT infrastructure, 5G is a game-changer [32]. It's not unusual to have trouble opening an e-mail at a hospital because of the huge volume of patients and visitors. Consider how critical it would be for healthcare systems to have access to more bandwidth to keep track of patient's health more often. It would be difficult to adequately monitor hospitals that only provide Wi-Fi on a 4G network.

However, most hospitals choose to utilize 5G networks since they need significantly less equipment and infrastructure than Wi-Fi six networks. As a result, 5G networks will revolutionize distant care and telemedicine, as they provide lower latency than 4G and far greater throughput than any prior cellular networks [17].

There are currently many organizations that are capitalizing on this new trend by developing networks that can handle and analyze patient information, enabling apps for training and administration reasons, as well as other functions. High frequencies of 30 to 300 gigahertz are possible in the next generation of wireless technology, but only six gigahertz are used by 4G networks.

11.2.14 TELEMEDICINE

People may be treated in the comfort of their own homes without having to go to a hospital thanks to decentralized healthcare provided via telemedicine [31]. Patients' personal information has already been elevated to the top of the priority list for 2019's cybersecurity efforts. Patients' private information must be protected since cyberattacks aimed against them are becoming more targeted and successful [43]. Currently, breaches can only be controlled after they have occurred, and the defensive side of cyber assaults comes at a significant cost.

11.2.15 BLOCKCHAIN

There are no intermediaries or security issues to deal with when using blockchain technology. Decentralized records may be connected using cryptography in the blockchain. Each of them is a "block," and it contains a mathematical procedure as well as a timestamp. Peer networks may keep track of digital transactions using the distributed ledger technology built into the blockchain. To avoid data breaches, healthcare organizations and providers may keep track of every exchange of a patient's medical records. Hackers who have been responsible for the vast majority of previous breaches involving medical data will be unable to do so any longer if blockchain technology is properly implemented [36]. Clear audit trails and enhanced privacy make clinical procedures more open and accountable.

11.2.16 BOTS

Natural language processing (NLP) and chatbots were widely accepted by healthcare experts and care providers as a means of diagnosing and treating patients. It is feasible to collect and analyze patient health data using patient-only apps, such as those already available on the market, as well as connecting patients and clinicians for diagnosis and treatment using patient-clinical apps.

11.2.17 VOICE SEARCH

Once a patient is discharged from the hospital, a virtual assistant may help them schedule appointments, provide transportation to the hospital and even walk them through the post-procedure process. When a patient asks a question, the system will search for relevant material based on their voice and provide it to them. For senior patients who may not be familiar with computers and iPads, this is a godsend. For the most part, all they have to do is use voice search to get the information they need. Siri, Alexa, Google Home, and Cortana are the most popular voice assistants.

A significant portion of their clients will leave a business that doesn't voice search ready. Companies with high listing accuracy and visibility will undoubtedly capture a portion of visitors while this is still in its infancy. When the accuracy standards for voice search are improved and perfected, they will be able to show up in voice search results.

11.3 THE PROPOSED FRAMEWORK ZORAWAR: BIG DATA AND VISUALIZATION ORIENTED LATENCY AWARE SMART HEALTH ARCHITECTURE

The proposed ZORAWAR: big data and visualization oriented latency AWARe smart health Architecture is designed with five primary divisions such as smart health sensors, low power short distance interface, fog division, cloud, and big data division and smart health community, as shown in Figure 11.2. The ZORAWAR's primary goal is to make remote health monitoring for a healthy way of life possible. In order to exchange data with the smart fog nodes or the cloud and big data division, this interface is necessary. For transmitting signals, it has Bluetooth, Zigbee, and Wi-Fi built in the next section named, "Low Power Short-Distance Interface." The next division comprises fog nodes. The citizens' health data is collected by the first division. The smart fog nodes are used in the second division. To estimate one's future health state based on gathered data, doctors employ a procedure called health status prediction. The outcome will be combined with the present data to improve the classification's accuracy. It's the smart fog nodes' job to categorize a user's current health state using sensor data and a forecast. It ends up in the cloud and big data divisions, where servers are located in both. Each step makes the healthcare system better by increasing its quality and accuracy. It is assumed that all users have the same number of sensors to collect health data for the proposed healthcare system.

The ZORAWAR was created with this premise in mind. The successive subsections of the proposed system explain the functions carried out in each of the divisions.

Wireless sensors installed around the home and senior patients' data may be used by this module to collect physiological and environmental data, which is then converted into analyzable form. This module cleans up the data by applying filters and preprocessing steps. Cellphones are becoming more used in healthcare solutions [18]. Even if IoT devices lack the processing power and networking capabilities, smartphones help by transmitting produced data to a cloud data center and then delivering apps from there [34]. The program allows users to choose how often they get data on their devices. Smartphones are equipped with sensors, such as a global positioning system (GPS), an accelerometer, and others, which collect data about their surroundings.

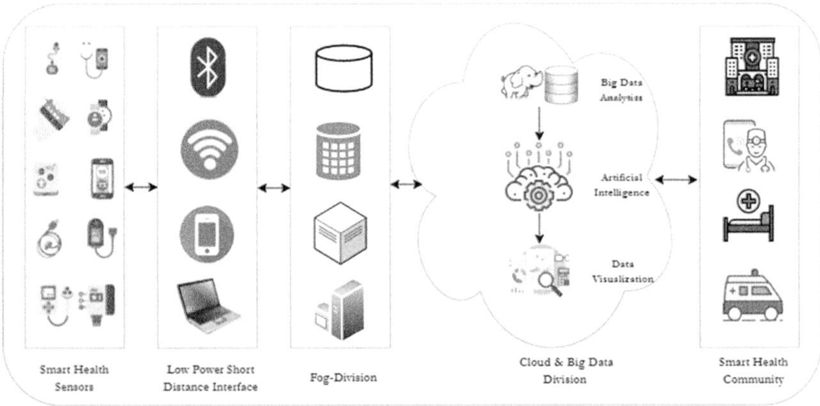

FIGURE 11.2 Smart ZORAWAR: big data and visualization-oriented latency aware smart health architecture.

11.3.1 LOW POWER SHORT DISTANCE INTERFACE

The ZORAWAR has numerous low-power wireless technologies are available, such as Bluetooth low energy (BLE), RF-based technologies such as ZigBee, RF4CE (receive four times as much energy as transmits), NFC, Nike+, and Wi-Fi, as well as infrared solutions backed by the Infrared Data Association (IrDA). However, having so many options make it more difficult to make a decision. Each technology makes trade-offs in terms of power consumption, bandwidth, and range for each of its features. However, some are built on open standards, while others are exclusive to their creator. To add to the confusion, new wireless interfaces and protocols are constantly being developed to meet the demands of the IoT [19].

11.3.1.1 WI-FI

Device networking and Internet access are the primary functions of Wi-Fi, a series of wireless network technologies based on IEEE 802.11

specifications. Digital devices may communicate with each other via radio waves. Computers, tablets, smartphones, smart TVs, printers, and smart speakers may all be connected to a wireless router using these networks. They're also common in public places like coffee shops, hotels, libraries, and airports, where people can use their mobile devices to access the Internet for free.

11.3.1.2 BLUETOOTH

Bluetooth is a short-range wireless communication protocol that makes it possible to send and receive data wirelessly. These two techniques appear to be very similar since they use the same radio frequencies. The three distinct kinds of Bluetooth technology that are often discussed are as follows:

1. **Bluetooth:** It is a bygone era remnant of a mobile era defined by the huge cell phone. Such Bluetooth technology is inefficient, unsafe, and often difficult to connect.
2. **BLE (Bluetooth 4.0, Low Energy Bluetooth):** Originally developed by Nokia, BLE is currently supported by the majority of major platforms, including iOS, Android, Blackberry, OS X, Linux, and Windows 8.
3. **iBeacon:** This is the trademark for an Apple-developed streamlined communication system based on Bluetooth technology. What it is: a Bluetooth 4.0 transmitter that sends a unique identifier called a UUID, which your iPhone recognizes. This significantly reduces the amount of work required to install for many suppliers before. Additionally, even users who are not technically savvy may readily utilize iBeacons such as Estimote.com or other options. Although technically distinct, iBeacon technology may be related to near-field communication (NFC) on an abstract level.

Bluetooth is integrated into a wide variety of goods, including smartphones, tablets, media players, and robotics systems. The method is especially advantageous for sending data between two or more devices that are located close to one another in low-bandwidth environments. Bluetooth is often used to send audio data between telephones and hand-held PCs (transferring files). Bluetooth standards make it easier for devices to find and configure services. Bluetooth devices may promote the whole range of services they provide. This simplifies the process of using services since, in comparison to other communication protocols, it provides better automation in areas such as Security, network address setting, and authorization configuration.

11.3.1.3 NEAR FIELD COMMUNICATION (NFC)

When two loop antennas are placed near together, electromagnetic induction occurs, creating an air-core transformer. It uses the ISO/IEC 18000-3 air interface to transmit data at rates ranging from 106 kbit/s to 424 kbit/s across the 13.56 MHz ISM band. A passive target may be powered by an NFC initiator, which generates an RF field to power the initiator (an unpowered chip referred to as a "tag"). Therefore, NFC targets might take the shape of tags, stickers, key fobs, or battery-free cards. NFC stands for NFC. Devices enabled by NFC can communicate with one other [20].

There are two ways of operation:

1. **Passive Communication Mode:** The initiator device sends out a carrier field, and the target device replies by making changes to its own carrier field. The initiator's electromagnetic field may be used to energize the target device in this mode, converting it into a transponder.
2. **Active Communication Mode:** The initiator and the target device exchange fields in order to communicate with one another. A device's RF field is off while it waits for data. This is the default setting for both devices.

11.3.1.4 FREQUENCY OF RADIO (RF)

Radiofrequency communications are arguably the simplest kind of device communication. ZigBee and Z-Wave protocols make use of a low-power radio frequency (RF) radio installed or retrofitted into electrical products and systems.

Z-Wave has a range of around 100 feet (30 m). Country-specific RF bands are being used. SRD bands are 868.42 MHz in Europe and 908.42 or 908.42 in the United States, respectively; in Israel, 916 MHz is used; in Hong Kong, it is 919.82, while in Australia/New Zealand, it is 921.82; in India, it is 865.2 MHz.

11.3.1.5 ZIGBEE

In accordance with IEEE 802.15.4, ZigBee is a low-power wireless protocol. However, transmission distances are restricted to 10 to 100 meters because of the low power consumption.

11.3.1.6 RFID

The use of EMWs to wirelessly identify objects is known as radiofrequency identification (RFID) [21]. Installing an active reader or tags that retain data, mostly authentication answers is standard procedure. Experts refer to this technique as an Active Reader Passive Tag (ARPT). RFID has a low range of around 10 cm but a long range of up to 200 meters. When Léon Theremin created the RFID in 1945 as an espionage weapon, many people were unaware.

11.4 FOG-DIVISION

The fog division is composed of several dispersed smart devices or nodes, referred to as fog nodes. A fog node is a computer, storage, and networking device that is dispersed close to the sensors that generate the data [39]. Fog nodes are responsible for four tasks: (i) collecting patient data from sensors; (ii) evaluating this data to determine the patient's health status; (iii) interacting with caregivers; and (iv) transmitting the data to the cloud. Local data processing is added to the fog to give it intelligence, enhancing the system's stability, reducing latency, overcoming internet disconnection, and speeding up decision-making in emergency circumstances.

11.4.1 FOG NODE

Fog nodes may operate on their own or alongside a cloud node to achieve their goals [34]. Fog nodes can perform predefined/dedicated services independently without cloud support since they contain computational power resources and data services capabilities (e.g., data processing and aggregations) on board. As an example, fog nodes may monitor and analyze real-time data from a pressure sensor and then initiate actions like opening or shutting a valve in response to a sensor pressure measurement. Fog, on the other hand, has limited hardware capabilities in comparison to the cloud. Therefore, it may rapidly get congested when data flow is excessive and surpasses the fog capacity. As a result, the tasks may need cloud or fog cooperation (refer Figure 11.3). It's evident that as the amount of service traffic on fog nodes grows, so does the risk of fresh arriving service requests being delayed due to the existing traffic. Low latency goes hand in hand with a large amount of traffic, which is related to fog node capacity and resource management.

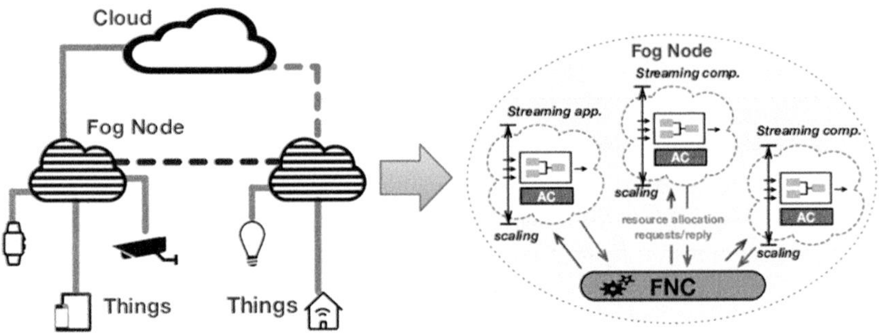

FIGURE 11.3 Container-based support for autonomic data stream processing through the fog.

Source: Reprinted with permission from Ref. [42]. © 2018 Springer Nature.

11.4.2 FOG REPOSITORY

Secure and analyzing data requires private, locally stored storage for the gateway to keep the data safe and accessible. Consequently, data that cannot be sent to the cloud is stored at the fog node for a short period of time.

11.4.3 DATA SECURITY MODULE

The safety and integrity of patient data are essential issues when designing healthcare frameworks since patient data is vulnerable during transmission from a gateway to the cloud. As a consequence, this module emphasizes the importance of encryption and decryption in ensuring the safety of healthcare information. Using encryption, data may be protected from unauthorized access by encoding it in an unreadable manner. When patient data is kept on fog nodes or communicated to a cloud server, the data security module in our proposed system strives for privacy.

11.4.4 DATA MANAGEMENT

Fog computing strategies retain the numbers of sensual close to reduce evocative information for announcements and operational reaction in addition to system plan adjustments. In healthcare situations, hesitation, and dormancy in decision-making may lead patients to incur eternal indemnities. The first sector component to build a cleaning strategy after obtaining numbers from the device network is the data dispensation sector component. Many bio-signals such as EEG, EMG, ECG, and others are used to construct the patient's situation utilizing the appropriate procedure. Using a lossy or lossless density technique, patient data may be compressed to save space when sent over a communication network. The majority of IoT healthcare applications utilize compression using a lossless technique since lost data might lead to incorrect interpretation of illness.

11.4.5 PATIENT IDENTIFICATION MODULE

Face, iris, and fingerprint biometrics are all included in this module's scope of responsibility for patient identification. Because technology can function with faces from many different angles and cameras and mobile phones are all over the place, face recognition is the most straightforward, fastest, cheapest, and most reliable method of identifying patients. Face recognition is a three-step process that begins with face detection, continues with feature extraction, and concludes with face recognition. The face detection process determines whether or not the acquired picture includes a face. If it includes a face, it recognizes and defines the position of the face inside the picture. Feature extraction is the process of extracting features from a recognized face using either appearance-based or geometric-based approaches. Finally, the recognition stage uses ML classifiers to compare the feature vector to all other vectors contained in the face database to determine the face's identification.

Pain detection and measurement are performed by this module as part of the patient monitoring system. There are two approaches for detecting pain: the first employs wearable sensors to gather vital indicators, including temperature, BP, and blood glucose, to determine the intensity of the sensation. When determining whether or not a patient is in pain, the second way looks at the patient's facial expressions to see how much. Because it represents a patient's health state and emotions, facial expression recognition is useful in healthcare settings. Pain from illnesses without a direct method to reflect it, such as headache, tooth pain, and internal medical aches may be defined by the system with the use of this equipment. Taking pictures of a patient's face regularly might help a caregiver gauge how much pain the patient is experiencing and whether or not assistance is needed. When treating and diagnosing disorders including schizophrenia, depression, autistic spectrum disorder (ASD), and bipolar disorder (BPD), the emotions of the patient are critical factors to consider. This module uses facial expressions and vital indicators to determine the pain level.

11.5 CLOUD AND BIG DATA DIVISION

11.5.1 CLOUD DIVISION

In the cloud division, resources, repositories, and servers are spread out around the globe. Using a cloud manager, these devices may receive processes and store patient data [22]. These data may be used by a clinician or caregiver to do a long-term study of the patient's health state and history.

11.5.1.1 CLOUD DATACENTER

A cloud data center is an ideal location for IoT in healthcare explanations. Computation on a big scale makes it possible to package service value while also increasing scalability and, ultimately, dependability cloud resources are virtualized and may be structured architecturally.

11.5.1.2 RESOURCE MANAGER

With IoT-enabled Healthcare information, this is in charge of arranging cloud resources. Resource scheduling may be done based on the amount of work that needs to be done, demand, and the current circumstances. It also confirms that the properties have very strict access controls. To run wrong demand, resource management defines interdependence between resources, so they are all run at the same time.

11.5.1.3 SERVERS

Cloud data centers are made up of servers that may be either homogenous or heterogeneous in terms of hardware configuration, depending on the cloud provider. There are two sorts of servers in this configuration: application servers and database servers. On the application server, the web services and backend requests are housed together with the database. Only the association operation and then the data source are changed in the database.

11.5.1.4 VIRTUAL MACHINES (VMS)

For each VM, the hardware resources are made available by the host server. They reduce the amount of available memory, processing speed, and storage capacity by using compression.

11.5.1.5 DATA ANALYSIS MODULE

This module is in charge of making judgments at the fog layer when a quick response is required. To make the right choice, information like the patient's age, weight, height, and any existing medical conditions is needed. The sensor data is received, and ML algorithms are used to determine whether an emergency scenario has arisen, and a quick alert is sent along with the gathered data. This is done in the cloud.

11.5.1.6 DATA COMPRESSION MODULE

Cloud storage is required for the long-term study of patient data, as well as the prediction of illnesses and quantification of risk. Because of this, networks get congested, latency increases, and storage space requirements balloon as more and more data is transmitted to the cloud over time. Compression of images is a solution to the prior issues. Two stages separate lossless compression from regular compression. To begin, convert the patient data into a different format. The second stage eliminates redundant coding by using an entropy encoder.

11.5.2 BIG DATA DIVISION

11.5.2.1 BIG DATA STORAGE

Sensor data is analyzed at the fog layer before being transferred to the cloud storage, which has a lot of room for storing large healthcare data files. It is possible for a variety of healthcare stakeholders to make use of patient data in various ways [23].

11.5.2.2 BIG DATA ANALYSIS

To help analyze all patient data, it makes all data accessible to assist and enhance long-term clinical decision-making processes, treatment, and medical research. This is a great service to the medical community. These data may be subjected to ML techniques, prediction algorithms, and data visualization algorithms to get more information.

11.5.2.3 DISEASES PREDICTION

Diseases that a patient is likely to develop in the future are predicted using characteristics including age, height, weight, and family history. Calculating the percentage of projected diseases using machine-learning techniques is the goal, which relies on the correlation of authorized vital indicators to do so.

11.5.2.4 DATA ACQUISITION

The initial step in the biomedical big data life cycle is data collecting. Structured, semi-structured, and unstructured formats are all used to acquire or gather data from diverse biological sources (medical images, medical devices, biomarkers, genomics, general health, and clinical data). Data cleansing, which identifies what is valuable to collect and what is pointless to discard, is a key difficulty in big data. The integrity and history of digital objects, where they originated from, and how they got to their current condition, or the current status of the data, is another key difficulty in big healthcare data. A software system's trustworthiness may be increased by using provenance in health datasets. Since processing pipeline records track the origin and transportation of data, this will help indicate the next processing stages. However, if a mistake occurs at a particular step, it will have an impact on all succeeding phases of analysis. An additional issue with big data in healthcare is the need to create metadata automatically. Information regarding data meaning, language, ideas, and connections may be found in metadata. It also tells you where your data came from. Data provenance must be understood, and huge streams of it processed before data is stored, which necessitates effective analytical algorithms.

11.5.2.5 DATA STORAGE

Large amounts of unstructured information in healthcare need greater storage capacity for big data analytics. To diagnose illnesses, patterns must be revealed, as well as actions that may have an impact on a patient's health. Furthermore, a solution for the growing number of digital instruments, such as wearable devices, apps, and electronic health records (EHR), carried by patients and physicians alike, is an absolute need. The human genome is so large that it creates hundreds of terabytes of data each year, and the sequence of that data doubles around every seven to nine months. The conventional techniques of data analysis are thus inadequate for managing such systems, and new, scalable, and trustworthy ones must be developed to assure high storage capacity.

11.5.2.6 DATA MANAGEMENT

The complexity of managing heterogeneous data of patients stems from its wide range. To access and visualize patient data in real-time and streaming, data management processes maintain their accuracy. At this point, it guarantees that the present data is validated even further. In addition to that, all areas of data management must be carefully managed. Some companies have rules in place to help with data management. Healthcare-related big data difficulties may be addressed with the help of the proposed Hadoop framework.

11.5.2.7 DATA ANALYSIS

To assist with heterogeneous data of patients acquired from many sources, large data analysis methods are necessary. This is where the code or algorithm that generates the final results is executed. Descriptive, diagnostic, predictive, and prescriptive analysis are some of the most common forms of analysis.

11.5.2.8 DATA VISUALIZATION

This step is supposed to deliver a variety of outputs, including reports on patient health monitoring and decisions. As a result, visualization must allow batch processing as well as real-time analytics to help medical systems make better decisions and avert crises.

11.5.2.9 NOTIFICATION MODULE

This module is in charge of delivering alerts to the patient's caregivers and family members in the event of an emergency. This informs the patient whether they need to take any action, such as when to take their medication, how much medicine to take, what food to consume, or if there are any additional instructions they should follow. Messages, voice calls, or voice over IP (VoIP) notifications are all voice telephone calls or VoIP.

11.6 CHALLENGES IN SMART HEALTHCARE ADOPTION

Smart healthcare systems that make use of EHRs and technology such as the IoT and big data are projected to seamlessly connect patients and clinicians across a wide range of healthcare systems in the future [35]. These systems are also becoming increasingly interconnected with various types of medical wearable technology, which are being worn for the purpose of real-time healthcare monitoring over the Internet. However, a number of problems must be overcome before digital healthcare can establish systems that are reliable, flexible, and compatible with one another. Following that, we'll go over some of the present roadblocks that are preventing mainstream adoption of digital healthcare services [24].

11.6.1 SECURITY AND PRIVACY

IoT devices might pose a security and privacy risk to their owners because of their ubiquitous nature. Access to the IoT devices by unauthorized individuals might have serious consequences for patient safety and privacy. There are medical and mobile devices, together with other linked devices in the cloud, that gather and transfer health records to and from their different locations. The device layer is subject to attacks such as tag cloning, spoofing, RF jamming, and cloud polling, among other techniques. Patients' safety can be jeopardized by denial of service (DoS) assaults against healthcare systems. Because of the sheer volume and complexity of new software and hardware vulnerabilities, it's difficult to quickly identify possible security concerns. Connected gadgets are multiplying, and this problem is only going to worsen. For example, insecure Web-based interfaces substantially expand the attack surface because of default authentication. Wearable technologies, on the other hand, have witnessed a significant increase in popularity in recent years. It's easy to find Internet-connected wearables because of the availability of strong search engines like Shodan, but there are no security guidelines in place to protect them.

11.6.2 AUTHENTICATION AND INTEROPERABILITY CONCERNS ACROSS MANY DOMAINS

Digital health transactions require the establishment of trust between entities operating in various domains, which can only be achieved through inter-realm authentication. An identity solution called Shibboleth supports the authentication of entities within and across organizational systems. Inter-realm authentication system Shibboleth has been successfully implemented and tested at the national level.

Authentication in a digital health system is made easier with Shibboleth-based systems, which are safe and reliable. The absence of facilities to host separate identity and service providers in a nation-based organization and to host them across all comparable digital health organizations means that not all digital health systems adopt Shibboleth implementations. Because of a lack of IT skills and money, systems like Shibboleth cannot be extensively used, particularly in underdeveloped nations.

Another issue that has to be addressed as a common concern for nations working on digital health ICT infrastructures is a lack of interoperability [25]. An inability to provide high-quality telemedicine services and a lack of international cooperation on sensitive medical data transfer are due to a lack of IT infrastructure and IT competence. As an example, the Catalan Digital Health System (CDHS) case study covers the following features:

- A user ID and password or an X.509 digital certificate is used to authenticate medical doctors against hospital systems;
- To ensure the safety of prescriptions, Catalan health services receive a Security Assertion Markup Language (SAML) assertion from each hospital;
- To authenticate against the Catalan Council of Pharmacies' database, pharmacies employ either X.509 digital certificates or user credentials (ID/password);
- Catalan Health Services receives a SAML assertion from the Catalan Council of Pharmacies for the purpose of granting them access to pending ePrescriptions;
- There must be an electronic signature from a doctor for each prescription supplied, and the Catalan Health Service must be informed.

11.6.3 HEALTH INFORMATION EXCHANGE BARRIERS

With the capacity to securely transfer healthcare information across a variety of healthcare organizations, the Health Information Exchange (HIE) serves to enhance the public's access to healthcare [25]. Consumer-mediated exchange directed interchange and query-based exchange is all viable solutions based on the IoT that are currently being used in HIE deployments.

It is much easier for patients to stay on top of their health-related issues when they have access to their own electronic records. To provide the best possible treatment for their patients, healthcare organizations often use "directed exchange" to pass along pertinent information about their patients to other medical professionals [26]. In the context of unexpected medical care, a query-based exchange is typically used when a healthcare organization requires a new patient's past health records from a previous healthcare organization. This is accomplished through the use of the HIE system to seek access to these records.

Security and privacy concerns are the biggest impediments to the implementation of HIE systems. The following are some of the problems with the present HIE systems: First and foremost, authorized insiders are abusing their privileges. Most often, this occurs when medical facilities, either carelessly or selfishly or in exchange for some sort of gain, give outpatient medical records to those who are not allowed to have access to them. Health Information Management Systems (HIMS) routinely provide media access to the medical information of celebrities and politicians. Insiders who have access to the system but not to the data may also be in violation of the rules. Former hospital personnel who haven't been barred from viewing their own medical information online, for instance. To get revenge on their former employers, the first group can exploit the already-existing access to the HIMS database to hack into the private information stored inside. An unauthorized intruder may try to get into the system either by attacking the system directly or by pretending to be a member of the medical team. HIMSs face a new and serious threat in the form of healthcare-related cybercrime. In terms of reputational damage, penalties, litigation, and so on, a security breach in a hospital can cost it as much as $7 million. A number of well-known companies, including Anthem, CareFirst, Premera, and UCLA Health, have suffered significant data breaches. In 2015, the Healthcare Information and Management Systems Society (HIMSS) conducted a cyber security assessment and found that 64% of healthcare businesses had been the target of external assaults in the previous year. During the past two years, Bloomer News reported that 90% of all healthcare institutions had been targeted. More than any other industry, the healthcare and medical industries are the most likely to suffer from data leaks.

11.6.4 DEVICE COMMUNICATION

In order to implement linked health solutions, it is necessary to have a solid communication system in place. To gather data, many devices now include sensors that speak the language of the device with which they are interacting [27]. As a result, many devices communicate with servers in this language. Sensors from different manufacturers may not be able to communicate with one another since they each have their own unique protocol. Important data is often isolated on data islands as a result of the IoT's central purpose being undermined by a fragmented software environment and privacy issues.

In addition, the existence of multiple devices raises questions about the safety of utilizing wireless network technology to connect medical devices. There are several scenarios in which the usage of a Wireless Personal Area Network (WPAN)-enabled device may result in collisions with other WPAN devices operating in the same frequency channel. Because it affects performance and can lead to dangerous scenarios, collision in WPANs disrupts healthcare delivery in various ways. When medical devices are connected via wireless communication, it is critical to ensure that they function effectively.

For physicians, smart health systems are not always simple to utilize. Some healthcare employees may be put off by the complexity of a system because of the many aspects there.

To be genuinely interoperable, a linked health system must be able to exchange data across many interfaces via both one-to-one and one-to-many connections, necessitating the cooperation of various systems. Authentication and encryption are critical in healthcare facilities, and devices must be able to communicate in a variety of formats and protocols. Directories containing device functionality, protocols, terminologies, and standard compliance are required for device management. Medical equipment is still unable to achieve the kind of "plug and play" interoperability that is already typical in non-health fields.

11.6.5 COLLECTION AND MANAGEMENT OF DATA

The use of IoT sensor devices in digital healthcare presents a number of data management difficulties. Formed or implanted inside the body, medical sensors are used to collect data. Because the human body is continually changing, there is a steady flow of new data. Furthermore, the data collected is a wide range of different types [28]. Electrocardiogram (ECG) data, for example, are often encoded in an XML format, whereas camera-based IoT devices typically capture data in a wide variety of image formats. When it comes to a connected health scenario, there is a wide range of components that are all interconnected: user devices, networks, systems, and vast amounts of data. The cyber-physical components of digital health systems must be developed using appropriate data-driven learning strategies in order to cope with their ever-changing nature. An in-depth understanding of a patient's health status can be gleaned from data analysis done correctly. This data isn't being used to its full potential since it's not being properly processed and mined, and collecting it also wastes computing resources. To help researchers, pharmaceutical companies, and healthcare providers quickly extract meaningful information from patients' data, several data analysis tools have been created during the past decade. The sheer volume and speed of data created in healthcare facilities, along with the lack of common data-gathering formats, creates a slew of problems. With regard to big data, the importance of integrity cannot be overstated.

It's difficult to get data in the healthcare system that's clean, structured, complete, and precise. In addition, the definitions of healthcare and the measurements used in the healthcare business are continually evolving. When it comes to financial metrics, physicians also report the length of stay (LOS) indicator. If consumers don't know which measure to use or don't know the definition of the metric presented, their decisions may be affected because LOS definitions can vary so much. The length of time a patient spends in bed is used by doctors to determine LOS. The 24-hour time frame is used to measure LOS for financial purposes. However, the 24-hour clock stops at midnight. If the meaning of "LOS" is not consistent, the collected data may be misinterpreted.

11.6.6 MULTI-DISCIPLINARY KNOWLEDGE IS USED IN THE DESIGN AND EXECUTION

A wide range of abilities, including those in embedded systems, network architecture, data analytics, and bioengineering, are required to achieve digital health. The development and implementation of such a complex system necessitate a wide range of expertise across many different fields [29]. The system must also be constantly updated to keep up with the ever-changing demands of the market. Ultrasound (US) and CAT scan imaging technologies, for example, have only a limited level of interaction with smart health systems at this time.

11.7 FUTURE VISION

Healthcare facilities are complex ecosystems with a wide range of difficulties and stakeholders to contend with and manage. An examination of how healthcare facilities grow and rely on teamwork as well as a mix of emerging technology, with a particular emphasis on the IoT, Big Data Analytics and AI [30]. With a complex ecosystem of diverse types of buildings, services, systems, and stakeholders, the healthcare industry operates in a highly regulated environment. Now that the digital revolution has begun, we must consider how it will affect the delivery of patient care in the future. The use of data to gain deeper insights into patient care and the condition of the facility can have a significant impact on the health of patients.

Having the capacity to deliver the correct data to the appropriate place and people at the right time is crucial in deciding the most effective treatment for patients, but it is also critical in assuring the facility's safe and proper operation and optimization. Because of the wide range of healthcare services available, there are several different facilities and divisions within a facility, each with its own set of requirements and challenges to address, ranging from significant effect to infection control to staff safety and patient experience and everything in between. All of these difficulties must be handled and solved in a realistic and collaborative manner, with the ultimate goal of tackling the primary priorities of the future health system, which are:

- Ensuring that the appropriate degree of access to services is provided to the people;
- Providing high-quality care and achieving positive outcomes;
- A cost that is reasonable for both the payee and the payer.

Patient-centered care has become more prevalent as a result of a greater focus on patient outcomes. Healthcare services such as outpatient, emergency, and specialty treatment are delivered closer to the patient's home in areas that are convenient for them under this new paradigm. In order to facilitate collaboration between care providers as well as the daily functioning of smart healthcare facilities and organizations, it is vital that we understand how data is maintained and used.

11.8 DRUG DISCOVERY

Drug discovery and development will be improved by the use of AI and big data in scientific research. All of the many stages of drug development, from target screening through the discovery of new drugs to testing in humans, are included. A common approach to identifying potential therapeutic targets is to compare established medications to a variety of bodily substances. Though time-consuming, this approach is often overlooked. To perform drug and target checks, AI has greatly expedited the procedure. To find RNA-binding proteins in ALS and cancer genomics, the Watson approach was used. It's also possible for the screening process to be adjusted or rectified at any moment through the use of AI.

Drug discovery relies primarily on high-throughput screening, which generates and tests a large number of compounds at a rapid pace. The cost and risk increase in tandem with the introduction of a new compound. Virtual drug screening utilizing AI may be an effective solution to this problem. The number of drug molecules that must be analyzed may be decreased by adopting computer pre-screening. The use of this approach may improve lead compound discovery, evaluate the expected activity of medicinal molecules, reveal potential compounds, and finally develop a library of compounds with relevant properties. Big data, AI, and the IoT are all being employed in pharmaceutical research. When it comes to screening for exclusion criteria, using AI to analyze and match a large number of instances may help reduce recruiting time and improve the accuracy of the target population. The real-time monitoring of lung disease clinical trials is now possible with smart wearable devices, giving clinicians access to more accurate and timely data [40]. Clinical trials may benefit from the use of new technologies like blockchain, which may assist ensure patient safety and test validity. To make sense of all the information they've gathered, scientists rely on a specialized platform.

11.9 CONCLUSION

The advancement of information and communication technologies is accelerating at an alarming rate right now. Overall, the adoption and deployment of these technical advancements in the healthcare industry provide enormous advantages to all stakeholders. Our research presents a technique to combine smart health applications with the connectivity of the IoT is presented which enables the usage of big data, AI, and reduces the complexity of context-aware designs. Digital healthcare technology adoption is now hindered by a number of significant barriers, both domestically and globally, which we address in this chapter, along with some possible solutions for accelerating adoption. A number of hurdles must be overcome before electronic healthcare becomes a reality in the healthcare business, which is growing more interested in the IoT and sophisticated analytics.

KEYWORDS

- **artificial intelligence**
- **digital healthcare**
- **global positioning system**
- **Internet of things**
- **natural language processing**
- **virtual reality**

REFERENCES

1. Hossain, M. S., Muhammad, G., & Guizani, N., (2020). Explainable AI and mass surveillance system-based healthcare framework to combat COVID-I9 like pandemics. *IEEE Netw., 34*(4), 126–132.
2. Muhammad, G., Alhamid, M. F., & Long, X., (2019). Computing and processing on the edge: Smart pathology detection for connected healthcare. *IEEE Netw., 33*(6), 44–49.
3. Miao, F., Liu, Z. D., Liu, J. K., Wen, B., He, Q. Y., & Li, Y., (2020). Multisensor fusion approach for cuff-less blood pressure measurement. *IEEE J. Biomed. Health Informat., 24*(1), 79–91.
4. Yang, F., Zhao, X., Jiang, W., Gao, P., & Liu, G., (2019). Multi-method fusion of cross-subject emotion recognition based on high dimensional EEG features. *Frontiers Comput. Neurosci., 13*, 53.
5. Gu, Q., Jiang, S., Lian, M., & Lu, C., (2019). Health and safety situation awareness model and emergency management based on multi-sensor signal fusion. *IEEE Access, 7*, 958–968.

6. Muzammal, M., Talat, R., Sodhro, A. H., & Pirbhulal, S., (2020). A multi-sensor data fusion enabled ensemble approach for medical data from body sensor networks. *Inf. Fusion, 53*, 155–164.
7. Van, S. T., Deschrijver, D., & Dhaene, T., (2019). *Sensor Fusion Using Backward Shortcut Connections for Sleep Apnea Detection in Multi-Modal Data*. arXiv:1912.06879.
8. Lin, K., Li, Y., Sun, J., Zhou, D., & Zhang, Q., (2020). Multi-sensor fusion for body sensor network in medical human-robot interaction scenario. *Inf. Fusion, 57*, 15–26.
9. Al-Shargie, F., (2019). *Fusion of fNIRS and EEG Signals: Mental Stress Study* (Vol. 2019, pp. 1–5). engrXiv. doi: 10.31224/osf.io/kaqew.
10. Mergin, A. A., & Premi, M. S. G., (2017). Pixel level fusion of medical signals using DCT, DWT and hybrid (DWT-DCT) transform based on maximum selection rule—A comparison. In: *Proc. Int. Conf. Comput. Methodologies Commun. (ICCMC)* (pp. 898–903). Erode, India.
11. Chen, J., Zhang, L., Lu, L., Li, Q., Hu, M., & Yang, X., (2020). A novel medical image fusion method based on rolling guidance filtering. *Internet of Things*.
12. Cabria, I., & Gondra, I., (2017). MRI segmentation fusion for brain tumor detection. *Inf. Fusion, 36*, 1–9.
13. Nathan, V., & Jafari, R., (2018). Particle filtering and sensor fusion for robust heart rate monitoring using wearable sensors. *IEEE J. Biomed. Health Informat., 22*(6), 1834–1846.
14. Simjanoska, M., Kochev, S., Tanevski, J., Bogdanova, A. M., Papa, G., & Eftimov, T., (2020). Multi-level information fusion for learning a blood pressure predictive model using sensor data. *Inf. Fusion, 58*, 24–39.
15. Fabiano, D., & Canavan, S., (2019). Emotion recognition using fused physiological signals. In: *Proc. 8th Int. Conf. Affect. Comput. Intell. Interact. (ACII)* (pp. 42–48). Cambridge, U.K.
16. Chen, C., Jafari, R., & Kehtarnavaz, N., (2016). A real-time human action recognition system using depth and inertial sensor fusion. *IEEE Sensors J., 16*(3), 773–781.
17. Zhang, W., Yang, J., Su, H., Kumar, M., & Mao, Y., (2018). Medical data fusion algorithm based on internet of things. *Pers. Ubiquitous Comput., 22*(5, 6), 895–902.
18. Du, J., Li, W., & Tan, H., (2019). Intrinsic image decomposition-based grey and pseudo-color medical image fusion. *IEEE Access, 7*, 56443–56456.
19. Qi, G., Wang, J., Wang, Q., Zhang, Q., Zeng, F., & Zhu, Z., (2017). An integrated dictionary learning entropy-based medical image fusion framework. *Future Internet, 9*(4), 61.
20. Baloch, Z., Shaikh, F. K., & Unar, M. A., (2018). A context-aware data fusion approach for health-IoT. *Int. J. Inf. Technol., 10*(3), 241–245.
21. Dautov, R., Distefano, S., & Buyya, R., (2019). Hierarchical data fusion for smart healthcare. *J. Big Data, 6*(1), 19.
22. Herrera-Luna, I., Rechy-Ramirez, E. J., Rios-Figueroa, H. V., & Marin-Hernandez, A., (2019). Sensor fusion used in applications for hand rehabilitation: A systematic review. *IEEE Sensors J., 19*(10), 3581–3592.
23. Haghi, M., Thurow, K., & Stoll, R., (2017). Wearable devices in medical internet of things: Scientific research and commercially available devices. *Healthcare Informat. Res., 23*(1), 4.

24. Abdel-Basset, M., Ding, W., & Abdel-Fatah, L., (2020). The fusion of internet of intelligent things (IoIT) in remote diagnosis of obstructive sleep apnea: A survey and a new model. *Inf. Fusion, 61*, 84–100.
25. Gravina, R., Alinia, P., Ghasemzadeh, H., & Fortino, G., (2017). Multi-sensor fusion in body sensor networks: State-of-the-art and research challenges. *Inf. Fusion, 35*, 68–80.
26. Sumithra, M., & Malathi, S., (2020). A brief survey on multi modalities fusion. In: *Emerging Trends in Computing and Expert Technology (Lecture Notes on Data Engineering and Communications Technologies)* (Vol. 35, pp. 1031–1041). Cham, Switzerland: Springer.
27. Li, C., & Zhu, A., (2020). Application of image fusion in diagnosis and treatment of liver cancer. *Appl. Sci., 10*(3), 1171.
28. Swayamsiddha, S., & Mohanty, C., (2020). Application of cognitive internet of medical things for COVID-19 pandemic. *Diabetes Metabolic Syndrome, Clin. Res. Rev., 14*(5), 911–915.
29. Yang, T., Gentile, M., Shen, C. F., & Cheng, C. M., (2020). Combining point of-care diagnostics and internet of medical things (IoMT) to combat the COVID-19 pandemic. *Diagnostics, 10*(4), 224.
30. Singh, R. P., Javaid, M., Haleem, A., & Suman, R., (2020). Internet of things (IoT) applications to fight against COVID-19 pandemic. *Diabetes Metabolic Syndrome, Clin. Res. Rev., 14*(4), 521–524.
31. Singh, P. D., Kaur, R., Singh, K. D., & Dhiman, G., (2021). A novel ensemble-based classifier for detecting the COVID-19 disease for infected patients. *Information Systems Frontiers*, 1–17.
32. Singh, K. D., & Sood, S. K., (2021). QoS-aware optical fog-assisted cyber-physical system in the 5g ready heterogeneous network. *Wireless Personal Communications, 116*(4), 3331–3350.
33. Achouri, M., Alti, A., Derdour, M., Laborie, S., & Roose, P., (2018). Smart fog computing for efficient situations management in smart health environments. *Journal of Information and Communication Technology, 17*(4), 537–567.
34. Singh, P. D., Kaur, R., Singh, K. D., Dhiman, G., & Soni, M., (2021). Fog-centric IoT based smart healthcare support service for monitoring and controlling an epidemic of swine flu virus. *Informatics in Medicine Unlocked, 26*, 100636.
35. Kaur, R., Singh, P. D., Kaur, R., & Singh, K. D., (2021). A delay-sensitive cyber-physical system framework for smart health applications. In: *Advances in Clean Energy Technologies* (pp. 475–486). Springer, Singapore.
36. Kaur, S., Singh, K. D., Singh, P., & Kaur, R., (2021). Ensemble model to predict credit card fraud detection using random forest and generative adversarial networks. In: *Emerging Technologies in Data Mining and Information Security* (pp. 87–97). Springer, Singapore.
37. Sood, S. K., & Singh, K. D., (2021). Identification of a malicious optical edge device in the SDN-based optical fog/cloud computing network. *Journal of Optical Communications, 42*(1), 91–102.
38. Angurala, M., Bala, M., Bamber, S. S., Kaur, R., & Singh, P., (2020). An internet of things assisted drone based approach to reduce rapid spread of COVID-19. *Journal of Safety Science and Resilience, 1*(1), 31–35.

39. Sood, S. K., & Singh, K. D., (2019). HMM-based secure framework for optical fog devices in the optical fog/cloud network. *Journal of Optical Communications*.
40. Seth, J., Nand, P., Singh, P., & Kaur, R., (2020). Particle swarm optimization assisted support vector machine based diagnostic system for lung cancer prediction at the early stage. *PalArch's Journal of Archaeology of Egypt/Egyptology, 17*(9), 6202–6212.
41. Sood, S. K., & Singh, K. D., (2019). Optical fog-assisted smart learning framework to enhance students' employability in engineering education. *Computer Applications in Engineering Education, 27*(5), 1030–1042.
42. Brogi, A., Mencagli, G., Neri, D., Soldani, J., & Torquati, M., (2018). Container-based support for autonomic data stream processing through the fog. In: *European Conference on Parallel Processing* (pp. 17–28). Springer, Cham.
43. Singh, K. D., & Sood, S. K., (2020). Optical fog-assisted cyber-physical system for intelligent surveillance in the education system. *Computer Applications in Engineering Education, 28*(3), 692–704.

CHAPTER 12

Signal Processing in Biomedical Applications in Present and Future Development

RAMAN CHADHA[1] and ROHIT KUMAR VERMA[2]

[1]UIE, Department of Computer Science and Engineering, Chandigarh University, Gharuan, Punjab, India, E-mail: dr.ramanchadha@gmail.com

[2]Assistant Professor, Department of Computer Science, Himachal Pradesh University Regional Center, Dharamshala, Kangra, Himachal Pradesh, India

ABSTRACT

The current region presents a short graph of several basic bits of the hypothesis of sign dealing with endeavoring, however much as could reasonably be expected to interconnect them. We start by examining the important considerations related to the tireless and discrete-time signs and designs with a commendation on the last decision. Then, at that point, a few basic musings of discrete-signal portrayal, signal quantization, channel plan, and execution are examined to survey the major things of the standard advanced sign dealing with the hypothesis. This fragment continues with a conversation on multi-rate signals dealing with nearby a summation of the perspectives and advantages of utilizing channel banks. From this information, we can see the value in the importance of the gadgets accessible for signal portrayal, for example, discrete changes in multiscale portrayals and edges. The last piece of the section rapidly

Computational Health Informatics for Biomedical Applications. Aryan Chaudhary and Sardar M. N. Islam (Naz) (Eds.)

concentrates on the chief pieces of sign appearance and adaptable sifting, which are major contraptions in learning the properties of emotional signs got in different appropriate applications. All contemplations examined in the current part are outlined through direct toy models. It is also open in the book Web-page a MATLAB code tending to the sifting and change musings. In the new years, the joint effort of biomedicine and arranging changed the assessment of the human body and agreeable clinicians with minute fortes of the human body. The current book part incorporates the likely gains of Biomedical Signal Processing. Biomedical signal processing joins strategies to tie down clinical or biochemical data to give work on clinical affirmation. The frameworks in the body are conceded through electrical or non-electrical signs called biosignals. Bio-signals are constantly investigated and analyzed. Biomedical signal processing is utilized to separate signs of revenue and perform all things considered examination. These records may be triumph over with bias that movement beat, circulatory strain, oxygen immersion categories, blood glucose, whim-whams conduction, mind advancement, and so on. Typically, similar opinions are taken at unambiguous twinkles and mentioned on an affected person's layout. Professionals surely see quick of what one chance of these traits as they get throughout the megacity — and treatment choices are made ward on those remoted readings. Biomedical symptom handling fuses the assessment of those appraisals to offer substantial records whereupon clinicians can choose. Masterminds are locating better styles of handling to manage these signs and symptoms using a collection of numerical formulae and checks. Working with standard-appraisal bias, the symptoms can be enrolled by means of programming to furnish specialists with regular information and more abecedarian particles of information to help medical tests. By making use of further refined means to insulate what our bodies are examining; we might also possibly choose the circumstance of an affected person's fulfillment via further inoffensive measures. Our bodies are generally passing on data roughly our substance. This information can be amassed through the operation of physiological instruments that pastime beat, circulatory stress, oxygen splashing degrees, blood glucose, whim-whams conduction, mind enhancement, and so forth. For the maximum element, similar opinions are taken at unambiguous extraordinary lighting institutions on a schedule and noted on a case's layout. Operating with commonplace bio-evaluation widgets, the sign may be figured through programming to offer experts constant records and lesser vital particles of statistics to help in medical appraisals. By using continually exercising present-day hopes to insulate what our bodies are communicating, we can also maybe pick the condition of a case's thriving through constantly royal measures.

12.1 INTRODUCTION

Biomedical signs are impressions of physiological sporting activities of residing matters, going from the best and protein plans to neural and coronary heart rhythms to tissue and organ pictures. Biomedical symptoms control goals disposing of large records from biomedical signs. With the manual of biomedical sign dealing with, specialists can discover every other technology, and specialists can screen-express problems.

Numerous years quicker, the vital purpose in the meeting of biomedical signs and symptoms dealing with turned into on secluding signs to forgo commotion [1–6]. Wellsprings of commotion ascend out of imprecision of instruments to obstruct electric connections. Exclusive sources are an instantaneous aftereffect of the standard frameworks themselves beneath observation. Dwelling animals are jumbled designs whose subsystems accomplice, so the cognizant signs and symptoms of a function subsystem commonly contain the signs of different subsystems. Doing away with troublesome sign elements can then underlie coming almost about biomedicine exposures. An important device for commotion cancelation exams out the signal spectra and covers undesired recurrent parts. Every other assessment structure receives quantifiable signs of coping. This development sees the data as abnormal signs and symptoms; dealing with, for example, Wiener sifting [6] or Kalman isolating [7, 8], uses true portrayals of the signs and symptoms to put off required symptoms parts.

The signs and symptoms are assessed and examined from the organs of the human body through the use of numerous devices. This type of signal taking care of is called biosignal dealing with. The essential troubles are to eliminate the commotion from the signs and symptoms, and the resulting statistics are extra beneficial for the clinicians [1–10]. X-rays, a puppy, CT, etc., make extra pics, and those photos are dealt with using synthetic Intelligence and AI computations. The biosignal managing and AI [12] primarily based scientific image assessment unequivocally tracking down the illnesses via trained experts. At the same time, as these denoising frameworks are grounded, the sphere of biomedical sign dealing with maintains on connecting as a result of the movement of various novel biomedical units. The development of medical imaging modalities, for instance, ultrasound (US), attractive resounding imaging (MRI), and positron radiation tomography (pet), connect with radiologists to picture the association and cutoff of human organs; as an instance, the department of organ structures evaluates organ perspectives [9]. Mobile imaging, as an example, fluorescence stamping and cell MRI facilitates researchers in seeing the automobile and headway of live cells [10]; following of mobile development maintains up with showing hydrodynamics [11].

The computerization of DNA sequencing helps geneticists with organizing DNA actions in chromosomes [12]; appraisal of DNA procedures secludes genomic data of residing animals [13]. The headway of massive worth chips draws in specialists to gage the outpourings of lots of attributes from numerous blood drops [14]; association research among verbalization levels and aggregates relaxes the parts of traits [15]. The models display indicators advantageous for accessing frameworks and make a contribution common to the development of biomedicine. Every other arising, first-rate want in various biomedical tests is shaped. In centers, the objective of solicitation is to see pathology from normal. For instance, seeing physiological facts, clinicians decide to anticipate that sufferers need to stumble upon the wise consequences of worry [16]; looking coronary heart MRI tests, cardiologists understand which locale the myocardium encounters disappointment [17]; investigating pleasant movements of a circle of relatives, geneticists amass the probability that the adolescents cozy the infection from their family members [18]. Those models show that an altered course may be a fundamental degree in scientific practice. Especially, stirred up defilement perceiving evidence will now not simply waste finding belongings yet, similarly, put off remedy or motive loss of patients' lives. Some other convincing impact with reference to the depiction in biosignal dealing with is proven with the aid of labs wherein examiners use demand frameworks to peer the portrayal of dim biomolecules. Via distinctive features of the remarkable throughput of modern biomolecular checks like crystallography, atomic appealing resounding (nuclear magnetic resonance (NMR)), and electron microscopy, everyday researchers have an instrumental manner of picking out the sub-atomic arrangement of tremendous measures of proteins. On the grounds that proteins with equal designs usually have close to limits, to see protein work, traditional logical professionals' motel to the entire evaluation of their essential tantamount characteristics [19].

Plan in biomedicine faces two or three challenges:

i. Specialists need to gobble up most of the day to accumulate sufficient facts to dependably see glaringly related cases, say, every day, and weird;
ii. Guide solicitation among these instances works real and drawn-out; and
iii. The most overpowering check takes place when the sign credits aren't recognizable and therefore not satisfactorily perceivable by knowledgeable government.

Mechanized systems for needs in signal dealing withhold the confirmation for beating a bit of this challenge and to help impartial biomedical bearing. A modified classifier can benefit from an enlightening file, the outright records, uproot human supervisors, and solicitation worked up highlights without tendency. This fragment rotates around modified depiction tests were given from sign managing techniques.

12.2 PROGRESS OF BIOMEDICAL APPLICATIONS

This part intends to build up numerous and same plans modern rising techniques that show new trends and usages contemporary signal and photograph taking care of in medical imaging. It'll help the two specialists and radiologists within the picture realize and help experts with exchanging the cutting-edge, unique advances.

12.2.1 COORDINATED ADVANCED TRUE SYSTEMS

A basic test for the execution of modern signal managing processes in actual automated structures (CPS) is the trouble trendy securing records from topographically streamed perception location focuses and dealing with/setting up the accrued information at the combined network [1, 2]. Considering the entirety, there was a nonstop flood today's electricity for the advancement brand new handed-on and shared sign preparing today's progress in which change, appraisal, in addition as manipulate are completed locally, and correspondence is obliged to neighborhoods. Computerized actual systems deliver vital today's and it has more practicable stepped forward, and real assaults via enemies on signal handling modules could have an effect on blend latest over the pinnacle effects along with patron records spillage, the demolition ultra-modern foundations, and jeopardizing human lives [3]. Obviously, the importance of collaboration among connecting focuses makes it basic to anticipate the publicity modern-day sensitive close by information throughout the passed on records combo step.

12.2.2 CYBER-REAL STRUCTURES

In signal taking care of Cyber-true structures give method and help to deal with prognostic problems in a plan latest supportive districts [4]. Device slanting computations (ML) are used to interrupt down the importance of cutting-edge clinical limits, as an instance, suspicion for infection development, medicinal brand new, affected person association, and many others [5]. ML is being utilized for information appraisal in the clinical discipline [6]. It fights that the beneficial execution of ultra-modern ML procedures can help the combination of trendy laptop-primarily based designs in the remedial organization's circumstance giving chances to assist and redesign made via supportive prepared specialists.

12.2.3 MULTIMODAL MIXED-MEDIA SIGNAL

It deals with Analysts in numerous fields using multimodal facts. Certainly, one of its maximum ordinary brand new is in the field cutting-edge human laptop affiliation (HCI). Here, a way is a logo name procedure for composing exertion: talk, imaginative, and prescient, face clarifications, penmanship, tendencies, head, and body upgrades [7, 8]. Multimodal interfaces assist the human pc interface [9] uproot the standard manage vicinity and mouse. Multimodal speaker confirmation perceives the awesome speaker in a valid video collecting, which incorporates several audio systems, considering the relationship between the sound and the headway in the video [10].

12.3 STATISTICAL SIGNAL

Statistical Signal looking after is a technique for handling sign managing, which treats indicators as stochastic cycles, utilizing their quantifiable houses to perform signal looking after endeavors. Verifiable strategies are for the most component used in sign dealing with packages [13, 14]. Sign dealing with strategies for data stowing with no end in sight and sound watermarking in signal taking care of methodologies for records disguising a unique device for embeddings and improving "included" facts in sturdy files. In this strategy, the hour modern picked components present day the host sound pennant is controlled in a way that might be perceived by using an expert with the fine "key" [15]. Without the important thing, the covered information is sick-defined; each aurally and via manifestly blocked electronic requirements managing assaults. The methodology portrayed in each aurally immediately and excited and can be associated with each important and modernized sound flag, the remaining including uncompressed and similarly filled sound file plans. Facts stowing away is a gift via relative level encoding and quantization document alternate stage encoding method [16].

12.4 OPTICAL SIGN

It deals with optical sign looking after joins specific fields modern-day optics and signal taking care of to be unequivocal, nonlinear gadgets and structures, important, and motorized signal, and pushed statistics alternate intends to perform speedy signal looking after limits that may perhaps paintings at the line speed ultra-modern fiber optic (FP) exchanges [17, 18]. Statistics can be encoded in plenitude, level, recurrence, polarization, and spatial highlights ultra-modern an optical waves to accomplish high-restrict transmission. Unmistakable optical nonlinearities and chromatic dispersing were appeared to connect with key sub-shape applications, for instance, recurrence change, multicasting, multiplexing, demultiplexing, and tunable optical postponements. Optical pennant arranging utilizing cognizant optical rehash researches ought to have assorted likely packages for optical correspondences. In the beginning, a method for identifying how to accomplish a tunable optical high-organize quadrature amplitude modulation (QAM) [19] age considering multichannel outright and an all-optical pilot-tone based self-homodyne-affirmation is used by situations:

- Specific wavelength division multiplexing (WDM) channels with the agreeable pressure modern pilot tones;
- Alone channel with a low-manage pilot tone; and
- At final, an apportioned statistics transmission task linked with through reconfigurable channel reducing and stitching.

12.5 VIRTUAL PHYSIOLOGICAL HUMAN POWER

The virtual physiological human is indivisible from software in computational biomedicine that expects to foster a sport plan contemporary strategies and degrees state-of-the-art progress to test out the human frame commonly [20, 21]. It's far predicated at the pivotal man or woman trendy data development, brought as a stable effect for that for the most component principle today's human concerns, our personal thriving and prospering. The VPH is a made get-collectively out contemporary computational plans and ICT-based contraptions for the lovely look and leisure ultra-modern human existence developments and body structure. As soon as sufficiently created, the VPN [22] will supply a giant imaginative foundation to the Physiome undertaking, to pathology-explicit activities in translational assessment, and to vertical answers for the biomedical enterprise.

12.6 ELECTROENCEPHALOGRAM (EEG)

Mind-computer interfaces research in EEG based totally frontal cortex laptop interfaces (BCIs) has been drastically linked at some point of the most current pretty a while. To such degree owes a gigantic diploma to the multidisciplinary and testing nature latest BCI get a few information approximately. Signal instruction and display certification definitely incorporate key regions of today's a BCI framework. Signal managing exams are associated with the EEG sign to translate mental states that are fitting for BCI activity. On this instructive exercise, the focal BCI musings, for example, thoughts development checking, BCI task, and the critical mental states for BCI, are provided. The critical contemporary basic intellectual states for BCI, to be unequivocal engine symbolism

(ERD/ERS), enduring kingdom visible evoked expected consequences (SSVEP) [23], and event-associated possible effects are brought close by affordable software perspectives. The EEG preparing for intellectual country unwinding is depicted for what it's really worth. The multivariate thought about the EEG was given at the side of the neuroscience latest on hemispheric cerebrum specialization is favorably taken into consideration to accumulate best mixes today's the precise signal making the EEG [24]. BCIs are named by using the modern-day mind motion applied for manipulating. Among two or three sets, cutting-edge EEG-based BCIs, which include P300, unflinching state visible evoked ability (SSVEP), event-associated desynchronization (ERD), and slight cortical capability-based totally signal getting ready.

12.7 NEURAL ASSOCIATIONS AND HANDLING

In people, coordinated efforts between nerve cell circuits, structures, and signs among little, meso, and big scope sizes of brain parts support the utilitarian relationship of the cortical area that maintains our everyday schedule development. Mathematical, process, and preliminary neuroscientists apply a group of methodologies, strategies, and estimations, each in animals and people going from single-cell records to whole mind imaging, to understand the center frameworks that manage the link among these scales. Despite the approach that our understanding into neural instruments, circuits, and associations' coated up cortical area parts and limits persistently creates, the connection of this knowledge to present a smart arrangement of emanant direct [25] and set up course of action occurring on totally different levels of the abstraction affiliation stays testing.

12.8 LARGE KNOWLEDGE IN BIOINFORMATICS

In medicine calculation, the steady inconveniences are: the board, assessment, and brink of the medical data. Streak coming up with engages America to form-fitting and valuable techniques to incorporate infinite photos for depiction, which may be patched up for each other [27]. In cure, the knowledge veteran is for the foremost half noninheritable from patients. This data is contained physiological signs, pictures, and records. They'll be handled or sent utilizing correct instrumentality and systems.

The associations used in an answer for the cut-off and transmission of image data is that the image recording and correspondence structure (PACS). The big data enhancements area unit is mentioned into four categories [28, 29]: (i) data storing up and recovery; (ii) bungle ID; (iii) data analysis; and (iv) stage combined to set up. These game plans area unit associated and should cover; as an example, most data input applications might uphold elementary data appraisal or a contrary approach around.

12.9 IMAGE REDOING AND EXAMINATION

The appraisal model has been really cuffed as a call instead of the quality meagerly mixed model for organizing image amusement techniques. Applying an associate degree, applicable analysis chief on the image of interest yields associate degree meagerly results [30] that empowers America to re-try the image from under-inspected data. Likewise, earlier within the assessment setting and hypothetically take into account the distinctiveness offers the degree that analysis chiefs once everything is claimed in done position and therefore the explicit 2nd restricted capability government. Considering the shot at unvarying co-build-up exposure (ICD) associate degree smart image re-trying model and a productive assessment, accomplishing basically higher amusement execution.

12.10 MEDICINE IMAGING

Utilization of laptop upheld headways in tissue coming up with artistic work has progressed the advance one a lot of field of laptop helped tissue coming up with (CATE). Three-layered (3D) printing is an additional substance-making method. This advancement outfits America with the dear probability to form 3D plans by recollecting material for a layer-by-layer premise, utilizing numerous kinds of materials, for example, ceramic ware creation, metals, plastics, and polymers.

Nowadays, tissue building assessments square measure happening on despite however you investigate it premise within the fields of recovery, recovery, or replacement of imperfect or hurt valuable living organs and tissues. 3D bioprinting is AN filmable making development that's finding its direction through all items of human existence. The flexibility of 3D printers will be mishandled in regions of medical specialty designing, as an example, key exploration, sedate movement, testing, and furthermore in clinical apply. Just about all current remedial nonorganic embeds, as an example, ear prostheses, a square measure created in doomed sizes and plans that square measure for the foremost half used for patients. This method permits additional precise redid aggregation of devices created to the patient's own specifics. Bio-printing is being employed to form additional precise nonbiologic and natural examinations. Depict a method for concealing data in sound documents that utilize the management of the amount of chosen transcendental components of the host sound record. We tend to portray a method for concealing data in sound documents that utilize the management of the amount of chosen apparitional components of the host sound record. For a model, in Figure 12.1, robotized measuring of growths stays to judge signal force changes in Mr photos, and this can be a hard issue on account of the curios influencing photos, for instance, incomplete volume impacts and power inhomogeneities. Low-level division methods, for instance, power thresholding, edge discovery, space developing, district mixing, and morphological activity, aren't applicable for the robotized measuring of the sign anomalies as these procedures rely on image directors that dissect force, surface, or form domestically in every voxel, and consequently, too effectively lead on by ambiguities within the image or need shopper association.

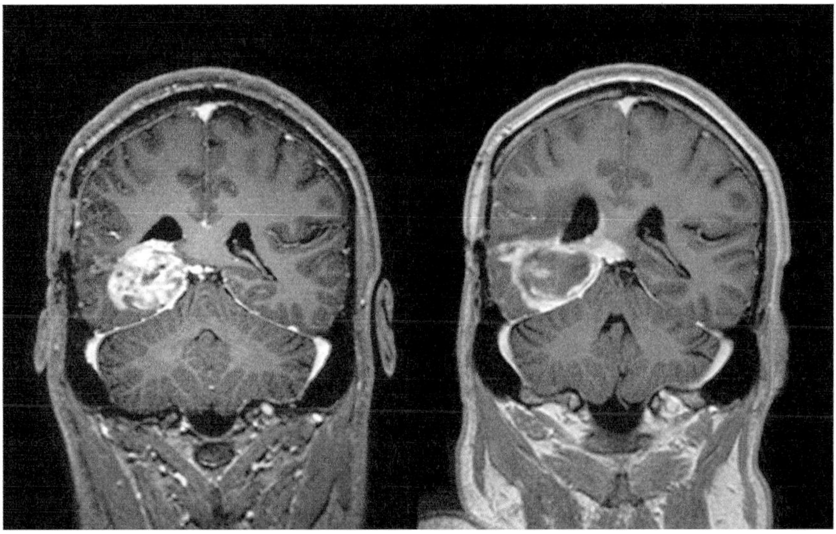

FIGURE 12.1 Image of a brain tumor before and after 10 days of treatment.

Source: Reprinted with permission from Ref. [31]. © 2020 Springer Nature.

12.11 INTELLIGENT IMAGING

Data-driven frameworks have gotten extended thought really to handle varied problems in medical specialty imaging. Data-driven models and methods thus forward. Offer promising execution in image redoing problems in appealing resonation imaging, handled picturing, and varied modalities in relevancy ancient philosophies victimization hand-made models, as an example, the distinct trigonometric function amendment, wavelets. During this term wraps the newest methods for creating all items of the imaging system data-driven, as well as information acquisition and inspecting image recreation, and taking care of assessment.

12.12 IMAGE COURTESY OF THE 3D SLICE

Smart imaging systems would endlessly acquire from mammoth data-sets and on-the-fly and adapt to hurry, efficiency, and movie execution or quality. The most effective illustration of shrewd imaging is that the ID of wedged and solid photos captivated with the separation skills in structure image surfaces. For this reason, a surface descriptor for retinal eye photos has been done, and therefore the region and time utilization has been diminished by the strategy for dilated paired examples (EBP). The first purpose is to decrease the dimensions and time utilization and what is more separate between age-related devolution (AMD) and diabetic retinopathy (DR), and normal structure photos with the tissue layer foundation surface by effort injuries past division stage with the planned strategy and effort promising outcomes. The most effective consequences of every investigation on the model set square measure featured in tables. This work utilizes the EBP administrator. Specifically, the presentation of EBP was contrasted and LPB as displayed in Figure 12.2.

12.13 PARTIAL DIFFERENTIAL EQUATION (PDE)-BASED IMAGE EXAMINATION

Image Examination is the most downside with the particular mode breaking down (EMD) estimation is its nonattendance of a speculative framework. Thusly, it’s troublesome to depict and appraise the 2-D case, the utilization of AN elective use to the algorithmic importance of the declared “sifting process” used as a chunk of the first EMD procedure. This technique, particularly considering fractional differential conditions

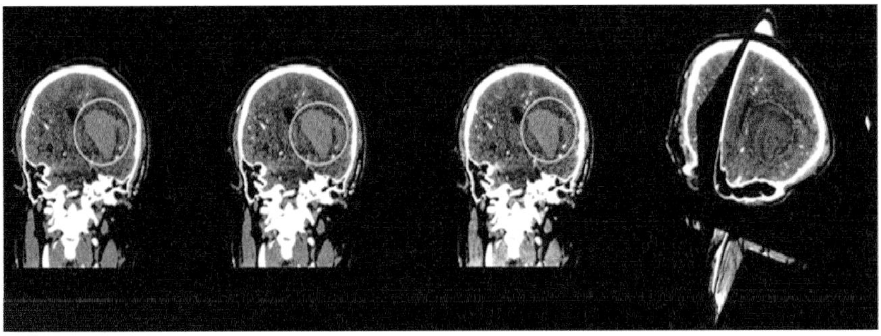

FIGURE 12.2 3D MRI image visual image victimization 3D slicer.

Source: Reprinted from Ref. [32].

(PDEs) and depends upon a nonlinear unfold primarily based filtering technique to handle the mean envelope assessment issue. Within the 1-D case, the potency of the PDE-based procedure, which appeared otherwise in relevancy to the principal EMD algorithmic interpretation, was additionally depicted in an exceedingly continuous paper. Lately, one or two 2-D developments of the EMD methodology have been planned. Even so many efforts, 2-D variations for EMD show up insufficiently acting and square measure terribly dreary. Thus AN extension to the 2-D area of the PDE-based methodology is mostly depicted. This technique has been associated with occasions of each banner and movie rot. The gained results ensure the price of the new PDE-based separating method for the breaking down of assorted forms of information. The ampleness of the methodology engages its use in numerous sign and movie applications, for instance, denoising, detrending, or surface investigation. Visual image of 3D MRI neural structure cancer image Figure 12.2 exhibits the illustration of 3D MRI photos utilizing 3D slicers wont to perform totally different examinations on mind growths in starting phases. Hyperspectral imaging (HIS) as these days, hyperspectral imaging (HSI) has ascended as a promising optical advancement for medical specialty applications, primarily continuously sciences investigate, nevertheless furthermore went for nonintrusive finish and movie directed operation. HIS advancements are used comprehensively in therapeutic exploration, zeroing in on varied natural marvels, and totally different tissue composes. Their high spooky assurance over a broad assortment of frequencies allows the obtention of spatial data contrasted with totally different lightweight human activity traditional blends. It's smart to give steady quantitative data to one or two of natural methodologies in each robust and debilitated tissue. Hyperspectral work and assurance square measure typically viewed because the key factors that understand HSI and MSI, whereas MSI supported distinct and for the foremost half isolated frequency gatherings, HSI basically utilizes terribly shut and connected supernatural gatherings over a relentless apparition vary, to breed the scope of every element within the image.

12.14 ARTIFICIAL NEURAL ORGANIZATIONS

In this image handling, healthful imaging strategies have usually been being used to find and identify proof of infection. Microcalcifications and lots more and plenty square measure the earliest signs of cancers that ought to be distinguished victimization current strategies. The issue within the portrayal of liberal and harmful microcalcifications equally causes a basic issue in auxiliary photography care. Machine-controlled classifiers are also necessary for radiologists in perceiving unselfish and risky models. Therefore, a counterfeit neural framework (ANN) [11, 26] which may be crammed in as an automatic classifier, is analyzed. In restorative image designing, ANNs are related to a range of information request and model affirmation tasks and rework into a promising portrayal instrument in chest damage. Thusly, uncommon judgments of image options can win the various course of action choices. These solicitations will be divided into 3 sorts: regardless, the strategy considering assessments, for instance, support vector machine (SVM); second, the procedure in context on administering, for instance, alternative tree and unforgiving sets; and third, counterfeit anatomical structure. Varied ANNs created rely on when increasing the real positive (TP) speech act rate and decreasing the sham positive (FP) and phony negative (FN) affirmation rate for the perfect outcome. Utilization of wavelets in ANNs, for instance, molecule swarm increased riffle neural organization (PSOWNN), biorthogonal spline riffle ANN, second-arrange faint side ANN, and Gabor wavelets ANN will work on the affectability and expresses that square measure gained in lots and microcalcification affirmation.

12.15 SPOKEN COMMUNICATION

In the field of medical specialty, sign, and movie taking care of involving specific interests within the instructional and examination field in medical specialty style. With the updated physiological data, a large strategy of ingenious works in clinical procedures makes use of this concept in useful applications. With the progress in medical specialty imaging, the extent of data created by multimodality image techniques, e.g., loosening up from patterned picturing (CT), engaging reverberation imaging (MRI), US, single-photon emanation registered to picture (SPECT), and antielectron outflow picturing (PET), engaging molecule imaging, EE/ magnetoencephalography (MEG), optical research and picturing, photoacoustic picturing, lepton picturing, and atomic power research, has developed dramatically and therefore the probability of such data has showing intelligence tense being very astounding. This addresses an exquisite check on the foremost effective thanks to developing new motivated imaging techniques and machine models for helpful information addressing, examination, and showing in clinical applications and in understanding the basic traditional cycle. Sign and movie handling is inevitable in gift day medical specialty imaging because it offers essential techniques to image improvement, redesign, coding, accumulating, transmission, assessment, understanding, and portrayal from any of an extending variety of various complicated distinctive modalities. To handle this bother, generally traditional image pre-processing strategy, as an example, feature extraction, image mix, gathering, and division would like adjusted shrewd procedures which will handle the mass and respectable assortment of the knowledge and often have the flexibility to fuse and handle data from non-imaging sources.

12.16 CONCLUSION AND FUTURE SCOPE

End This half preponderantly targeted around signs and most up-to-date methods in clinical image handling, which can build additional interest in medical specialty exploration fields. With the foremost recent patterns in data procurance, a large game-plan of ingenious works in clinical procedures square measure applied in useful applications. In medical specialty imaging, the knowledge securing frameworks like patterned picturing (CT), engaging reverberation imaging (MRI), US SPECT, antielectron discharge picturing (PET), optical research, then forth, catches photos of the patients. These frameworks develop dramatically and manufacture huge data that have additional useful information. The elite presentation registering (high power command (HPC)) methods break down the photographs and envision pictures in 3D as a pixel-wise investigation with very little handling time. The many difficulties within the mind growth discovery square measure to research the particular space, shape, and numerous cancer tissues and nontumor tissues. Unreal reasoning (AI) and AI (ML) address these difficulties that uphold radiologists and what is more for patients.

KEYWORDS

- **artificial intelligence**
- **diabetic retinopathy**
- **electroencephalogram**
- **high power command**
- **hyperspectral imaging**
- **machine learning**

REFERENCES

1. Jin, J., Allison, B. Z., Wang, X., & Neuper, C., (2012). A combined brain-computer interface based on P300 potentials and motion-onset visual evoked potentials. *Journal of Neuroscience Methods, 205,* 265–276. doi: 10.1016/j. jneumeth.2012.01.004.
2. Katzenbeisser, S., & Petitcolas, F. A. P., (2000). *Information Hiding: Techniques for Steganography and Digital Watermarking.* Norwood MA: Artech House.
3. Kohl, P., & Noble, D., (2009). Systems biology and the virtual physiological human. *Molecular Systems Biology, 292,* 1–6. doi: 10.1038/msb.2009.51.
4. Amini, S., Veilleux, D., & Villemure, I., (2010). Tissue and cellular morphological changes in growth plate explants under compression. *Journal of Biomechanics, 43*(13), 2582–2588.
5. Saini, S., & Vijay, R., (2015). Back propagation artificial neural network. In: *Proceedings of the 5th International Conference on Communication Systems and Network Technologies* (pp. 1177–1180). Gwalior, India.
6. Behr, J., Choi, S. M., Grosskop, S., Hong, H., Nam, S. A., Peng, Y., et al., (2000). Modeling, visualization, and interaction techniques for diagnosis and treatment planning in cardiology. *Computers & Graphics, 24*(5), 741–753.

7. McInerney, T., & Terzopoulos, D., (1996). Deformable models in medical image analysis: A survey. *Medical Image Analysis, 1*(2), 91–108.
8. Lustig, M., Donoho, D. L., Santos, J. M., & Pauly, J. M., (2008). Compressed sensing MRI. *IEEE Signal Processing Magazine, 25*(2), 72–82.
9. Wang, L. V., (2009). Multiscale photoacoustic microscopy and computed tomography. *Nature Photonics, 3*(9), 503–509.
10. Beard, P., (2011). Biomedical photoacoustic imaging. *Interface Focus, 1*(4), 602–631.
11. Ghesu, F. C., et al., (2016). Marginal space deep learning: Efficient architecture for volumetric image parsing. *IEEE Transactions on Medical Imaging, 35*(5), 1217–1228.
12. Wang, G., (2016). A perspective on deep imaging. *IEEE Access, 4*, 8914–8924.
13. Esteva, A., et al., (2017). Dermatologist level classification of skin cancer with deep neural networks. *Nature, 542*(7639), 115–118.
14. Okada, M., (1979). A digital filter for the QRS complex detection. *IEEE Transactions on Bio-Medical Engineering BME, 26*, 700–703.
15. Ergun, E., & Batakçı, L., (2009). Audio watermarking scheme based on embedding strategy in low frequency components with a binary image. *Digital Signal Processing, 19*(2), 277–286.
16. Kadambe, S., Murray, R., & Boudreaux-Bartels, G. F., (1999). Wavelet transform-based QRS complex detector. *IEEE Transactions on Biomedical Engineering, 46*, 838–848.
17. Awad, E. S., (2015). Data interchange across cores of multi-core optical fibers. *Optical Fiber Technology, Part B, 26*, 157–162.
18. Hamilton, P. S., & Tompkins, W. J., (1986). Quantitative investigation of QRS detection rules using the MIT/BIH arrhythmia database. *IEEE Transactions on Biomedical Engineering BME, 33*, 1157–1165.
19. Kahn, J. M., & Ho, K. P., (2004). Spectral efficiency limits and modulation/detection techniques for DWDM systems IEEE. *Journal of Selected Topics in Quantum Electronics, 10*(2), 259–272.
20. Chen, S. W., Chen, H. C., & Chan, H. L., (2006). A real-time QRS detection method based on moving averaging incorporating with wavelet denoising. *Computer Methods and Programs in Biomedicine, 82*, 187–195.
21. He, B., Li, G., & Lian, J., (2002). A spline Laplacian ECG estimator in realistic geometry volume conductor. *IEEE Transactions on Biomedical Engineering, 49*(2), 110–117.
22. Perrin, F., Pernier, J., Bertrand, O., Giard, M. H., & Echallier, J. F., (1987). Mapping of scalp potentials by surface spline interpolation. *Electroencephalography and Clinical Neurophysiology, 66*, 75–81.
23. Kawakatsu, H., (2015). Methods for evaluating pictures and extracting music by 2D DFA and 2D FFT. 19th international conference on knowledge-based and intelligent information and engineering systems. *Procedia Computer Science, 60*, 834–840.
24. Kawakatsu, H., (2014). Fluctuation analysis for photographs of tourist spots and music extraction from photographs. In: *Lecture Notes in Engineering and Computer Science: Proceedings of the World Congress on Engineering 2014; WCE 2014* (Vol. 1. pp. 558–561). London, UK.
25. Manandhar, P., Ward, A., Allen, P., Cotter, D. J., Mcwhirter, J. G., & Shepherd, T. J., (2016). An automated algorithm for measurement of surgical tip excursion in ultrasonic vibration using the spatial 2-dimensional Fourier transform in an optical image. 44th annual symposium of the ultrasonic industry association. *Physics Procedia, 87*, 139–146.

26. Bhateja, V., Patel, H., Krishn, A., Sahu, A., & Lay-Ekualille, A., (2015). Multimodal medical image sensor fusion framework using cascade of wavelet and contourlet transform domains. *IEEE Sensors Journal, 15*(12), 6783–6790.
27. Mjahad, A., Rosado-Muñoz, A., Bataller, M. M., & FrancésVíllora, J. V., (2017). Ventricular fibrillation G-MJF. Tachycardia detection from surface ECG using time-frequency representation images as input dataset for cyber-physical systems. *Computer Methods and Programs in Biomedicine, 141*, 119–127.
28. Arenja, N., Riffel, J. H., Djioko, C. J., Andre, F., Fritz, T., Halder, M., et al., (2016). Right ventricular long axis strain-validation of a novel parameter in non-ischemic dilated cardiomyopathy using standard cardiac magnetic resonance imaging. *European Journal of Radiology, 85,* 1322–1328.
29. Mavratzakis, A., Herbert, C., & Walla, P., (2016). Emotional facial expressions evoke faster orienting responses, but weaker emotional responses at neural and behavioral levels compared to scenes: A simultaneous EEG and facial EMG study. *NeuroImage, 124*, 931–946.
30. Vuilleumier, P., & Pourtois, G., (2007). Distributed and interactive brain mechanisms during emotion face perception: Evidence from functional neuroimaging. *Neuropsychologia, 45*(1), 174–194.
31. Tabatabaei, P., Asklund, T., Bergström, P., Björn, E., Johansson, M., & Bergenheim, A. T., (2020). Intratumoral retrograde microdialysis treatment of high-grade glioma with cisplatin. *Acta Neurochir (Wien), 162*(12), 3043-3053. doi: 10.1007/s00701-020-04488-2. Epub 2020 Jul 14. PMID: 32666378.
32. Zhang, X., Zhang, K., Pan, Q., & Chang, J., (2019). Three-dimensional reconstruction of medical images based on 3D slicer. *Journal of Complexity in Health Sciences, 2*(1), 1-12. https://doi.org/10.21595/chs.2019.20724

CHAPTER 13

Emerging Trends in Healthcare and Drug Development

CHINJU JOHN, AKARSH K. NAIR, and JAYAKRUSHNA SAHOO

Department of Computer Science and Engineering,
Indian Institute of Information Technology, Kottayam, Kerala, India,
E-mail: chinjuj.phd201001@iiitkottayam.ac.in (C. John)

ABSTRACT

In recent times, artificial intelligence (AI) and its subdomains, such as deep learning (DL), machine learning (ML), natural language processing (NLP), and so on, have been undergoing massive developments increasing its applicability extensively. This had led it to be widely employed in healthcare based and its allied domains for a wide array of tasks through this work. We aim to provide an insight into the applications of AI and DL in the healthcare context. We perform an empirical analysis on the evolution of AI in the healthcare sector from a subdomain-based perspective. We use a systematic approach to present various subdomains of AI and their effect in various healthcare applications over time. Additionally, we also discuss the challenges faced during real-life implementations and areas with scope or developments as well. Finally, we also focus our work on the role of AI in rural healthcare management and prospective roles as well. We try to present the relevance of AI in rural healthcare scenarios as a creative alternative for the issues faced by the same. The inferences derived from our study point to the fact that in the near future, the healthcare sector holds to key to most of the unresolved issues in the healthcare

Computational Health Informatics for Biomedical Applications. Aryan Chaudhary and Sardar M. N. Islam (Naz) (Eds.)

sector, and solutions can be generated only when we are able to harness the complete potential of AI.

13.1 INTRODUCTION

Artificial intelligence (AI) has been one of the most fiercely debated topics for the last couple of decades simply due to its high versatility in application scenarios and functionality. AI empowers machines and prompts them to perform cognitive tasks analogous to what humans do. In previous days, the only factor differentiating humans and machines was their cognitive abilities, but with the advent of AI, machines were capacitated with high-level cognitive abilities empowering them to perform all those tasks that were solely bounded to humans. For generating such Intelligence, AI systems usually make use of training data from which it infers specific patterns or results, thus helping it to perform deputed functions. AI can also be termed one the most versatile technology with a vast range of applicability across multiple domains. Prior to being termed as intelligent, a system undergoes certain steps through which it attains its aforementioned cognitive abilities, such as the reasoning phase, analysis phase, and adaptive phase [1]. The system achieves various capabilities as it undergoes each phase, i.e., with the first stage, it performs data collection and algorithm generation through which the system will get an idea as to how the designated tasks can be performed. In the second phase, the system performs the idealistic path selection for its particular applications from a group of eligible candidate algorithms. In the final stage, it does the fine-tuning of the system where they undergo self-updation and try to attain better performance.

Due to the current pandemic situation, healthcare and related service sector personnel are undergoing unprecedented pressure in multiple aspects, including labor scarcity, managerial demands, and so on. Ultimately, a fruitful solution for the same can only be devised by assimilating effectual technologies into the systems. The last decade has witnessed a huge hike in the number of AI-based applications being employed in the healthcare sector on a gradual basis. Looking into the current trend, it is estimated that the future of healthcare will be taken over by AI directly or its allied applications and several human-oriented tasks will gradually undergo a shift into technology-oriented mode.

With all this being said, the applicability of AI for such extremely sensitive tasks is highly questionable, and an active debate has been taking place for a long time over the same. Heath care tasks require a high level of intellectual skills, and the addition of such technology will make it devoid of rational aspects as well. Along with all the positive sides, AI also comes with a set of negative aspects as well. AI-based applications possess the ability to alter the essential features of the domain it is applied to. In the case of the healthcare domain, it poses a high threat as such changes may lead to several ethical issues corresponding to regularization, learning, and so on, gradually leading to the development of systems that is fragmented or inexplicable, thus resulting in the generation of inequitable results. To tackle these concerns, a systematic approach is needed to be devised prior to the enactment of such an AI-based system, especially in the context of healthcare, thus providing the users with a way to identify challenges and rectify them with ease.

The base of such approaches should never be derived from strict managerial policies as they usually lack the ability to be flexible on problems and stick on to the proposed framework only. Such directives only lay out a path to understand the procedures and remain futile when it comes to related issues. The only solution is to adopt a user-oriented approach aiming for their speculations, requirements, and benefits. The biggest stumbling block for the same is the absence of a method that is rationally capable of conducting such a study. Anyway, the rational and analytic issues concerned with the employment of healthcare-based AI systems also occur at a higher level with reference to various levels of relationship with end-users as well. This may bring about multiple issues as for every healthcare-related task in general and diagnostic task specifically, the paramount concern is always for accuracy. Thus, whenever an ethical analysis of an AI-based healthcare system is done, the evaluators should always examine the reason for the ethical issues at several levels and different phases of the system lifecycle. Thus, a gradational evaluation is the basic requirement as the chances of the ethical structure of the algorithm is always open to changes through several ways at any stage of development [2]. For instance, a diagnostic algorithm designed with an ethical perceptive and tested meticulously could still destroy a system once it gets put into instances where it undergoes immense data overload, which it was not designed for. Thus, performing such an ethical review of medical AI technology is highly required, and only when the system meets the basic requirement constraint can it be made use of to the fullest by the community as well as the medical industry. If not, it will always persist as an unresolved puzzle failing to unravel its full caliber. The discussions we have done so far describe the task in a futuristic perspective only and lack to give an awareness of the current issues related to the implementation of medical AI. We need to undergo an in-depth analysis to get a better angle on that, going through their pros and cons in detail. Figure 13.1 provides a basic idea of various subdomains of AI in healthcare scenarios.

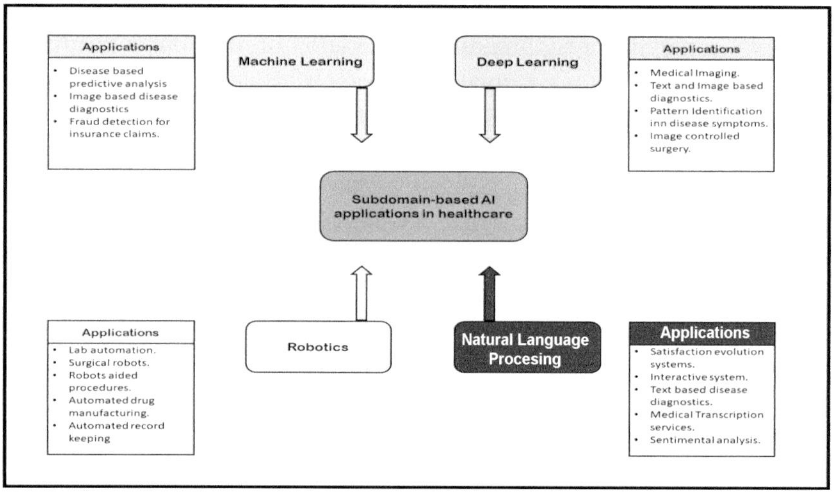

FIGURE 13.1 Subdomain-based AI application in healthcare.

AI is the terminology used to refer to a broad group of technologies rather than a single entity. Even though their working methodology, as well as their motto, maybe the same, each technology has its own areas of application where they excel at. Some of the component technologies that make up the term AI are ML, DL, NLP, Expert systems, and so on [3]. The majority of these domains have direct applicability in the medical context, although the task they perform may vary accordingly. Through this study, we strive to provide a detailed analysis of AI and DL in the context of healthcare. We will be primarily discussing their path of development and evolution. Later on, we move on to their applicability in tasks highly related to the current scenarios, such as drug discovery and rural health management. Towards the later stages, some of the common challenges

AI faces in the medical scenario will also be discussed, along with the emerging opportunities as well.

The contribution we aim to provide through this study are the following:

- Perform a detailed study on the evolution of AI and DL in healthcare;
- Conduct an in-depth analysis of the most recent literature on DL in healthcare over the last few years;
- Present the growth of DL technology in the pharmaceutical industry, especially in drug discovery and allied tasks;
- Additionally, we identify the various challenges faced by the technology in real-life implementation in rural communities along with possible futuristic opportunities.

13.2 TRANSFORMATION OF AI IN HEALTHCARE

The role of AI has been evident in the healthcare sector for a very long time. The first surge was under the name of ML, where models such as Artificial Neural Network (ANN) played a huge role in familiarizing and making the technology popular.

They were the sole controller of AI in healthcare for almost 60 years and gradually began to fall out of place by the end of the 20th century. The 21st century has witnessed the taking over of healthcare tasks by DL methodology, which itself was an improvised version of the basic ANN models thus also known by the name Deep artificial networks. They possessed models with higher specifications such as layers and parameters, thus making them capable of performing tasks of higher complexity compared to the basic ANN model, thus increasing their applicability compared to the latter [4]. Deep Neural Networks (DNN) are specifically designed to be able to make use of system constraints to the maximum, making them capable of dealing with large volumes of data. This enables them to perform various tasks based on Computer Vision, NLP, and even gaming theory problems. DNN is built on a layer-based architecture in which each layer acts as a basic processing unit which is termed as 'neurons,' and forms an interlinked structure with the aid of several weighted links. Such models are trained using techniques such as backpropagation which primarily makes the system understand the desirable changes to be made with the initial training weights or parameters that are used to perform training at a particular layer with respect to the input from the previous layer. Generally, DL based models are said to be highly autonomous once training is done and deployed, as such models are able to do high-level inferencing of patterns of different levels from varying types of datasets with continual training and necessary updation.

AI generally and ML and DL specifically have transformed the current day healthcare industry vastly. As discussed earlier, the role of AI is applicable in multiple facets of healthcare. The effect of the application of AI on healthcare-based tasks can be evaluated in two main aspects: First, for making use of healthcare big data, thus adding higher values to the data, and secondly, as direct aid to medical personnel for healthcare service delivery at various levels. The application of ML-based technology has already gained high popularity when it comes to information extraction for inferring treatment patterns and medical diagnosis in electronic health record (EHR)-based big data applications. Such tasks have also opened up opportunities for taking a data-driven approach when it comes to medicinal result prediction, identification of various subgroups of diseases, and so on. A real-life example of such a technology can be seen at the IBM Watson Health cognitive computing system, where the system employs ML-based techniques for running a decision support system for cancer-based diagnostic and treatment assistance for medicos. The main motto of the system was to develop a system with higher accuracy and lower cost by making use of huge amounts of real-life medical data and several research articles.

In the following portion of the section, we will be presenting the evolutionary stages that AI has gone through over the last decade in the healthcare sector from a subdomains-based perspective. We also present a few application-based classifications of AI as well.

13.2.1 MACHINE LEARNING

The direct applications of ML were one of the primary ventures of AI in the healthcare sector. The high applicability and fast developments in ML had facilitated the introduction of newer terminologies and algorithms to the healthcare context as well. One of the most traditional applications of ML in healthcare was related to precision medicine, where it made use of ML to perform prediction of the future treatment plan with the aid of previous treatment history of the patients. The applicability of ML in healthcare varies highly from applications related to radiology, cardiology, ophthalmology, and so on. ML-based tasks function best for oral diagnostic procedures and image-based diagnostics. In cases of dermatological applications, ML plays a vital role in vision-based diagnostics [5]. Melanoma is an example of one such disease that is easily identifiable using ML-based techniques. On a wider aspect, dermatology-based ML applications are mainly employed for tasks such as disease diagnostics and classification via oral procedures, skin disease evaluation, and so on. In cardiological applications as well, ML has found a strong foothold. The combination of ML techniques in echocardiography for machine-driven calculation of aortic valve areas in aortic stenosis is a very famous application of such type. Similar to such applications, ML can also be found on several other diagnostic task-related to cancer, ophthalmology, and so on. Apart from the basic image-based knowledge extraction operations, ML has the complete potential to alter the way several healthcare tasks are performed, such as decision support systems, detection, and interpretation of findings, post-processing, and dosage estimation, examination quality control, and so on. Some of the most commonly employed techniques for the same are Naive Bayes, SVM, and CART, which have proven their validity via high accuracy for all the above-mentioned tasks, respectively [6].

Applications of ML aren't just confined to diagnosis and therapy. Additionally, they play an important part in ancillary jobs as well. Fraud detection in cases of insurance claims is one such instance. By constructing user contentment evaluating models using techniques such as regression, ML can also be utilized to improve the quality of life of patients and medical professionals. Even though ML catered to most of the basic needs, higher-level applications needed higher-level solutions [7]. Thus, DL started to be introduced into the healthcare context.

13.2.2 DEEP LEARNING

DL is a subdomain of ML that makes machines capable of performing high-level human decision-making functionalities via processing data and inferencing patterns within. DL procedures are highly versatile when it comes to data as they are capable of processing both labeled as well as unlabeled data. DL techniques have completely altered the way tasks such as speech recognition, object detection, image processing, and so have been executed over time. In the context of healthcare services, DL applications can be seen as being highly applied in image-intensive tasks such as radiology, pathology, ophthalmology, dermatology, cancer detection, and image-controlled surgical procedures [8]. It does similar jobs with more accuracy than its human counterparts.

Medical imaging is a well-known arena where DL has been proving its usefulness which has even led to the expectations that DL applications will eliminate the requirement for human interaction in image-related diagnostic activities. In the current scenario, DL has been frequently utilized to make certain inferences from MRI and CT scan images. The use of CNN and neural autoregressive distribution estimation (NADE) for brain tumor diagnosis is one such application [9]. In the light of the COVID pandemic, DL has also found ways to make use of CT scan images for COVID detection procedures as well. Cancer diagnostic is yet another stronghold of DL techniques along with many other similar applications such as cardiology based, radiology based, and so on.

Additionally, DL has proved its mettle in text-based applications as well. The Recurrent Neural Network (RNN), a type of DL model, is an ideal choice for sequence modeling issues, such as predicting the next word from health record data. Such model work with the aid of weighted connections that makes it possible for them to produce outputs via combined intermediate results generated over a period of time [10]. The Long Short-Term Memory (LSTM) is yet another DL model that has been highly applied in text-based tasks. Missing data prediction is a major application of LSTM models, and in the healthcare sector, it has been made use of in instances related to ECG and EEG signal processing. It easily outperforms Linear Regression (LR) and Gaussian Process Regression by utilizing the capacity to understand long-term dependencies to locate missing data. DL models have been employed in several other use cases, with the rest to text data as input which will be presented in detail under the NLP system [11]. Summing up, these are some of the very basic applications of DL in the healthcare context, and this list expands to Medical Robotics, Interactive systems, Expert systems, and so on.

13.2.3 NATURAL LANGUAGE PROCESSING (NLP)

Natural language processing (NLP) is a subdomain of AI which makes use of DL-based techniques to make human languages understandable to computers. Some of the prime functionalities of NLP systems include task-related to speech recognition, text data analysis, language translation, and any other tasks which are directly related to languages. When it comes to text-related applications in the healthcare domain, NLP provides the solution to a vast range of tasks, some of which includes analysis of unorganized clinical records of patients, preparing records from scientific data (scanning/test reports), deciphering treatment-related conversations, and even organize interactive AI applications [12].

With the outreach of EHR systems, one of the most prominent applications of NLP is knowledge mining from unstructured EHR and other health-related data [13]. When discussing such an application, the employment of NLP for data retrieval from diagnostic reports should be presented as well. A combination of NLP along with DL models have been employed in some applications where they empower the system to generate inferences from the reports and generate a system capable of doing oncology-based disease predictions related to cancer outcomes [14]. Similarly, when it comes to cardiology-based applications as well. NLP has found its place in the system performing report evaluation and diagnostics with higher expertise than any of the existing technologies or human counterparts. The scope of NLP is not just limited to the few applications discussed above but is far and varying. Most of the applications make use of DL or ML models to form a complete system which automatically brings them under the scope of DL and ML as well. Thus, some of the other instances of NLP can be seen during the discussions of other technologies also.

13.2.4 ROBOTICS

With the ongoing technological advancements, robotics is becoming an integral part of healthcare mechanisms. The systems are evolving to be highly intelligent, thus enabling them to achieve a certain level of autonomy with respect to the task they perform. Robotics can't be termed as an individual domain as they make use of DL models to achieve the aforementioned discriminative power and Intelligence. Generally, Robotics is used to refer to applications undertaken by both physical robots as well as robotic process automation procedures. The role of physical robots in healthcare has been well defined for a long period of time. From basic applications such as clinical assistants and basic surgical assistants to high-end applications such as robot-aided surgeries, the range of applications has constantly been evolving [15]. As surgical assistants, they perform an array of tasks varying from high precision invasive dissection, improving vision for remote locations to medication control and so on.

Robotic process automation presents an entirely different set of applications where the system mainly focuses on performing structured works on managerial grounds. Even though such tasks are categorized under robotics, they don't make use of physical robots at any level. Moreover, they employ computer programs to perform the functionalities, thus making the system semi-intelligent. Such systems are usually developed with certain points in mind, such as cost efficiency and functional transparency, along with prime preference to performing monotonous tasks such as advanced authorizations, amendment of patient details, and so on [2]. When properly combined with other domains, their task-performing capability can be made diverse, and the system can be made to perform NLP-based tasks such as billing, receipts, or image-based task such as signature verification, identity verification, and so on. Extensive research predicts the growth of the domain solely lies in its capability of integrating with other technologies and constantly undergoing evolution.

13.2.5 AREAS OF APPLICABILITY OF AI

In the preceding sections, we had performed a detailed discussion on the various roles of AI and DL, specifically in healthcare scenarios. In this section, we will be analyzing the domain-based applicability of DL and its applications in healthcare in a broader perspective. Some of the major applications where AI-based technologies have already carved out their space can be listed as in subsections.

13.2.5.1 ADMINISTRATIVE LEVEL APPLICATIONS

To cope up with the changing user expectations throughout the world, the healthcare sector is also on a path of evolution in the way it delivers its services. The newly employed methodologies are more oriented towards ensuring the quality of service alongside ensuring the quality of treatment as well. Tedious works relating to data entry or result analysis can be easily taken care of by AI-driven systems. For the last few years, the healthcare sector has witnessed an amalgamation of ML and NLP techniques for EHR-based data mining and report-based diagnostic functionalities, thus reducing the need for clinicians [2]. Similarly, NLP-based applications for performing "voice over text" functionalities are also being employed for transcription services and healthcare record analysis. Some of the other common functionalities that AI-based systems are used for include patient scheduling, insurance claim verification, medic-patient interactive chatbots, medicine stock prediction using predictive analysis, and so on. There are various other applications that have undoubtedly transformed the administrative sector and the perception of healthcare administration.

13.2.5.2 DIAGNOSTIC SUPPORT

AI has currently moved to a stage where it has started to take over human tasks up to a level. Providing diagnostic support is one such sensitive area where AI is trying to establish itself strongly. The arsenal of AI for diagnostic tasks is wide and varying. They perform the task with such high efficiency that they serve as additional support to human-aided diagnostic procedures even though not as a total alternative. Due to the prudence with which research is being conducted, this field is fast increasing its applicability and achieving excellent usefulness. Previously, diagnostic applications were primarily performed by ML programs for ailments such as coronary artery disease, liver diseases, and so on, whereas as of now, it has been replaced by high-end DL programs. Such DL models easily outperform older methods in terms of accuracy as well as overall performance [16]. Even when such system considerably surpasses human excellence, it's better to think of them as support systems and a hybridized approach that incorporates both human expertise and AI's discriminative power as we're still far away from attaining complete autonomy for such delicate use cases.

13.2.5.3 AI IN PHARMACEUTICS

AI has recently started shifting its concentration towards the pharmaceutical sector, especially drug discovery-related procedures for the last few years. The incorporation of AI-based techniques for the generation of pharmaceutical goods has opened up the scope for thrift development and deployment of technologies from research centers to real-life situations.

AI in the pharmaceutical industry plays a wide variety of roles, all the way from basic assistance in decision-making procedures to determining the ideal therapeutic process for a person to generating personalized medications and even for managing medical data, thus adding to its value. As one of the premiers in medical AI technology, Eularis had introduced a platform known as E-VAI, particularly for providing analytical and decision-making assistance using AI techniques for medical tasks. The introduction of such a platform sparked the research community for developing better alternatives as to the formerly made use of ML algorithms and was highly user friendly, allowing the end-user to access a whole set of data related to various aspects. The applications of AI on pharmaceuticals are so vast that they can't be covered through a single work. Our focus is mainly on the applications of AI in drug discovery which we will be presenting in the coming section.

13.3 AI IN DRUG DEVELOPMENT

Drug development is mainly about chemicals and their combinations, and with a vast expanse of chemical space, it is always possible to incubate and develop a huge number of drugs. A major hurdle with such a task is the deficiency of ample technological support, thus limiting the produces turning them into tedious and costly tasks. Such limitations can easily be addressed with the introduction of AI-based systems into the pharmaceutical industry. AI can help identify elements and also help in facilitating easier validation of the target as well as the drug structure modifications. Figure 13.2 provides a basic overview of the various applications of AI in drug discovery and development. Even though AI has a set of advantages, it also comes across some major challenges related to data features such as its volume, heterogeneity, unpredictability, and so on. The usual data sets used for drug development procedures are high in volume containing huge amounts of elements and compound details, which are not best suited for traditional ML algorithms. An effective alternative for such an issue is to make use of quantitative structure-activity relationship-based models that are capable of making better predictions when it comes to such datasets containing chemical terminologies.

Computer-aided drug discovery (CADD) is an evolutionary term associated with the drug discovery paradigm that came as a subsidiary to the traditional drug discovery approach with the invention of computers. With the advent of biological data made available as records stored in the databases in sequence or structure formats, it has laid a new milestone in the discovery of new drugs. The interdisciplinary branch called Bioinformatics deals with the storage and manipulation of such data. When AI waves were introduced in 2006 by Geoffrey Hinton, it was indeed the beginning of another dimension of intelligence [4]. Now it has shown its strength in changing the conventional pipelines of the healthcare industry, especially in the phases of new drug identification and drug repurposing. The power of AI was even utilized to identify the drugs for COVID-19 through AI-enabled drug repurposing strategies [17].

The conventional drug discovery approaches can be mainly classified into two. When the sequence or structure of proteins or small peptides are used for target identification and lead identification, it will be classified as Structure-Based Drug Discovery (SBDD), and if ligands are used for the same, the name will be changed to Ligand Based Drug Discovery (LBDD). However, AI can be effectively accommodated in both of the approaches, especially the DL architectures [18]. The design and development of a novel drug from scratch are more like searching for a needle in a haystack. Even if we succeed in finding such a potential lead, the possibility that it can complete all the clinical phase trials and be the winner is still questionable. Years of hard work and effort could be marked as 'rejected' in a trial outcomes sheet. AI could come as a savior in such a scenario as we could identify the potential compounds with required Absorption, Distribution, Metabolism, Excretion (ADME) values, minimal adverse effects, and having a lesser chance of failure during trials [19].

The first phase or stage of a drug discovery process is always the target identification, i.e., the destination on which our drug compound must act for the results. Once the target is known, choosing the right lead molecules which has the "Drug-like" characteristics is another pivotal task that is achieved with the aid of combinatorial chemistry and high-throughput screening in the traditional drug-discovery paradigm [20]. Once the "lead compound" is identified with the desired features, the next phase is "lead optimization," where the molecular properties are hyper tuned to fit the requirements. Structure-Activity Relationship (SAR) studies are used for optimization purposes and have been proven to be a success over the years. The rule-of-five designed by Lipinski [21] for evaluating the drug-likeness of a compound is always a benchmark for considering the compound that possesses the drug's potential. With the help of Molecular Dynamics, these compounds can be simulated in the organic environment to study how they behave with time and how effectively they can bind to the target site [22]. Once the compounds are prepared and filtered as per the criteria, they will be moved to real-life testing phases called the clinical trials, which is again a well-stipulated procedure with strict monitoring under the supervision of authorities, with prior approvals. These are the standard approaches undertaken during the drug discovery and testing phase. The rest of this section deals with the areas which could be enhanced with the application of AI for the Drug discovery process along with research progress on those domains by performing the same.

13.3.1 TARGET IDENTIFICATION

Drugs are designed in such a way that they can interact with proteins at their molecular level. The agonist or antagonist mechanism of the drug molecule can control the protein that is responsible for the disease by controlling its activity. Protein structure prediction is always a trending area of research in the Bioinformatics community. DeepMind's AlphaFold [23] was conceived from the aforementioned AI concepts to predict the structure of a protein from the amino acid sequences with a homology-based mechanism. The tool was able to predict the de novo protein structure at a promising accuracy. ML paradigms such as SVMs and a Naïve Bayesian (NB) classifier have also been employed for the identification of potential therapeutic targets by mining the large datasets collected from the normal and diseases states of molecules as well [24].

13.3.2 LEAD COMPOUND DESIGNING

Drugs are like 'keys' that can lock or unlock the proteins which are responsible for an illness. Molding the key or choosing the keys which could match the proteins is always the golden step in the drug discovery process. The future drugs in their pre-processed molecular versions are called 'ligand' initially. These ligands can be identified using SBDD approaches like virtual screening using protein structure so that the compounds can be chosen specifically to the binding cavity. If there are approved drugs or ligands which are available from the previous studies but failed in clinical trials due to their physical-chemical features can be considered as template molecules for identifying the similar ones from the databases, which is known as Ligand Based Virtual Screening (LBVS). AI platforms such as SOM, along with the vast databases available, can be used to link several compounds to numerous targets and off-targets. Bayesian classifiers and SEA algorithms can be used to establish links between the pharmacological profiles of drugs and their possible targets [25]. Popova et al. proposed a method called ReLeaSE for generating chemical compounds and focused chemical libraries with desired physical, chemical, and bioactivity properties that are based on deep reinforcement learning (RL) [26].

13.3.3 HIGH THROUGHPUT VIRTUAL SCREENING (HTVS)

Virtual screening of compounds that can show affinity to the target protein structure can be done via AI algorithms such as Nearest-Neighbor Classifiers, Random Forest, SVMs, and DNN [27]. DL models trained and tested with sufficient datasets can screen the drug-like molecules with required physical-chemical properties. When protein structure is used for virtual screening, docking of the ligands at the binding sites will be scored on parameters to rank them based on docking score. And in Ligand-based virtual screening, the similarity index of compounds is calculated based on the distance, and closer ones will be identified. The DL model called PyRMD is claimed to be a completely AI-powered LBVS module that uses Random Matrix Discriminant (RMD) algorithm [28].

13.3.4 STRUCTURE-ACTIVITY RELATION (SAR) PREDICTION

Quantitative Structure-Activity Relationships of compounds are focused areas of study in drug discovery approaches. These *in silico* studies could reveal the contributions of chemical groups responsible for the interactions associated with biomolecules. DeepSnap, a DL-based method, was developed by Uesawa at the Meiji Pharmaceutical University in 2018 [29]. In this DeepSnap-DL approach, the 3D-optimized molecular structures, which can be rotated at any arbitrary angle on the x, y, and z-axes, were photographed as a ball-and-stick model with different colors to represent the corresponding atoms to automatically input as much structural information as possible into the DL models.

Random forest and SVM-based approaches were already used for SAR studies, and a visible transition from the latter to DL models is observed these days. The approach proposed by Bouhedjar et al. has adopted the word embedding approach for representing the SMILE strings for predicting the activity of corresponding structures [30]. They have used a hybrid DL architecture composed of CNN and LSTM. These promising results suggest that their approach can be applicable for predicting any physicochemical, biological, or pharmacological properties of interest (Figure 13.2).

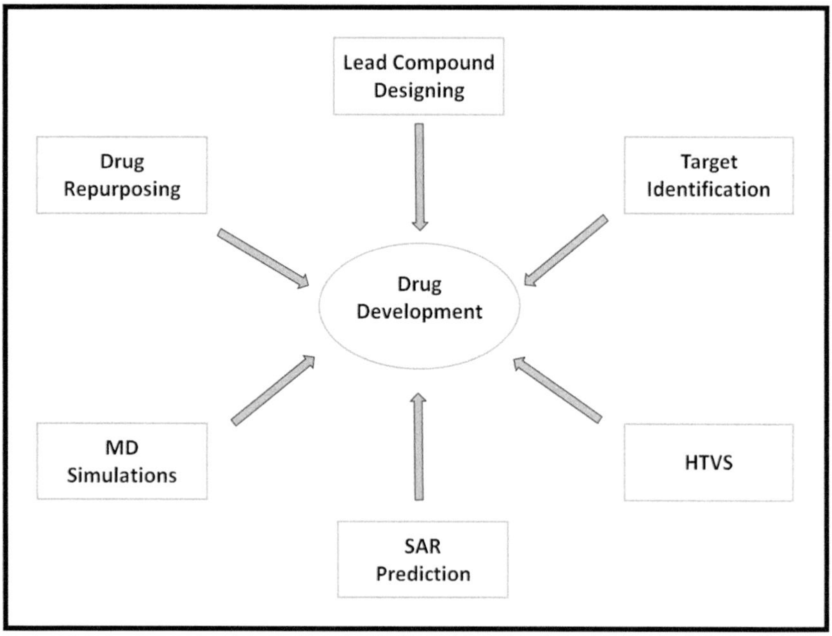

FIGURE 13.2 Application domains of AI for drug discovery.

13.3.5 MOLECULAR DYNAMICS SIMULATIONS

Molecular dynamics is another major application of AI mainly performed to reveal the changing nature of a biomolecule in the virtual ambiance created in silico to resemble the original environment to which they belong. They play a key role in identifying the binding power of protein-drug complexes and how they are attaining the equilibrium state. A variational approach for Markov processes (VAMP) was also used to develop a DL framework for molecular kinetics using neural networks, dubbed VAMPnets [31]. They have used VAMPnets to learn molecular kinetics from simulation data of a range of model systems. VAMPnet was capable of implicitly learning the feature transformation from Cartesian coordinates to backbone torsion with high efficiency.

13.3.6 DRUG REPURPOSING

Repurposing is the process of making use of the FDA-approved drugs for any particular disease for treating against any other disease. Identifying such drugs as an immediate requirement was observed in the pandemic time. DL algorithms have proved their applicability even for identifying the repurposable drugs to treat COVID-19 which has served as a huge bonus considering the labor and resource-constrained situation during the pandemic [17].

13.4 CHALLENGES AND FUTURE SCOPE

13.4.1 CHALLENGES

Even though hopeful results are being obtained via DL architecture-based systems, DL-based healthcare application still needs to go along before becoming false proof. Even though there are many such issues, we will be only discussing some of the major issues which we felt are of utmost significance—in subsections.

13.4.1.1 DATA QUANTITY

DL is a term used to refer to a combination of extremely completive computational tasks. One such example is an ANN where a huge number of network parameters are required to be configured precisely for generating a functional model. The foundation to achieving such a task is primarily the presence of vast volumes of data. Even though there are no rules stating the minimal volume of data needed for a functional model, the generally followed procedure is to make use of data having a minimum of at least 10 multiples of the number of samples as parameters used in the model. The following feature is one of the prime reasons behind the success of DL in applications making use of vast amounts of data. Such DL techniques are also employed in various AI subdomains such as computer vision, NLP, and so on. Even so, when it comes to the healthcare scenario, the whole dynamics change rapidly. Statistics state that the number of people who are able to access high-end healthcare facilities is very few and little compared to people who are not able to even afford basic healthcare [32]. Thus, generating enough data for highly complex medical tasks are a real matter or concern in such systems as with a lesser amount of data, the system performance also gets compromised, which is not an ideal situation when it comes to sensitive application such as in healthcare.

13.4.1.2 DATA CHARACTER

When compared to other domains, the major peculiarity of healthcare data lies in their inherent properties of being highly unstructured, heterogeneous, and deficient. Performing model training with such data with huge volumes and varying characteristics will only result in an incomplete model with deficient characteristics. Thus, various issues need to be tacked prior to making data fit for training, such as data inadequacy, redundancy, and data insufficiency.

13.4.1.3 DATA TRANSIENCE

Diseases are continually evolving and converting themselves in unpredictable ways over time. Currently, most DL models in use or in their implementation stage in the healthcare domain are ignorant of this fact as they are only made capable of dealing with vector-based inputs and are incompatible when it comes to temporal data. Since major instances of healthcare data happen to be dynamic and time-related, developing DL models suited for such applications is a highly relevant aspect of the development of DL-based applications in healthcare.

13.4.1.4 DOMAIN INTRICACY

Compared to other applications such as image processing or NLP tasks, healthcare-based, and biomedical applications are highly intricate and sensitive tasks. The tasks involved in the healthcare domain closely relate to disease identification or allied tasks, which are extremely heterogeneous, and when it comes to certain diseases, a thorough understanding of their genesis or progression is still void. Furthermore, in a real-life clinical scenario, the number of patients is restricted, which also adds to the issue.

13.4.1.5 ACCOUNTABILITY

Despite the fact that DL models have proved their mettle in a variety of applications, they are frequently considered black boxes due to the presence of hidden layers. Even though such an issue may not be lethal in domains where the end-user also plays a role in determining final decisions, in a healthcare scenario, it is ideal to have a clear understanding of both the quantifiable execution of the algorithm as well as their reason for the adopted approach. In reality, model accountability is highly critical for persuading medical practitioners to follow the prediction system's recommendations.

13.4.2 FUTURE SCOPE

These aforementioned challenges have opened up several prospective research opportunities for the overall improvement of the domain. In the following section, we will be having a brief decision about some of the research directions which we feel will pave the way for the development of DL-based applications in healthcare in the near future.

13.4.2.1 FEATURE ENHANCEMENT

Due to the fact that the amount of healthcare data available for research is highly limited, we should devise measures to utilize the same data in a much more efficient way by trying to extract more features from it and also identifying better techniques to efficiently work upon them. Such feature enhancement can be done to data generated from various sources such as EHR, wearable systems, body area networks, genetic data, online media, and so on. An essential and difficult research problem would be to perform appropriate integration of such extremely varied data and make it fit into DL models. The security aspect of such sensitive EHR and allied data should also be taken into consideration [33]. With all this being said, the reality is that little effort has been taken to accomplish the following objective, and works concentrating on approaches for heterogamous medical data interpretation are scarce in the context of DL. A possible solution that could be identified for the issue is to make use of the hierarchical schema of DL and process the data from each source individually with a suitable model and then combine back the resultant data into a stacked model, thus representing the data as a whole (Autoencoders or Bayesian networks can be employed).

13.4.2.2 DISTRIBUTED INFERENCING

Each medical environment has its unique set of patients. Thus, an individual site will not be possessing enough data about multiple medical conditions, which calls for either data transfer or collaborative training. When it comes to medical data transfer, the sensitive nature of the data holds many restrictions making it hard to be done in real life. Therefore, the optimal alternative is to develop a model that can be trained in a distributed manner that ensures minimal data leakage and extensive privacy protection making Federated DL an ideal contender for the same. Federated DL models are highly secure alternatives when compared to the traditional methodologies and also have immense scopes for development and research.

13.4.2.3 PRIVACY PRESERVATION

Privacy preservation is an essential feature in the DL system as concerns arise when we start scaling the system to higher levels. In recent times, the vulnerability-related issues of ML models deployed as services in various contexts have been discussing in detail in multiple works [34]. Even though such attacks stick to the network protocols and functionalities of the model completely, they try to infer parameters and data through various approaches such as Application Program Interface (APIs), thus inflicting serious privacy breaches. Such attacks are highly potent, and multiple preventive mechanisms have been made available. And some of the most highly popular ones include techniques such as Differential Privacy, Homomorphic Encryption and so on. When it comes to Differential Privacy, they make use of noise generating mechanism to embed noise into these data, thus masking it and making the adversary not able to identify the exact data or match it with a particular context or user. The major issue with such an approach is that the noise value needs to be kept in check, or the data may become so distorted that it will turn useless, and this makes the real-life implementation of differential privacy significantly complex. The complexity of the same scenario increases manifold when we shift to DL paradigms as the number of parameters to be privacy preserved also increases exponentially. Thus, taking into consideration the volume, fields, and sensitivity of data to be processed via DL models, these issues need to be addressed prior to their deployment, especially in the healthcare industry.

13.4.2.4 AMALGAMATING WITH EXISTING KNOWLEDGE

As of now, the existing healthcare-based systems have gathered expert knowledge over a period of time which is priceless when compared to newly developed systems still configuring their cognitive capabilities. Already, medical data is highly limited, and expert data comprises of valuable insights, and the key to the future development of DL-based healthcare systems lies in their capability of amalgamating the existing knowledge along with the newly acquired ones. Platforms such as PubMed provides a huge stock of medical data, which opens up opportunities for further learning when efficiently mined and the extracted data can even be directly used for DL model training procedures to obtain high-level output from the system. Another alternative is to make use of a semi-supervised learning system in which the end-user always has a grip over the processes, and the system itself is capable of doing a hybrid learning mechanism, meaning that it can make use of labeled as well as unlabeled data equally well for model training thus bringing down the need for data and helping the system to attain the same performance as with higher amounts of data in conventional learning methods.

13.4.2.5 DYNAMIC MODELS

Since healthcare is a highly dynamic domain undergoing a huge number of changes in one aspect or the other, technologies involved with the domain should also be capable of adapting to the change. When it comes to DL models, applications involving the management of EHRs, health monitoring devices, and wearable devices, such temporal changes are clearly evident. Thus, one of our developmental aims should be in developing a dynamic and temporally compatible DL model as it will play a huge role in adapting our model with respect to patient needs and conditions, thus improving system performance and user experience as well [35]. For this purpose, we consider RNNs and DL architectures with embedded memories as idealistic solutions capable of playing a major role in the betterment of medical DL applications.

13.4.2.6 ACCOUNTABLE MODELS

Model execution and accountability are two critical factors in healthcare-based applications. It is highly doubtful that medical personnel prefer to work upon a system that they are not able to figure out. Even though the exceptional performance of DL models makes them highly desirable, it is hard to give a clear understanding of how the performance was obtained or what decisions had led to the result generation in such systems. Thus, figuring out a way to give a better explanation of the system procedures is a major hurdle preventing the widespread of DL into a higher number of applications in the healthcare sector. Our study proposes that the research direction should also be specific in generating approaches or algorithms which makes DL models more explainable to the user, thus reducing the smokescreen that exists between the model and the end-user as well as methodologies to aid the systems already in place for explaining the estimations of data-driven systems.

13.5 AI FOR RURAL HEALTH DEVELOPMENT

Particularly in developing nations, attaining equal opportunities with respect to healthcare service delivery to all fractions of the society is a major part of building an inclusive society. In most countries, government-run agencies along with semi-government and other welfare organizations play a huge role in trying to implement it in rural areas, even though the ground reality is still far from being ideal. When we use the term rural, we are referring to geographical locations which tend to be isolated from the major cities and will be having a limited population along with limited infrastructural developments as well. For developing countries in ideal situations also, the healthcare infrastructure will be limited and life expectancy period less compared to other developed economies. This situation worsens exponentially when we look into rural communities. Poverty can be as one of the prime reasons for the same, and other factors such as lack of employment, low level of infrastructural development and everything just adds to it. Decreased accessibility to qualified healthcare resources is also one of the most direct reasons related to substandard healthcare conditions, which is highly contributed to by the formerly mentioned factors as well [36]. Lesser public health funding, lesser number of healthcare insurance users, deficiency of qualified medical personnel, lack of ample transportation resources, and so on also contribute their part to the diminishing healthcare services in developing economies, especially towards their rural sector.

As previously discussed, most of the rural communities in developing countries have a high deficiency of highly skilled medical practitioners and a vast majority of patients need to be attended to by nurses or other paramedics. The only favorable thing in such situations was that in the majority of the instances, the medical conditions will be uncomplicated, mundane, and could be easily cared for with a bare minimum of resources and thus enabling healthcare-based AI systems to be employed as an effective alternative for the skilled labor shortage. In the preliminary stages, such technology was referred to as "Computer-assisted medical technology." In India, the primary efforts of computer-aided healthcare systems were developed in the form of a disease detection system referred to as the Early Detection and Prevention System (EDPS) that facilitated a platform where the needy could undergo medical diagnostic procedures even without the need for a doctor [37]. The system required nurses and any skilled paramedics who could understand medical procedures recommended by the system through the course of the treatment. The system was highly accepted due to its high-performance accuracy as well as its interactive nature, thus laying a strong foundation for a technical revolution in the healthcare industry in rural scenarios.

Apart from its applications in basic healthcare scenarios, AI-based techniques have also been used for developing disease-specific or task-specific systems aiming for rural societies. The development of comestible endoscopic capsules can be considered as a real-life example. They are primarily cheap and do diagnosis tasks, especially for cancer-related diseases, in turn becoming an alternative for highly complex and expensive test procedures. Such devices are particularly suited for low-cost treatment plans that are ideal for rural communities where the chances of occurrence of cancer are high. Leukemia is another potent disease whose treatment needs extensive diagnosis, including the type, stage, and all, which requires highly skilled doctors who specialize in the same. Such procedures are usually expensive, and due to that reason, it is not widely available as well [38]. Making use of AI-based principles, several methodologies had been proposed as an alternative solution for the same, and acute leukemia classification from born marrow images based upon their morphological properties was one of the most prominent breakthroughs. Thus, such a method can be considered as an effective alternative to their expensive counterparts, especially in developing countries. Another similar condition arises with peripheral neuropathic cases as well. The complex nature of the diseases makes it highly tricky to be diagnosed in the absence of experienced neurologists, thus limiting its diagnosis and treatment in rural scenarios. For such instances, Decision support systems are needed, and works are already available discussing such novel AI-based support systems that aid in diagnostics and in generating medical prescription reports as well. Experts evaluate the model to be highly effective with a very high value of accuracy and tag it as an ideal solution for rural communities having peripheral neuropathies cases.

In short, the above-discussed instances are just a few real-life examples through which we put forward as methods to overcome the accessibility-related issues for rural communities in developing countries by making use of cost-efficient new-generation AI-driven technology as a substitute for the traditional medical systems that are expensive and highly complex. Thus, it can also be stated that AI technology applied in healthcare does not only limit itself to the enhancement of services delivered by physicians but also ensure improved delivery of services as well along with reduced medical expenses. It also proves to be efficient in empowering paramedical workers to cover up for instances with deficiency or lack of doctors as well. With all this being said, we should also note that as of now, only very few reliable systems or technologies are available, and they alone are not sufficient to be termed as a complete alternative for skilled human resources [37]. In order to achieve an improvement in the overall degree of healthcare services in rural communities, and efficient AI system will surely go a long way. Prior to installing the system, some of the basic necessities of the system should be met as well, which include things such as adequate infrastructure, including electricity and network connection, timely updation and continual training, ample financial support, and so on.

When discussing healthcare needs in rural environments, the system should also be specifically suited to cater to the needs of the situations. Thus, general healthcare system turns out to be not well suited where application-based variance is needed. In the following section, we will be discussing an AI-based hierarchical healthcare system that can be employed in the context of rural healthcare developmental scenarios:

1. **Primary Level Applications:** Such systems can be employed at the most fundamental level of rural healthcare scenarios, such as primary healthcare centers or small clinics. The prime motto of such systems should be to tackle issues related to economic constraints, inaccessibility, the efficiency of skilled staff, and so on. Thus, the strong points of the system should be easy and efficient diagnostic capability, cost efficiency, compactness, and mobility, connectivity, and poor efficiency, and user-friendly interface as well. Such a system can be employed for various tasks such as running tests on blood or urine, generating ECG or EEG reposts, monitoring PG or glucose levels, and similar tasks that relate to mundane diseases requiring basic examinations. The system will comprise of a computer system enabled with AI functionality that is capable of performing diagnostic operations as well.
2. **Intermediate Level Applications:** Such systems are usually employed in bigger hospitals or clinics with a higher number of doctors and medical departments. The primary role of these systems includes probing basic training to elementary healthcare personnel, keeping a check of AI-based healthcare systems, maintaining, and keeping them updated, and also generating reports and maintaining data-based regarding diseases, their controls, and similar things. Apart from all this, such clinics may also possess specialized AI-based systems that are capable of nursing people with potent and complex ailments.

3. **High-Level Applications:** The primary role of such systems is to synchronize various stages of healthcare-based AI systems such as developmental tasks, promotional woks and facilitate cooperative functioning. For ensuring the long-term existence of these systems, such cooperative measures are a high necessity. A contiguous rapport between multiple parties such as fund providers (e.g., government organizations), fundraisers (e.g., NGOs), and developers (e.g., technical people) is needed for the widespread promotion of such systems. As an end line, we can say that continuous collaboration is the key to keeping the system streamlined and be capable of adjusting to the ever-changing healthcare requirements.

To summarize, it is evident from the study performed that promoting healthcare-based AI technology in rural communities will be one of the baby steps toward eradicating the long-lasting disparity between rural and urban communities when it comes to even basic living necessities. The introduction of such technology will surely take society a far way when it comes to building an inclusive society.

13.6 CONCLUSION

The standard perspective for data storage and overall management of healthcare data evolves to be highly circulative which brings out the lesser quality and decreases the value of data. Thus, there should be a new approach for using the data in future. With AI becoming more popular, the healthcare sector also started employing AI-based techniques, gradually satisfying the requirements of data aggregation, and also resulting in the development of procedures with increased performance and accuracy. This aids in the development of robust models capable of performing high-level activities such as automated disease diagnosis, as well as high precision methods for maximizing resource utilization for specific tasks in a well-timed and dynamic way.

AI has entirely transformed the way healthcare tasks are being performed. From basic tasks such as image or text-based disease diagnostics tasks to high-end applications such as surgical robots and medical support systems, the domain has spread itself over a vast area. The evolution of AI in the healthcare context was thrift but sturdy. The gradual introduction of various subdomains unto the scene has only brought more support to the following techniques to be added. The AI primary technique to be introduced was ML for basic diagnostic tasks followed by DL for high-end applications of the same. Later on, technologies based on NLP and even combinations of AI and IoT were implemented with high efficacy in healthcare systems. Along with all the merits, applications of AI do face certain shortcomings as well, some of which are related to data features such as quantity, character, transience, intricacy, and so on. The domain has an extensive scope of development if future works are centered on the challenges faced in real-life scenarios. AI also can play a vital role in improving the healthcare scenario in rural economies. Rural medical communities face several issues for which AI can be seen as a prospective and effective solution. Skilled medics shortage, lack of extensive diagnostic equipment, and so on can be easily overcome with effective implementations of AI-based systems. Thus, we conclude that applications of AI in the healthcare sector have a never-ending scope if it can be harnessed efficiently.

KEYWORDS

- **artificial intelligence**
- **artificial neural network**
- **deep learning**
- **deep neural networks**
- **healthcare systems**
- **machine learning**

REFERENCES

1. Kun-Hsing, Y., Andrew, L. B., & Isaac, S. K., (2018). Artificial intelligence in healthcare. *Nature Biomedical Engineering, 2*(10), 719–731.
2. Davenport, T., & Ravi, K., (2019). The potential for artificial intelligence in healthcare. *Future Healthcare Journal, 6*(2), 94.
3. Jorge-Martinez, D., Shariq, A. B., Edeh, M. O., Chinmay, C., Qaisar, S., De-La-Hoz-Franco, E., & Ariza-Colpas, P., (2021). Artificial intelligence-based Kubernetes container for scheduling nodes of energy composition. *International Journal of System Assurance Engineering and Management*, 1–9.

4. Bengio, Y., & LeCun, Y., (2007). Scaling learning algorithms towards AI. *Large-Scale Kernel Machines, 34*(5), 1–41.
5. Callahan, A., & Nigam, H. S., (2017). Machine learning in healthcare. In: *Key Advances in Clinical Informatics* (pp. 279–291). Academic Press.
6. Toh, C., & James, P. B., (2021). Applications of machine learning in healthcare. *Smart Manufacturing: When Artificial Intelligence Meets the Internet of Things*, 65.
7. Futoma, J., Morgan, S., Trishan, P., Doshi-Velez, F., & Leo, A. C., (2020). The myth of generalizability in clinical research and machine learning in health care. *The Lancet Digital Health, 2*(9), e489–e492.
8. Chen, Y. W., & Jain, L. C., (2020). Deep learning in healthcare: Paradigms and applications. Springer.
9. Jia, Z., & Deyun, C., (2020). Brain tumor identification and classification of MRI images using deep learning techniques. *IEEE Access.* pp. 1–1.
10. Miotto, R., Fei, W., Shuang, W., Xiaoqian, J., & Joel, T. D., (2018). Deep learning for healthcare: Review, opportunities and challenges. *Briefings in Bioinformatics, 19*(6), 1236–1246.
11. Esteva, A., Alexandre, R., Bharath, R., Volodymyr, K., DePristo, M., Katherine, C., Claire, C., et al., (2019). A guide to deep learning in healthcare. *Nature Medicine, 25*(1), 24–29.
12. Iroju, O. G., & Janet, O. O., (2015). A systematic review of natural language processing in healthcare. *International Journal of Information Technology and Computer Science, 7*(8), 44–50.
13. Juhn, Y., & Hongfang, L., (2020). Artificial intelligence approaches using natural language processing to advance EHR-based clinical research. *Journal of Allergy and Clinical Immunology, 145*(2), 463–469.
14. Po-Hao, C., Hanna, Z., Galperin-Aizenberg, M., & Tessa, C., (2018). Integrating natural language processing and machine learning algorithms to categorize oncologic response in radiology reports. *Journal of Digital Imaging, 31*(2), 178–184.
15. Porkodi, S., & Kesavaraja, D., (2021). Healthcare robots enabled with IoT and artificial intelligence for elderly patients. *AI and IoT-Based Intelligent Automation in Robotics*, 87–108.
16. Han, C., Leonardo, R., Kohei, M., Takafumi, N., & Hideki, N., (2020). Bridging the gap between AI and healthcare sides: Towards developing clinically relevant AI-powered diagnosis systems. In: *IFIP International Conference on Artificial Intelligence Applications and Innovations* (pp. 320–333). Springer, Cham.
17. Zhou, Y., Fei, W., Jian, T., Ruth, N., & Feixiong, C., (2020). Artificial intelligence in COVID-19 drug repurposing. *The Lancet Digital Health.*
18. Kit-Kay, M., & Mallikarjuna, R. P., (2019). Artificial intelligence in drug development: Present status and future prospects. *Drug Discovery Today, 24*(3), 773–780.
19. Dana, D., Satishkumar, V. G., Luce, G. St. S., David, L., Farha, N., Quaisar, A., Latha, P., et al., (2018). Deep learning in drug discovery and medicine; scratching the surface. *Molecules, 23*(9), 2384.
20. Pina, A. S., Abid, H., & Ana, C. A. R., (2010). An historical overview of drug discovery. In: *Ligand-Macromolecular Interactions in Drug Discovery* (pp. 3–12). Humana Press, Totowa, NJ.
21. Ming-Qiang, Z., & Barrie, W., (2007). Drug discovery beyond the 'rule-of-five.' *Current Opinion in Biotechnology, 18*(6), 478–488.

22. Borhani, D. W., & David, E. S., (2012). The future of molecular dynamics simulations in drug discovery. *Journal of Computer-Aided Molecular Design, 26*(1), 15–26.
23. AlQuraishi, M., (2019). AlphaFold at CASP13. *Bioinformatics, 35*(22), 4862–4865.
24. Fang, J., Ranyao, Y., Li, G., Dan, Z., Shengqian, Y., Ai-Lin, L., & Guan-Hua, D., (2013). Predictions of BuChE inhibitors using support vector machine and naive Bayesian classification techniques in drug discovery. *Journal of Chemical Information and Modeling, 53*(11), 3009–3020.
25. Achenbach, J., Pekka, T., Lutz, F., & Ewgenij, P., (2011). Computational tools for polypharmacology and repurposing. *Future Medicinal Chemistry, 3*(8), 961–968.
26. Popova, M., Olexandr, I., & Alexander, T., (2018). Deep reinforcement learning for de novo drug design. *Science Advances, 4*(7), eaap7885.
27. Durrant, J. D., & Rommie, E. A., (2015). Machine-learning techniques applied to antibacterial drug discovery. *Chemical Biology & Drug Design, 85*(1), 14–21.
28. Amendola, G., & Sandro, C., (2021). PyRMD: A new fully automated AI-powered ligand-based virtual screening tool. *Journal of Chemical Information and Modeling, 61*(8), 3835–3845.
29. Matsuzaka, Y., & Yoshihiro, U., (2021). A molecular image-based novel quantitative structure-activity relationship approach, DeepSnap-deep learning and machine learning. *Current Issues in Molecular Biology, 42*(1), 455–472.
30. Bouhedjar, K., Abdelbasset, B., Abdelmalek, K. N., Anouar, B., Amine, B., & Abdelhafid, D., (2020). A natural language processing approach based on embedding deep learning from heterogeneous compounds for quantitative structure-activity relationship modeling. *Chemical Biology & Drug Design, 96*(3), 961–972.
31. Mardt, A., Luca, P., Hao, W., & Frank, N., (2018). VAMPnets for deep learning of molecular kinetics. *Nature Communications, 9*(1), 1–11.
32. Gerke, S., Timo, M., & Glenn, C., (2020). Ethical and legal challenges of artificial intelligence-driven healthcare. In: *Artificial Intelligence in Healthcare* (pp. 295–336). Academic Press.
33. Bartoletti, I., (2019). AI in healthcare: Ethical and privacy challenges. In: *Conference on Artificial Intelligence in Medicine in Europe* (pp. 7–10). Springer, Cham.
34. Ragesh, G. K., & Baskaran, K., (2016). Cryptographically enforced data access control in personal health record systems. *Procedia Technology, 25*, 473–480.
35. Bi, H., Jiajia, L., & Nei, K., (2021). Deep learning-based privacy preservation and data analytics for IoT enabled healthcare. *IEEE Transactions on Industrial Informatics, 18*(7), 4798–4807.
36. Iliashenko, O., Zilia, B., & Alissa, D., (2019). Opportunities and challenges of artificial intelligence in healthcare. In: *E3S Web of Conferences* (Vol. 110, p. 02028). EDP Sciences.
37. Strasser, R., Sophia, M. K., & Sophie, M. R., (2016). Rural health care access and policy in developing countries. *Annual Review of Public Health, 37*, 395–412.
38. Guo, J., & Bin, L., (2018). The application of medical artificial intelligence technology in rural areas of developing countries. *Health Equity, 2*(1), 174–181.
39. Vijai, C., & Worakamol, W., (2021). Rise of artificial intelligence in healthcare startups in India. *Advances in Management, 14*(1), 48–52.

CHAPTER 14

Future Directions in Healthcare Research

KIRANDEEP SINGH,[1] PRABHDEEP SINGH,[2] and MOHIT ANGURALA[3]

[1]*Chandigarh University, Punjab, India*

[2]*Graphic Era (Deemed to be University), Dehradun, Uttarakhand, India, E-mail: ssingh.prabhdeep@gmail.com*

[3]*Khalsa College of Engineering and Technology, Amritsar, Punjab*

ABSTRACT

By using digital technologies, the healthcare industry is always evolving toward a better health system for the patients. The healthcare industry has brought many opportunities in the growing business as a result of digitization. Now, hospitals are now connected to the Internet and gather useful information from a variety of sources to collect information about the patient's health status through digital gadgets. The primary aim behind this chapter is to present information about the present and future technology, as well as how they might aid in the prediction of diseases before they occur. This chapter mainly focuses on recent technology innovations utilized in computational health informatics, such as artificial intelligence (AI), machine learning (ML), and signal processing. Furthermore, the existing standards' various challenges in the existing healthcare industry have been discussed in an effective manner [11]. The future of computational Health Informatics technologies is an opportunity for developing and growing countries using the recent developments in Health Informatics Technologies.

Computational Health Informatics for Biomedical Applications. Aryan Chaudhary and Sardar M. N. Islam (Naz) (Eds.)

14.1 HEALTH INFORMATICS

As a field, health informatics is on the rise. It's all about making the best use of data and technology to improve patient care. It's at the crossroads of public health, information technology, and healthcare. As it pertains to the gathering, storage, retrieval, and use of information in public health and healthcare, this is an important topic. It is through the use of health informatics technology that we may all benefit from better health outcomes and lower healthcare expenditures, as well as increased administrative effectiveness and more access to excellent medical treatment [4]. Following is a short discussion on Health Informatics Technologies.

14.2 mHEALTH

Health informatics and technology in the field of mobile health is mHealth, or the use of mobile technology in healthcare is making it easier for patients to get treatment and reducing costs. Patients are able to manage their health better, contact healthcare professionals, plan appointments, and obtain health information thanks to an expanding number of mobile applications. In addition, remote monitoring of some health issues and medical equipment is being made easier for patients thanks to technological advancements. These firms are just a few of the many that are developing mobile apps to revolutionize how physicians and patients communicate in the current day, such as Pager, Oscar Health, and Vesta Healthcare (Figure 14.1).

14.3 THE USE OF REMOTE MEDICAL CARE THROUGH TELECOMMUNICATIONS

Telemedicine, like mHealth, is concerned with establishing a virtual relationship between a patient and a doctor. Medical professionals are freed from the inconvenience of having patients travel long distances only to meet them in person with the advent of telemedicine. Telemedicine is a godsend to the thousands of individuals who are unable to get medical treatment due to their distant location, lack of transportation, or inability to walk (Figure 14.2).

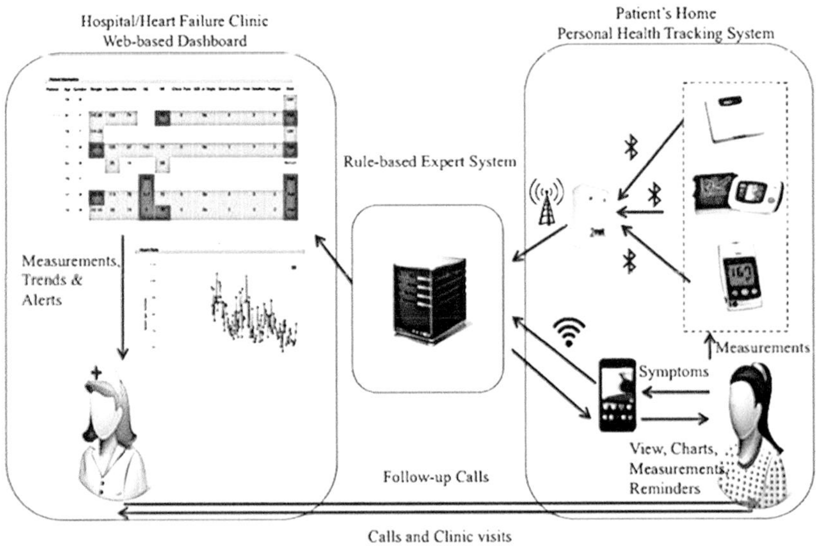

FIGURE 14.1 Design and usability of a heart failure mHealth system.

Source: Adapted from Ref. [14].

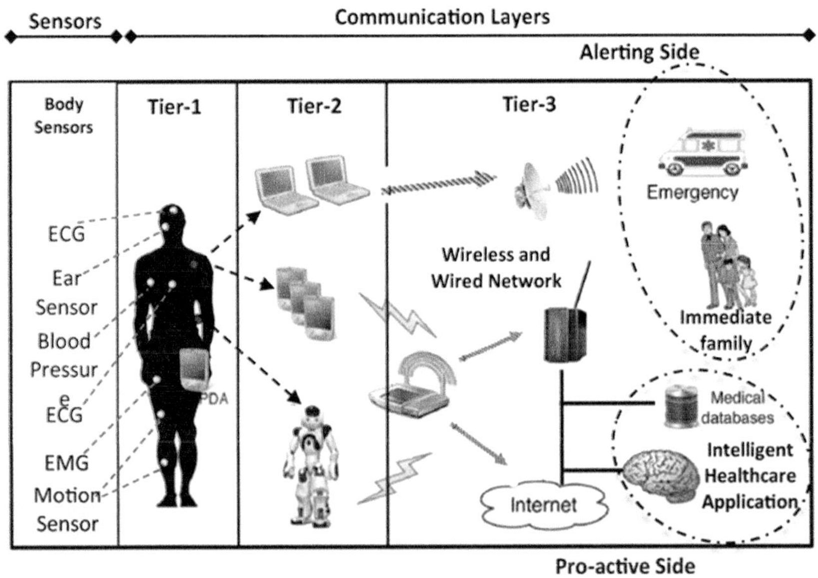

FIGURE 14.2 Real-time remote health-monitoring systems in a medical center.

Source: Reprinted with permission from Ref. [15]. © 2018 Springer Nature.

14.4 ELECTRONIC MEDICAL RECORDS ARE BECOMING MORE COMMON

Improved patient outcomes may be achieved via the use of electronic health records (EHRs) because clinicians can diagnose and treat patients using a comprehensive picture of their past and present health. Healthcare inequalities and e-prescribing and telehealth procedures have also been enhanced by the use of EHRs (Figure 14.3).

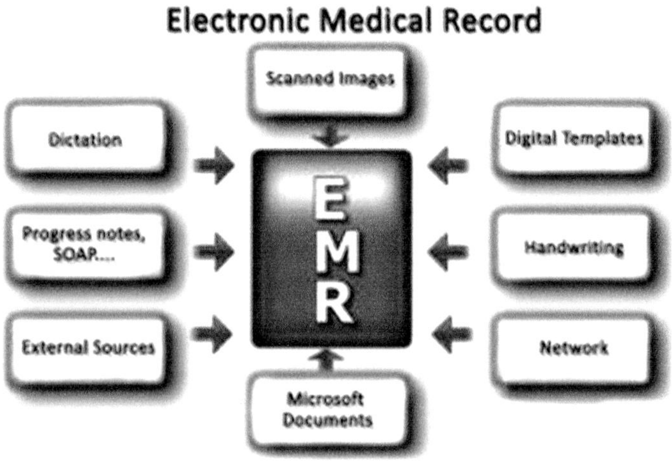

FIGURE 14.3 Electronic medical record.

Source: Reprinted with permission from Ref. [16]. © 2018 Elsevier.

EHRs enable physicians to coordinate patient care and assure accuracy in today's digitally linked world, but they also provide patients with the ability to be their own advocates. Instantaneous access to all of their medical records is available to them. Patients may view their records and test results via patient portals at many clinics and hospitals, and they can even connect with their primary care physician using these portals.

14.5 HEALTHCARE INFORMATION TECHNOLOGY THAT IS INTEROPERABLE (HEALTH IT) SYSTEMS

It is the capacity to run a secure system that provides allowed stakeholders, and only approved stakeholders, rapid access and exchange of health data and information no matter where they are situated interoperability. The Healthcare Information and Management Systems Society provides a more in-depth description (HIMSS). When multiple information systems, devices, and applications can access, share, integrate or utilize data together in a coordinated way, inside, and outside the limits of a particular area or country, in order to create a smooth flow of information, it is called "interoperability."

To find a new doctor, patients had to either search down all of their medical data and provide them to them or just depend on an oral history of their past treatment before interoperable Health IT systems were available. Even if a patient's medical records are spread across several divergent computer systems located in various places, healthcare practitioners may now access them all through Health IT solutions. In order to maintain continuity of treatment, it is essential that each medical institution has a complete record of each patient's medical history.

In addition to electronic medical records (EMR), wearables provide another way for users to gather data, promote prevention, and improve health outcomes. However, there has been a massive increase in wearable technology since the smartwatch and Fitbit initially came onto the scene [3].

Patients and their physicians may benefit from wearables because they can alert them to emerging medical concerns. Wearables capture real-time data, which is then evaluated by a system that may alert physicians if anything is wrong with a patient. Using this method, physicians may be proactive and reach out to people who may be in need of acute medical assistance without even seeing it. Patients with asthma, for example, maybe put on an Automated Device for Asthma Monitoring and Management (ADAMM). Wearers and physicians may be alerted to an impending asthma attack or other medical emergency by using an app that links to this monitor (Figure 14.4).

14.6 BIOPRINTING IN 3D

Drugs, prosthetics, and even human tissue and organs might be created using this cutting-edge medical technology, which is still in its early stages. An important step forward in the development of 3D printing was made in 2018 when scientists produced human ears and successfully glued them to mice's skins. In Australia, scientists have successfully implanted 3D-printed vertebrae into a human patient with chordoma cancer, which may be even more thrilling. Improved, longer-lasting, and more effective orthopedic implants are being created using 3D orthopedic implants, which have revolutionized the field of joint and bone replacement. More than 4 million orthopedic implants are expected to be implanted on patients by 2027.

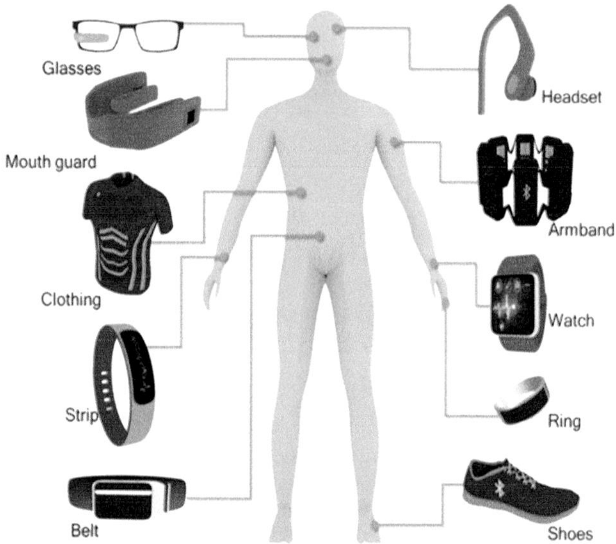

FIGURE 14.4 Evolution of wearable devices with real-time disease monitoring for personalized healthcare.

Source: Reprinted from Ref. [17]. © 2019 by the authors. Licensee MDPI, Basel, Switzerland. This article is an open access article distributed under the terms and conditions of the Creative Commons Attribution (CC BY) license (http://creativecommons.org/licenses/by/4.0/).

According to the Kidney Project, a bioartificial kidney is being developed to be used in the event of kidney failure. There were no notable safety issues when they implanted a kidney bioreactor prototype containing functioning human kidney cells into pigs. Every patient who is eligible for a transplant might benefit from this innovation, not only those at the top of the list.

14.7 ROBOTICS

The use of robots in healthcare administration and facility maintenance is increasing as artificial intelligence (AI), and ML capabilities improve [4].

Surgical assistants and delivery and transportation assistance are being assisted by robots with powerful AI skills. One of two robots used by the Avera McKennan Hospital in Sioux Falls, S.D., to sanitize operating rooms and eradicate superbugs is dubbed Xena, and it has become a crucial instrument in the battle against the COVID-19 epidemic [1, 5]. Zimmer Biomet Holdings' ROSA robotically assisted complete knee replacement surgical technology was approved by the FDA in 2019 (Figure 14.5).

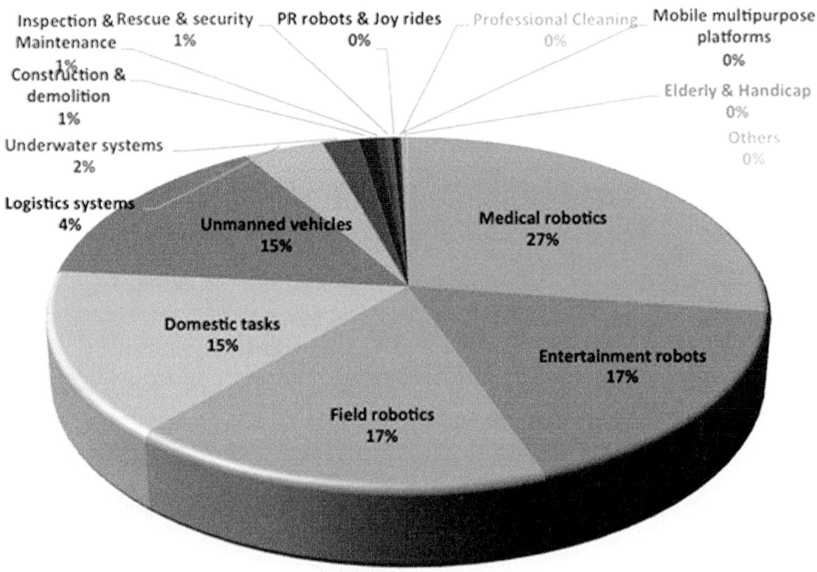

FIGURE 14.5 Vitality of robotics in the healthcare industry: an internet of things (IoT) perspective [18].

14.8 BLOCKCHAIN

Many businesses, including the healthcare industry, are benefiting from blockchain technology. Transactions and assets in a corporate network may be recorded and tracked more easily with the help of blockchain, a decentralized, unchangeable ledger [7]. There are two types of assets: tangible and intangible. On a blockchain network, almost anything of value may be recorded and sold, reducing risk and costs for all parties. When a transaction is placed into a ledger, it cannot be changed, making it both ultra-secure and accessible (Figure 14.6).

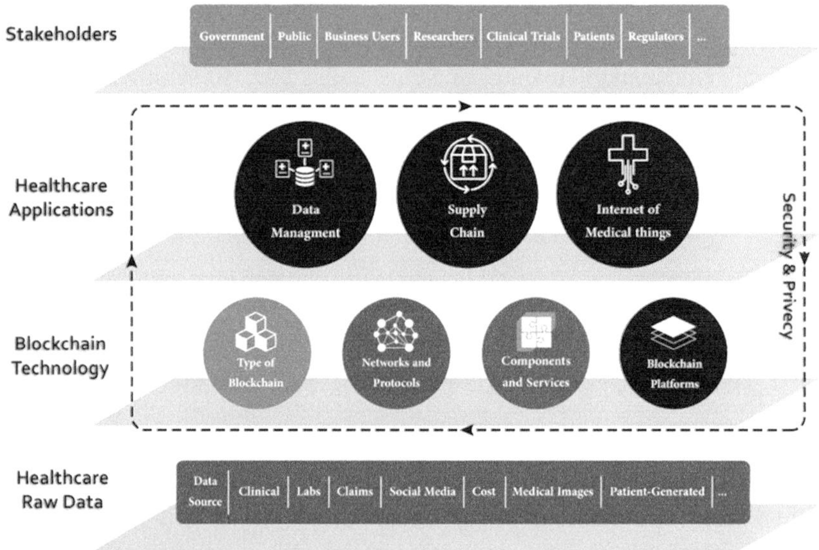

FIGURE 14.6 Blockchain technology in healthcare.

Source: Reprinted from Ref. [12]. © 2019 by the authors. Licensee MDPI, Basel, Switzerland. (http://creativecommons.org/licenses/by/4.0/).

14.9 STANDARDS HEALTH INFORMATICS

Despite the availability of communication technology to assist data sharing, it is still difficult and expensive to send data from one computer to another for clinical treatment, patient safety, and quality improvement. This potential has been hindered by the random adoption of data standards for organizing, describing, and encoding clinical information so that the data may be understood and accepted by the receiving systems. When it comes to healthcare organizations, there is a lack of common data standards that are preventing information from being shared between commercial clinical labs and healthcare institutions, pharmacies, and healthcare practitioners addressing prescriptions. Healthcare and regulatory organizations are making a lot more effort than they need to in order to create, send or utilize unique reports since there is no standard way to describe data for any of these datasets or criteria. In order to address this issue, the federal government is integrating its safety-related systems. A large number of the necessary data standards have already been developed; the development of the remainder is still ongoing.

- Data exchange, terminologies, and knowledge representation are the three main areas where standards for healthcare data need to be defined. These data standards' implementation is addressed in the concluding portion of this section.
- Standards for the transfer of data:
  - Message formats, document architectures, clinical templates, user interfaces, and patient data linkage are all in need of standardization in order to be used in data exchange.
- Standards for message formatting:
  - A common encoding specification, information models for establishing the connections between data pieces, document architectures, and clinical templates for arranging data as they are transferred enhance interoperability via the usage of message format standards Consolidated Health Informatics announced in March 2003 that all federal healthcare services agencies must adopt the primary clinical messaging format standards. The IEEE is now working to implement these standards, and the International Organization for Standardization will do so as well.
- 90% of big hospitals already use the HL7 V2.x series as the principal data exchange standard for clinical communications. A number of technical issues with the standard have made it difficult to implement the changes. First, the framework was designed to enable a variety of terminologies to describe a data element without being explicit about the exact codes within the language they use. When it comes to optionality, there have been variations since various vendors have distinct information models. This has caused confusion and misunderstandings about how the standard is applied. V2.x may be communicated over the Internet and represented as an extensible markup language (XML) syntax standard even if it does not support web-based protocols. V2.x, on the other hand, does not provide an information model that is required for the more sophisticated conveyance of clinical data.
- The completion and implementation of HL7 Version 3.0, in which only a few data fields are subject to interpretation, would be necessary to resolve these concerns, although they might be readily achieved. The scope of the V3 standard remains the same as that of V2.x, although the first version of V3 did not cover the domains of

patient referrals, patient care, or laboratory automation, all of which are critical to patient safety. As a first step, it is necessary to speed up the development of V3 and produce implementation guidelines for the use of both V2.x and V3, as well as compatibility between the two standards. To ensure interoperability, both V2.x and V3 call for a standardized nomenclature to be supplied at the data element level. Since V2.x is likely to be extensively used for some time, it is vital to fix any issues with compatibility between this standard and others. For example, inpatient, and outpatient pharmacies utilize V2.x for drug order communications, whereas retail pharmacies use NCPDP Script. For high-performance systems, compatibility between these two standards is essential.

- While HL7 V2.x used a Reference Information Model (RIM) to build up the communications format, HL7 V3 does not use one. Requirements-oriented modeling (ORIM) is used to define the types of data that are needed and the qualities of those types, such as attributes and relationships. The RIM's information needs are defined in an organized manner, reducing the amount of room for error. Increased semantic interoperability is achieved by populating data fields with explicit controlled vocabulary. Data may be exchanged fast and easily using XML, but semantic interoperability requires the RIM.
- Entity, role, participation, and act are the four high-level classes at the heart of the RIM. Figure 14.1 is a simplified picture of the structural interactions that make up the RIM, which should help you better comprehend the model's foundation.
- High-risk operations may be identified by information modeling, which has a direct influence on patient safety. Retrospective analysis of a patient safety occurrence and active patient care safety are intertwined, and both need adequate information linkages. As a result, the information model should make it easier to establish links between organizations and the purposeful activities they do in order to analyze better patient safety concerns and bigger problems of cost and quality. In the case of an aircraft disaster, comparing data from a flight data recorder with data from the cockpit's voice recorder is as critical as examining the relationship between a precipitating event and an unfavorable occurrence.

14.10 CHALLENGES

14.10.1 DEVELOPING PROACTIVE MODELS, METHODS, AND TOOLS TO ENABLE RISK ASSESSMENT

Health IT-based clinical applications might pose hazards to patients, their caregivers, and even their creators if they are used in an unorthodox manner. An overall proactive risk can be derived for an error class when severity and likelihood estimate of a potential error are combined. Clinical staff or expert opinion are the most common sources for current estimations of severity and probability. Inaccuracies and under-reporting in such event data make them a poor foundation for frequency assessment. Our understanding of the potential danger of these catastrophes necessitates new, proactive, data-driven models, approaches, and tools. Employees of healthcare organizations and health IT companies must also "have the knowledge, experience, and competencies needed for completing the clinical risk management duties given to them." Prioritizing efforts to establish compensatory controls for these mistakes will assist in prioritizing attempts to avoid or at least lessen the risk of these errors happening. For many situations, more accurate frequency estimations should be attainable because of the use of digital systems like the EHR. Standardizing the design elements and functionalities of user interfaces. Errors in data input and understanding are common when user interfaces are poorly designed. Since different EHRs, intensive care units (ICU) and infusion devices present patient identification information in various ways, users may need to acknowledge their acceptance of entered data in a variety of ways before proceeding with the next step in the process. Providers are forced to frequently swap mental models about how each interface works because of the absence of acknowledged and implemented standards in this area. Improved and more uniform approaches to enable users to submit data are needed, as is automated verification that the supplied data are valid for a specific patient. Last but not least, the sector must adhere to well-established standards for the development, creation, and testing of safety-critical Health IT-related patient safety concerns are characterized according to the stage of the health IT lifecycle in which they arise.

Challenges in both design and development Developing risk assessment models, methodologies, and instruments. Developing standard features and services for user interface design Third, ensuring the safety of software in a clinical setting that is network-enabled [2]. Unambiguous patient identification is essential at this step. Challenges in both implementation and use Developing and deploying safety-enhancing decision support systems figuring out the best ways to handle IT system changes securely Keeping tabs on, evaluating, and optimizing the performance System performance and safety may be monitored automatically using real-time technologies that are being developed. A single health IT vendor's product offering may be comprehensive, but new and stand-alone applications will always be produced that must be interfaced with the current system (s). There should be no errors throughout the whole process of building the software, implementing, patching, and upgrading it. Software development and testing approaches for isolated and self-contained systems have not yet been established to support the massively networked systems that will be needed for seamless patient data exchange across EHRs, enterprises, communities, and ultimately countries [6]. Healthcare should be considered a safety-critical business, and IT components employed in it should be given the same priority as those in aerospace, nuclear, and military industries. As an example, Scandinavian nations and the United Kingdom are moving forward with the development of health IT guidelines and even mandating some processes, whereas the United States has not yet recommended a stringent regulatory environment for health IT that is led by the industry or the government. In any case, the FDA's recent announcement of a software developer "pre-certification" program, which certifies software developers rather than particular projects, is a good step forward that strives to combine safety and innovation. Putting a system in place that makes it possible to identify a patient with absolute certainty. Accurate patient matching across EHRs, businesses, communities, and countries is one of the biggest threats to patient safety. The United States, Germany, Italy, and Canada are just a few countries that have yet to implement national standards for patient identification that use ambiguous, non-unique, temporary, and changeable data to determine a match. The most common methods of patient matching still rely on these ambiguous, temporary, non-unique, and changeable data to determine. We need a way to correctly connect patients across companies, places, and time periods. It's possible that failing to identify the same patient's data in two distinct places might be as critical as improperly matching two separate patients' data. Challenges in both implementation and use creating and applying for safety-enhancing decision assistance. Errors and blunders will continue to be made by busy clinical application users. A "safety net" and a "cockpit" are needed to help people do the correct thing as well as to catch mistakes. Today's computer-based clinical decision support systems heavily depend on "alerts" and "reminders" to physicians, although they are often disregarded by them. There are times when a computer proposes a course of action that the physician ignores, resulting in yet another kind of mistake. Clinical decision assistance must be created, implemented, and controlled in such a way that it has the most effect. Successful health IT relies on providing the right amount of and assuring the safety and dependability of AI-driven automation, as well as making sure that humans are kept informed and "in the loop" about what is occurring.

How can a machine know when to interrupt a human being? When is it acceptable to override the machine and let humans have the final say? To continue working, users must respond to current interruptive notices, which appear and ask for a response from that user. Although these alerts are intended to help clinicians, they are often rendered useless by the lack of comprehensive clinical information reflected in the computer or by clinicians' inability to comprehend their own thinking in the context of a given situation. The answers to each of these questions must be sought out. Methods for managing IT system changes in a secure manner must be identified and put into practice. Safety hazards are present when a new system is implemented when a commercial off-the-shelf EHR replaces an in-house built one or when substantial improvements are made to an existing EHR. Managing various forms of system transitions, such as partial implementation, record migration, software upgrades, and downtime, calls for a variety of best practices. Is there a need for an anomaly detector? Identifying and resolving problems reported by users is what? Staff have never worked without health IT before, so how can we prepare them for downtimes? Most healthcare systems are dependent on their health IT systems. Health services research has shown us that even if standards and best practices are clear and accessible, adapting them is still a huge barrier. Keeping tabs on, evaluating, and optimizing the performance improving the efficiency and security of computer systems by creating real-time solutions for automated monitoring and surveillance [13]. Retract-and-reorder tool, possible data sources, data collection strategies for each measurement topic, and entities that were accountable for performance were all recommended to advance the scientific path to measuring health

IT safety in a scientific manner. To expand real-time measurement and make surveillance more automated, researchers would need to cooperate closely with computer scientists, health IT vendors, and healthcare organizations to generate new scientific understanding as well as methodologies and tools. The establishment of a cultural and legal framework/safe harbor to facilitate the exchange of information regarding dangers and adverse occurrences [9]. Unless this changes, it will be unable to collect enough data to identify common failure mechanisms and predict the probability of similar future catastrophes. A national or worldwide health IT reporting system that collects and examines major patient safety concerns with the support of committed professionals is what we advocate for as a required, blame-free solution. Similar to the current aviation industry "near-miss reporting system," which has developed a list of "near-misses" that must be recorded and includes extra information on incidents, such a system might be patterned after. On top of all of this, we must start looking into ways to combine data from existing registries for equipment failures and hazards, medical record reviews, 13 user complaints, and medico-legal investigations, for example, to better understand the nature, origins, consequences, and outcomes of IT problems in healthcare. Analysis of patient safety event records has already shown positive results in identifying IT safety issues. Creating models and approaches for patients/consumers to enhance the safety of health information technology. It's important to ask what role consumers and their caregivers may play when it comes to spotting and resolving IT-related issues. There are a number of ways in which people may raise concerns about the quality of care they receive. With the introduction of activity monitoring and personal/shared health data, will they be asked to take on new responsibilities and assume greater responsibility for their own healthcare? Allowing for easy access to patient progress notes and other clinical data would need a culture transformation, which is a significant "non-technical" obstacle in and of itself.

14.11 RECOMMENDATIONS HEALTH INFORMATICS

Technological advances in healthcare are nothing new, and this should be no surprise to anybody. When it comes to medical care, the days of being treated by a real country doctor are long gone, replaced by machinery and software more akin to those used by Dr. McCoy in "Star Trek."

Technology's influence on healthcare has evolved, but it hasn't been all rainbows and sparkles. Even while their more mundane appearance may make them more difficult to notice, the most significant improvements in healthcare are sometimes just as revolutionary as the tiny robots zipping through veins. Medical informatics is at the forefront of the current technological revolution in medicine because it blends communication, information technology, and healthcare to enhance patient care. Here are six ways in which it's already reshaping the healthcare industry.

14.11.1 SAVINGS THAT ARE STARTLING

Healthcare is not only costly, but it is also inefficient. As many as half of all medical costs might be lost due to unnecessary or redundant procedures, the costs connected with more conventional means of exchanging information and delays in treatment. Much of the waste may be reduced by using an electronic and networked system. Health informatics eliminates mistakes, improves communication, and boosts efficiency where before there was expensive ineptitude and blockage [8].

14.11.2 KNOWLEDGE SHARING

There's a reason doctors refer to their work as a "practice," and that's because they are always expanding their knowledge and improving their abilities. The field of health informatics makes it possible to transmit information more quickly regarding patients, ailments, treatments, medications, and the like. The practice of medicine improves when more information is shared between healthcare professionals and patients, which benefits everyone along the way, from hospital administrators and doctors to pharmacists and patients.

14.11.3 PARTICIPATION OF THE PATIENT IN THE DECISION-MAKING PROCESS IS A THIRD

Patients are better able to take responsibility for their own healthcare when they have electronic access to their own health history and treatment suggestions. Patients who have access to care portals can better understand their diagnosis and treatment options, as well as keep track of their prescriptions and other health-related symptoms. They are also better able to communicate with medical professionals, resulting in improved results. Individuals might feel as if they are a vital member of their own healthcare team thanks to health informatics because they really are.

14.11.4 INHUMANITY IN THE DELIVERY OF CARE

One of the drawbacks of using information and technology to improve patient care is that treatment is becoming less personal. Rather than a doctor getting to know a patient in-person to provide the best treatment, data, and algorithms are doing the "knowing."

Patients' health data may be sorted using algorithms to discover what's wrong and what treatment is needed. While the long-term implications of this data-driven strategy need to be explored, it is essential that patients and their care providers have easy access to a true and accurate record.

14.11.5 COORDINATION HAS BECOME BETTER

Because medical treatment is becoming more specialized, patients may expect to see up to a dozen different doctors during a single hospital stay. Health informatics is the answer to the problem of coordinating a growing number of experts. It's mind-boggling how many different interactions a single patient may have with a team of individuals discussing treatment, and unless those talks and efforts are made in unison, issues will occur, and care will suffer. The essential coordination is made feasible through health informatics.

14.11.6 AIMS ACCOMPLISHED

Improved outcomes are the most significant manner in which informatics is altering healthcare. Coordinated teams that use EMR enable better diagnoses and reduce the likelihood of mistakes, resulting in greater quality and safer treatment. In addition to saving time and money for hospitals, clinics, and healthcare providers, such as patients, insurance companies, and state and federal governments, automation allows doctors and nurses to become more efficient. This frees up more time to spend with patients.

14.12 FUTURE DIRECTIONS

14.12.1 THERE'S A LOT MORE TECHNOLOGY OUT THERE

In the next 10 years, the amount of knowledge and tools accessible to clinicians in the field of genetic diagnosis and therapy will likely rise in an unimaginable way. Improvements in minimally invasive imaging and less invasive treatment with catheters will lead to better functional results and an earlier return to activity. The use of DNA chips or genetic fingerprinting will greatly enhance risk assessments in the future. A better understanding of the hazards will allow other technologies to lengthen life. However, new tools like these will need that confront and address a variety of new ethical problems.

The use of electronic technology is expected to enhance productivity as well. The need for elaborate compliance systems would be much reduced if invoicing was linked directly to the content of the medical record. No of what "billing form" a plan uses, software should eventually be able to automatically charge that plan for services.

14.12.2 ADDITIONAL DETAILS

The quality of patient care data will rise along with it, thanks to advancements in technology. The electronic medical record will not only be able to store patient information, but it will also be able to provide instantaneous information on "best practice," whether it is derived statistically from the practice of the physicians in that AHC or based on health plan data or nationally generated practice guidelines. Online clinical research has several advantages. Quality criteria from the patient viewpoint may be enhanced by the capacity to question large numbers of patients and wide sectors of the community.

In addition, we will improve our knowledge of illness severity. Furthermore, the "risk" of a future year's disease costs will be better recognized when this information is available. AHCs will profit from this knowledge since they generally care for patients who are more unwell.

14.12.3 THE PATIENT WILL BE THE ULTIMATE CONSUMER IN HEALTHCARE

Patients will become the ultimate customers as more and more people use the Internet, and businesses may no longer pick the health plan for their workers but instead give them a "defined contribution" to purchase their own healthcare. The value of patient satisfaction surveys and other forms of patient feedback will rise in the coming years. Customers who buy their own health insurance policies will be able to choose doctors and hospitals near their homes, reducing the need for health plans that are sold to employers that cover a vast area.

14.12.4 CHANGES IN THE MODE OF DELIVERY

Process and results will improve when data is made more widely available to the public. They will either improve or cease if they aren't capable of getting the greatest results. In the next 10 years, a considerable number of people with relatively common illnesses will benefit from improved treatment processes and results. Care for these patients will become more regularized, allowing for a better knowledge of the appropriate delivery model for healthcare services. Measurement of outcomes of nurse practitioners, generalist doctors and specialty physicians in the care of particular illnesses, for example, would allow better "hand-offs" and more efficient use of resources.

There will be an increased need for practitioners as the number of patients rises, and the focus will shift away from arguing over who will treat which patients to optimizing the care model. There will be an increasing demand for experts in the areas of illness that now affect the elderly and also in areas of developing diseases that are today uncommon but will become more frequent as other more common diseases become prevented. Nonphysician practitioners will be particularly useful in places where treatment is offered on a regular basis in the next 10–20 years due to a potential shortage of doctors. As the population ages, the need for hospital beds will continue to fall, but it is expected to rise again in the future owing to the elderly.

Patient access to more information on the Internet will lead to increased self-diagnosis and self-care. There will be less need to visit a practitioner's office thanks to this information as well as direct Internet video conversation with the practitioner's office. Patients in the most critical condition will be cared for in hospitals, while those in better health will communicate with one another over the Internet. However, as we understand from adult education, whether or not technology is accessible, individuals desire to engage with other humans, and although visits may reduce, they will not decline as much as technology may enable.

14.12.5 THERE'S A CHANCE FOR NEW IDEAS

More and more patients are being treated in the same way, and the competition among practitioners will shift from the best outcomes for common diseases toward the ability to innovate: developing the best care delivery models for patients with common diseases or developing new strategies to treat patients with uncommon diseases. AHCs will benefit from this once again.

14.12.6 COSTS ARE LIKELY TO RISE

It is apparent that greater information on appropriate treatment and more effective care models will emerge with technology for patient home care, in addition to more efficient billing, fewer wasteful tests and procedures, but these advances will be overshadowed by growing expenses. Increasing the number of people with chronic cardiovascular disease by only one-third would result in a 13% increase in the cost of healthcare services. People will be forced to take a costly prescription for the rest of their lives in order to avoid atherosclerosis, according to a new study on the so-called "magic bullet."

The number of people who are uninsured is expected to rise. When the cost of healthcare rises, businesses will try to lower their expenditures by limiting coverage and putting the burden on their employees. Because of rising prices, fewer people will be able to afford even minimal health insurance, which will lead to a rise in the number of people without coverage. There will be an increase in the number of uninsured as these numbers rise, which will lead to an increase in the cost of healthcare and an increase in the number of uninsured in a vicious cycle.

14.12.7 PROVIDERS WILL GET PAID, LESS

The bulk of payers, including the government and commercial insurance, pay the same amount to all providers, regardless of the kind of service. As expenses rise, health plans will pay those who must and limit payments to those who don't, resulting in higher costs for everyone. Providers will be handed what is left unless methods are devised to show value and expand market share in addition to their own administrative expenses, pressures on the bottom line for private plans, and payments for new pharmaceuticals and equipment.

14.12.8 HEALTHCARE REFORM IS NECESSARY

Within the next 5 to 10 years, the situation will deteriorate to the point of crisis for the United States. Employers' healthcare expenditures will continue to climb, and the number of uninsured individuals will continue to rise as the gap between what can be afforded and what is accessible widens; this will lead many to consider leaving the healthcare industry. The uninsured population will grow to include members of the middle class. The healthcare system might be reformed by the disaffected and their employers. We previously presented a mechanism that might provide universal healthcare coverage by the year 2010 in this scenario.

14.13 FUTURE HEALTHCARE TECHNOLOGY TRENDS

The healthcare business places a high value on excellent patient experiences, and as a result, many providers are moving away from volume-based treatment and toward value-based care. Value-based care compensates providers by assessing the cost of a service in relation to the quality of treatment, patient outcomes, and other criteria instead of following a typical fee-for-service approach. According to Forbes, value-based care will account for up to 15% of global healthcare expenditure by the end of 2019. Value-based care will become a more realistic alternative as technology advances, enabling medical providers to monitor the quality of treatment, bundle packages, and convert to EHRs.

For patients, this is wonderful news since the quality of their particular experience takes precedence. The next generation of hospital managers will assist their facilities to adapt to this new approach via the use of new technological tools. In order to lead the implementation of value-based care solutions, professionals might benefit from completing an online bachelor's degree Program in general studies with an emphasis in health services management.

In the future, healthcare workers will be required to combine conventional talents in care, communication, and leadership with modern technology and analytic abilities. The following developments and trends will complement the development of these new skills.

14.13.1 CENTERS FOR SPECIALIZED OUTPATIENT CARE

Healthcare professionals are providing more timely and specialized care in an effort to enhance the patient experience. Colonoscopies and cataract removals are examples of outpatient operations, which are often less invasive than inpatient procedures. As an alternative to typical hospital settings, several treatments are increasingly available at outpatient surgical facilities.

In order to lower total medical expenses, ASCs, and other specialist outpatient care clinics move specialty services to sites where they may be provided exclusively. There is less need for space and people at bigger medical facilities, while the ASC provides more concentrated treatment for patients. For example, experts are allowed to concentrate on their area of expertise, which improves the quality of treatment. ASCs also have reporting and quality-measurement requirements that are almost two times as high as those for outpatient procedures performed in comparable hospitals.

The utilization of quality data comparing ASCs and hospitals for outpatient treatment by healthcare providers may help educate patients about the advantages of these specialized facilities. ASC educators play a critical role in educating the general public about ASCs and assisting patients in their transition to an ASC setting when necessary.

Personalized healthcare may be achieved via precision medicine. Instead of treating everyone the same, precision medicine considers a person's genetics, environment, and way of life when developing individualized treatment plans and preventative measures. In addition to its usage in cancer, precision medicine has a wide range of applications in pharmaceutical research and healthcare technology, such as monitoring tumor cells' genetic profiles to determine the most successful therapies. There are over 6,000 genetic illnesses that may be identified, so healthcare companies need the correct technology to keep track of patient data and use insights to improve patient care. Around 65% of patients administered the world's most popular medications failed to react to therapy, according to a survey by Scipher Medicines. Predicting a patient's response to a specific pharmacological therapy may be made easier with precision medicine, which reduces healthcare expenditures associated with wasteful prescriptions and poor patient treatment while also preventing unwanted side effects and drug interactions.

Patients may now use precision medicine to make good lifestyle choices and keep tabs on their physical well-being thanks to the growing field of preventive healthcare provided by the precision medicine movement [10]. Healthcare professionals will use current technologies like wearables to gather consumer health data and analyze it in a manner that patients can comprehend and exploit with ease as the future of healthcare technology develops a more patient-centered emphasis.

A population health management approach using predictive analytics. Data is becoming more important in the healthcare business. Through population health management, healthcare practitioners may use analytics to enhance health and detect at-risk patients (PHM).

Data from numerous sources, including payers, hospitals, main, and specialty physicians, specialists, and more, is compiled into a single patient record known as a PHM. Using this data to establish a single patient record improves both clinical and economic results for healthcare providers. As of 2019, Forbes expects that half of all healthcare businesses would have dedicated resources for accessing, sharing, and analyzing real-world data.

Healthcare administrators and data scientists are able to utilize this data in real-time to discover and fix patient care gaps in the population. Using this data, they may create programs that improve outcomes and save costs. Healthcare administrators who can develop and execute policies, manage, and monitor budgets, and assure adherence to procedures and laws will be required to use PHM systems intelligently in the future of healthcare.

14.13.2 IN-THE-MOMENT DATA VISUALIZATION

The use of big data in medicine is not a new phenomenon, and developments in predictive analytics will be part of healthcare in the future to boost response efficiency, give real-time reporting, and identify at-risk groups. It's predicted that predictive analytics will become an industry worth more than $9 billion by 2020. Predicting health-related events such as how the weather affects patients with respiratory diseases is done utilizing historical and transactional data. The gathering and production of bigger datasets will boost accuracy and help businesses to recognize patterns as healthcare organizations share more data. The flu season can be predicted more accurately if a network of hospitals pooling their data works together. When preparing for a shift to real-time data visualization, healthcare managers must keep up with the rapid pace of technological change. When implementing modern technologies, administrators must work with physicians and technicians as well as medical facilities, IT teams and government agencies in order to ensure their success.

14.14 TELEHEALTH

It is also expected that healthcare advancements will change how we handle various medical conditions in the future. Modern medicine has undergone a sea change with the advent of telehealth, or the use of mobile devices, computers, and streaming services to access healthcare. People who reside far from a medical center may use telehealth to communicate more often with their doctors. People who need therapy for diseases like communication problems or mental health issues may now get it through a video chat from the convenience of their own homes.

While telehealth is still in its infancy when it comes to dealing with communication disorders, degree programs in communication sciences and disorders are training future speech-language pathologists and other professionals for the hybridization of conventional and telehealth therapy.

Increasing the quality of life for patients. Healthcare revolves around the well-being of patients. Healthcare professionals are increasingly turning to the newest developments in technology and data to help them provide better care to their patients. A job in the healthcare field may be satisfying since medical professionals play a role in improving the health of the general public and reducing the cost of healthcare.

14.15 DEVELOPING AND UNDER-DEVELOPING COUNTRIES

In poorer nations, healthcare delivery is often subpar and subject to vast variation. A substantial amount of research from industrialized nations consistently indicates variances in the procedure, and these results have revolutionized how the quality of treatment is evaluated today. Only eight of the 306 U.S. hospital areas studied in 2002 research adhered to evidence-based standards for at least 80% of patients. These changes seem independent of availability or cost of care: neither more supply nor increased expenditure resulted in better treatment or better survival. Similar findings have been seen in studies from poor nations. When it comes to treating the same conditions, the treatment offered by specialists in tertiary and teaching hospitals may be preferable to that provided by general practitioners in primary care institutions.

The lack of resources in underdeveloped countries may be a factor in the heterogeneity and poor quality of treatment. It's possible to deliver high-quality treatment despite very limited resources, according to limited evidence. Jamaican researchers found that improved procedure alone was connected to considerably higher birth weight in a cross-sectional investigation of government-run primary care facilities. Only 37% of perinatal fatalities were attributable to financial restrictions, according to Indonesian research.

The widest range of clinical practice in developing nations may be seen in comparisons across systems or countries. It was revealed that 75% of cases were improperly identified, treated, or monitored and that 61% of patients received inadequate antibiotics, fluids, feedings, or oxygen treatments by researchers in seven countries studying clinical practice. In another study, researchers used vignettes to evaluate physicians' knowledge and practice in California and FYR Macedonia. FYR Macedonia is a poor nation, yet the top 5% of Macedonian physicians performed as well as or better than the typical Californian doctor in the whole procedure.

Using the identical clinical vignettes across five developing nations, an international team conducted research commissioned specifically for this chapter. According to internationally accepted standards based on current scientific research, the team assessed the method for common ailments. Only a little variation in quality existed among nations. The variation in the quality of physicians within a nation was 10 times greater than the variation across countries. Given the vast range of outcomes, it's clear that initiatives to enhance health status must include measures that alter clinical treatment.

14.16 CONCLUSION

Digital transformation in healthcare is becoming increasingly important for both academics and practitioners. It has an impact on many aspects of healthcare businesses, including the purchase of digital resources, the creation of digital growth plans, the transformation of internal organizational structure, and the establishment of appropriate metrics and objectives. In this chapter, an integrated view of the state of the art of digitalization in the Healthcare Informatics literature is presented to highlight the important aspects and commercial uses of the latest technologies in the healthcare system. Furthermore, various existing health informatics domains are explored to identify the potential advantages of existing healthcare systems so that the latest digital technology can be utilized in healthcare sectors for the betterment of the various stakeholders. Finally, deliberate use of digital technology and data-driven and predictive care will enable a move toward digital health informatics models.

KEYWORDS

- **artificial intelligence**
- **digital health informatics models**
- **electronic health records**
- **extensible markup language**
- **internet of things**
- **machine learning**

REFERENCES

1. Singh, P. D., Kaur, R., Singh, K. D., & Dhiman, G., (2021). A novel ensemble-based classifier for detecting the COVID-19 disease for infected patients. *Information Systems Frontiers*, 1–17.
2. Singh, K. D., & Sood, S. K., (2021). QoS-aware optical fog-assisted cyber-physical system in the 5g ready heterogeneous network. *Wireless Personal Communications, 116*(4), 3331–3350.

3. Achouri, M., Alti, A., Derdour, M., Laborie, S., & Roose, P., (2018). Smart fog computing for efficient situations management in smart health environments. *Journal of Information and Communication Technology, 17*(4), 537–567.
4. Singh, P. D., Kaur, R., Singh, K. D., Dhiman, G., & Soni, M., (2021). Fog-centric IoT based smart healthcare support service for monitoring and controlling an epidemic of swine flu virus. *Informatics in Medicine Unlocked, 26*, 100636.
5. Kaur, R., Singh, P. D., Kaur, R., & Singh, K. D., (2021). A delay-sensitive cyber-physical system framework for smart health applications. In: *Advances in Clean Energy Technologies,* (pp. 475–486). Springer, Singapore.
6. Kaur, S., Singh, K. D., Singh, P., & Kaur, R., (2021). Ensemble model to predict credit card fraud detection using random forest and generative adversarial networks. In: *Emerging Technologies in Data Mining and Information Security* (pp. 87–97). Springer, Singapore.
7. Sood, S. K., & Singh, K. D., (2021). Identification of a malicious optical edge device in the SDN-based optical fog/cloud computing network. *Journal of Optical Communications, 42*(1), 91–102.
8. Angurala, M., Bala, M., Bamber, S. S., Kaur, R., & Singh, P., (2020). An internet of things assisted drone based approach to reduce rapid spread of COVID-19. *Journal of Safety Science and Resilience, 1*(1), 31–35.
9. Sood, S. K., & Singh, K. D., (2019). Hmm-based secure framework for optical fog devices in the optical fog/cloud network. *Journal of Optical Communications*.
10. Seth, J., Nand, P., Singh, P., & Kaur, R., (2020). Particle swarm optimization assisted support vector machine based diagnostic system for lung cancer prediction at the early stage. *PalArch's Journal of Archaeology of Egypt/Egyptology, 17*(9), 6202–6212.
11. Sood, S. K., & Singh, K. D., (2019). Optical fog-assisted smart learning framework to enhance students' employability in engineering education. *Computer Applications in Engineering Education, 27*(5), 1030–1042.
12. Khezr, S., Moniruzzaman, M., Yassine, A., & Benlamri, R., (2019). Blockchain technology in healthcare: A comprehensive review and directions for future research. *Applied Sciences, 9*(9), 1736.
13. Singh, K. D., & Sood, S. K., (2020). Optical fog-assisted cyber-physical system for intelligent surveillance in the education system. *Computer Applications in Engineering Education, 28*(3), 692–704.
14. Alnosayan, N., Chatterjee, S., Alluhaidan, A., Lee, E., & Feenstra, L. H., (2017). Design and usability of a heart failure mHealth system: A pilot study. *JMIR Human Factors, 4*(1), e6481.
15. Albahri, O. S., Zaidan, A. A., Zaidan, B. B., Hashim, M., Albahri, A. S., & Alsalem, M. A., (2018). Real-time remote health-monitoring systems in a medical centre: A review of the provision of healthcare services-based body sensor information, open challenges and methodological aspects. *Journal of Medical Systems, 42*(9), 1–47.
16. Aldosari, B., Al-Mansour, S., Aldosari, H., & Alanazi, A., (2018). Assessment of factors influencing nurses acceptance of electronic medical record in a Saudi Arabia hospital. *Informatics in Medicine Unlocked, 10*, 82–88.
17. Guk, K., Han, G., Lim, J., Jeong, K., Kang, T., Lim, E. K., & Jung, J., (2019). Evolution of wearable devices with real-time disease monitoring for personalized healthcare. *Nanomaterials, 9*(6), 813.

18. Patel, A. R., Patel, R. S., Singh, N. M., & Kazi, F. S., (2017). Vitality of robotics in healthcare industry: An internet of things (IoT) perspective. In: Bhatt, C., Dey, N., & Ashour, A., (eds.), *Internet of Things and Big Data Technologies for Next Generation Healthcare; Studies in Big Data* (Vol. 23). Springer, Cham. https://doi.org/10.1007/978-3-319-49736-5_5.

Index

C

M

Q

R

T